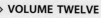

VOLUME TWELVE

ADVANCES IN NEUROTOXICOLOGY

Stem Cells in Neurotoxicology

VOLUME TWELVE

ADVANCES IN NEUROTOXICOLOGY

Stem Cells in Neurotoxicology

Series Editors

MICHAEL ASCHNER

LUCIO G. COSTA

Edited by

AARON B. BOWMAN

School of Health Sciences, Purdue University, West Lafayette, IN, United States

MICHAEL ASCHNER

Albert Einstein College of Medicine, Bronx, NY, United States

LUCIO G. COSTA

University of Washington, Seattle, WA, United States; University of Parma, Parma PR, Italy

ELSEVIER

ACADEMIC PRESS

An imprint of Elsevier

Academic Press is an imprint of Elsevier
125 London Wall, London, EC2Y 5AS, United Kingdom
50 Hampshire Street, 5th Floor, Cambridge, MA 02139, United States
525 B Street, Suite 1650, San Diego, CA 92101, United States

First edition 2024

ISBN: 978-0-443-13254-4
ISSN: 2468-7480

For information on all Academic Press publications
visit our website at https://www.elsevier.com/books-and-journals

Publisher: Zoe Kruze
Acquisitions Editor: Leticia Lima
Editorial Project Manager: Palash Sharma
Production Project Manager: Abdulla Sait
Cover Designer: Gopalakrishnan Venkatraman

Typeset by MPS Limited, India

Contents

Contributors

Aaron B. Bowman
School of Health Sciences, Purdue University, West Lafayette, IN, United States

Thomas Hartung
Center for Alternatives to Animal Testing (CAAT), Bloomberg School of Public Health and Whiting School of Engineering, Johns Hopkins University; Doerenkamp-Zbinden Chair for Evidence-based Toxicology, Baltimore, MD, United States; CAAT Europe, University of Konstanz, Konstanz, Baden-Württemberg, Germany

Ellen V.S. Hessel
National Institute for Public Health and the Environment, Bilthoven, The Netherlands

Hyunjin Kim
School of Health Sciences, Purdue University, West Lafayette, IN, United States

Ribhav Mishra
School of Health Sciences, Purdue University, West Lafayette, IN, United States

Maren Schenke
Center for Alternatives to Animal Testing (CAAT), Bloomberg School of Public Health and Whiting School of Engineering, Johns Hopkins University, Baltimore, MD, United States

Lena Smirnova
Center for Alternatives to Animal Testing (CAAT), Bloomberg School of Public Health and Whiting School of Engineering, Johns Hopkins University, Baltimore, MD, United States

Xueqi Tang
School of Health Sciences, Purdue University, West Lafayette, IN, United States

Jack R. Thornton
Center for Alternatives to Animal Testing (CAAT), Bloomberg School of Public Health and Whiting School of Engineering, Johns Hopkins University, Baltimore, MD, United States

Anke M. Tukker
School of Health Sciences, Purdue University, West Lafayette, IN, United States

David Yi
School of Health Sciences, Purdue University, West Lafayette, IN, United States

Victoria C. de Leeuw
National Institute for Public Health and the Environment, Bilthoven, The Netherlands

Preface

Stem cells are fundamental to development, including development and maturation of the nervous system. As such, their discovery created as well realization of their potential to create in vitro model systems along their developmental trajectory. This volume explores the expanding utilization of stem cells in neurotoxicology.

The volume begins with a historical context beginning with the idea that organisms are made up of cells, and that the creation of new progeny occurs via the genesis of a stem cell which drives the development of the entire adult body. Hyunjin Kim, doctoral student with Aaron Bowman maps out the road from the first use the term 'stem cell' with dualistic definitions in evolutionary theory and developmental theory to the present understanding of cellular identity, progressive restrictive potency, but developmentally resistant yet now observed bi-directionality that now allows the creation of pluripotent stem cells from somatic cells of an organism. In chapter two, a discussion of the advantages and disadvantages of pluripotent stem cell based neuronal models versus immortalized cell culture models in the context of neurotoxicological research is led by Xueqi Tang, a doctoral student with Aaron Bowman.

The volume continues with a deep dive into how stem cells models have been utilized and hold great potential for discoveries of the mechanisms of neurotoxic agents upon the developing and mature nervous system. Renowned stem cell scientists Thomas Hartung, Lena Smirnova, Maren Schenke and Jack Thornton discuss brain organoid models in the context of developmental neurotoxicity as well as their use to study neurotoxicological outcomes on learning and intelligence. Internationally recognized environmental health scientist Ellen Hessel and her colleague Victoria de Leeuw both experts in the use of stem cell based neurotoxicological testing for regulatory and mechanistic work lay out of conceptual framework, as well as critical factors for robust, reproducible and translatable neurotoxicological assessment with stem cell models. Ribhav Mishra, postdoctoral fellow with Aaron Bowman, explores both the role of stem cells in neurodegenerative disease and their being target cells for neurotoxicological agents, and as well, how stem cells may be used both in the study of pathogenesis as well as their therapeutic potential. Anke Tukker, senior postdoctoral fellow with Aaron Bowman, lastly discusses how human stem cells are providing insight into the mechanisms of persistent toxicity, long-after the chemical agent is no longer in the system.

The chapter ends with a methodological exemplar for the creation of a human cortical neural cell model by directed differentiation for human induced pluripotent stem cells. This chapter provides a step-by-step level of detail for investigators new to the field to enable production of non-transformed neural systems. Representative images and troubleshooting guidance is included by Hyunjin Kim, David Yi for this method currently in practice in the laboratory of Aaron Bowman, guest editor for this volume and co-author of this and other chapters. The method is placed into the context of the neurodevelopmental biology foundation from which it was developed and the breadth of scientific knowledge that it is built upon.

We express our sincere gratitude to all the authors who took time from their busy schedules to prepare their chapters with excellent content and relevant information on the intersection of stem cells with the study of neurotoxicology and brain health, as well as the role of stem cells in the development and maintenance of the brain. The contributors are highly qualified and considered authorities in this field. Editors greatly appreciate their dedication and hard work to this series. The chapters in this volume should be of interest to students, academicians, governmental employees and industrial researchers. Editors also appreciate the efforts made by the editorial project manager, Mr. Palash Sharma, and other Elsevier editorial staff for their various roles in the preparation of this book series.

Aaron B. Bowman
School of Health Sciences, Purdue University,
West Lafayette, IN, United States

Michael Aschner
Department of Molecular Pharmacology,
Albert Einstein College of Medicine, Bronx,
NY, United States

Lucio G. Costa
Department of Environmental and Occupational Health Sciences,
University of Washington, Roosevelt, Seattle, WA, United States

CHAPTER ONE

The long road to the development of stem cells as a model for central nervous system health and disease

Hyunjin Kim and Aaron B. Bowman*
School of Health Sciences, Purdue University, West Lafayette, IN, United States
*Corresponding author. e-mail address: bowma117@purdue.edu

Contents

Abstract

The advent of human induced pluripotent stem cells (hiPSCs) has revolutionized the field of biological and biomedical sciences by pushing the frontiers in drug discovery, disease modeling, and regenerative medicine. In less than twenty years since its discovery, hiPSC technology has established itself as a powerful tool to tackle some of the most complex questions in understanding human brain health and disease. Going back to the heyday of classical embryology and evolutionary developmental biology, we describe some of the pioneering discoveries that formed the conceptual basis for the development of hiPSCs, along with historical anecdotes to provide additional context. Further, we highlight how principles of in vivo neurodevelopment such as

Advances in Neurotoxicology, Volume 12
ISSN 2468-7480, https://doi.org/10.1016/bs.ant.2024.08.003

neural induction and regional patterning are recapitulated during in vitro neural differentiation. Finally, we provide a brief overview of recent technological advances that further support our efforts in realizing the full potential of hiPSCs.

1. Introduction

This chapter aims to introduce the reader to pluripotent stem cells with a primary focus on neural lineage differentiation. We provide a historical overview of how the field of stem cell research evolved over time by describing some of the pioneering works done in classical embryology and their culmination in the development of human pluripotent stem cells, along with occasional anecdotes for additional contextual information. Furthermore, we describe how this versatile model system elegantly recapitulates core aspects of in vivo neurodevelopment including neural induction and regional patterning to serve as an efficient platform to study human central nervous system (CNS) health and disease. Finally, we provide a brief overview of recent technological advances and tools that can be further utilized to promote impactful translational research. We hope that this introductory chapter provides a better understanding of this exciting model system and facilitates further integration of stem cell-based models in toxicological studies.

2. Stem cells: A historical perspective

Stem cells are unspecialized cells that are defined by their ability to proliferate, self-renew, and to give rise to other cell types. These characteristic properties that define stem cells, however, were not established until the 1960s and the term "stem cells" has gone through a series of conceptual transformations over the course of more than a hundred years. In this section, we describe the conceptual origin of stem cells by going through some of the pioneering works conducted in evolutionary biology and classical embryology from which several key properties relating to stem cell biology were formulated.

2.1 In the beginning

Since antiquity, a fundamental question in biology was on how a complex organism can arise from a single starting cell. By approximately the seventeenth century, two opposing schools of thought had developed around this age-old topic: "preformationism" and "epigenesis". The former, which itself consisted of two competing views (spermism and ovism), postulated that germ cells harbored a fully preformed organism. According to this idea, development was nothing more than a growth in size and unfolding of the animalcule or in the case of humans, homunculus (essentially a miniature human being), that dwelled curled up in the head of a sperm or an egg (Poczai & Santiago-Blay, 2022). As microscopic embryology started to gain momentum, preformationism gradually declined in popularity and gave way to epigenesis. Conceptualized in large part by Caspar Friedrich Wolff's astute observation and insightful interpretation of the progressive development of form in the chick embryo, this principle posited that development is not merely a period marked by growth in size, but also a concomitant increase in complexity (Wolff, 1759). The ascendancy of epigenesis at the expense of preformationism was further fueled by discoveries in descriptive embryology in the nineteenth century. Several key concepts and terms relating to development were established during this era, including the identification of the three embryonic germ layers by Christian Heinrich Pander and Karl Ernst von Baer, the former of whom made the initial discovery by microscopically studying two thousand domestic chick embryos that he harvested at fifteen-minute intervals (Akhurst, 2023; Pander, 1817). Another important milestone of this era is the development of the cell theory by Rudolf Virchow, Robert Remak, and Theodor Schwann and the notion that every cell stems from another cell via cell division—*"omnis cellula a cellula"* (Lendahl, 2022; Virchow, 1861). Collectively, the replacement of preformationism by epigenesis and the pronouncement of the cell doctrine can be seen as defining milestones as they establish the key conceptual basis that physiological complexity and phenotypic diversity of somatic cells arise from progressive cell division during embryonic development.

2.2 On the conceptual origin of a "stem cell" and its properties

The term "stem cell" was initially used in the context of evolutionary biology and makes its first appearance in scientific literature as early as 1868 in the works of the German zoologist Ernst Haeckel, who had previously studied medicine with Virchow (Maehle, 2011; Ramalho-Santos & Willenbring, 2007). It is interesting (perhaps even ironic) to note that Virchow himself was a fervent anti-evolutionist, even referring to his former student as a "fool" (Masic, 2019). Referred to by his contemporaries as the "German Darwin", Haeckel was one of the biggest proponents of the Darwinian principle and is recognized to have dubbed the terms "phylogeny" and "ontogeny", wherein the former's prefix, phylo-, is derived from the Greek word *phylon* and can be translated into *Stamm* in German to mean "stem" (Dayrat, 2003; Levit & Hossfeld, 2019). Haeckel used this prefix in the context of his phylogenetic trees, or as he called them, "Stammbäume" (family or stem trees), to refer to terms like "Stammorganismen" (stem organisms). As such, his coinage of the term "stem cell" or "Stammzelle" was of a figurative nature, rather than experimentally derived, with his evolutionary viewpoint serving as its linguistic foundation (Dröscher, 2014). Under his conceptual framework, the term stem cells or "Stammzellen" carried two contextual meanings. First, it was used to describe the evolutionary (phylogenetic) unicellular ancestor from which all multicellular organisms emerged (Haeckel, 1868). Subsequently, in proposing his controversial biogenetic law of "phylogeny to ontogeny", he applied the term to embryological development (ontological) to refer to the fertilized egg as stem cells or the ancestral cell of origin for all other cells of an organism (Haeckel, 1877).

In 1885, August Weismann formulated his influential "germ plasm" theory to reconcile how species preserve their characteristics between generations and yet display enough variation to transform over evolutionary time (Bline et al., 2020). In an era before the advent of modern genetics, Weismann, with remarkable insight, contemplated that the nucleus played a central role in unifying heredity, development, and evolution under his theoretical construct. He postulated that the only cells capable of harboring hereditary information in its totality (i.e. germ plasm) and transmitting them to the next generation were germ cells (Weismann, 1885). According to his notion of continuity of the germline, germ cells were segregated from somatic cells during embryonic development to replicate and maintain the full set of germ plasm across generation. In contrast, somatic cells were allocated different combinations of the constituents of the germ plasm, or "determinants", through progressive unequal nuclear divisions (also known as "qualitative divisions") that instructed them to differentiate into a particular somatic lineage, say a muscle cell for example (Weismann, 1892). Importantly, the full complement of the unmodified germ plasm was partitioned in germ cells, shielding them from any transfer of heritable information from somatic cells and/or the environment, in effect falsifying the Lamarckian view of inheritance of acquired characteristics (Surani, 2016). The germline-soma segregation and the notion that heritable information is only transferred from germline to soma and not in reverse is also known as the Weismann barrier. Weismann went blind later in life rendering him to primarily focus on theoretical questions in biology, but his monumental work went on to inspire a multitude of researchers in his generation and beyond.

Influenced by Weismann's work, Theodor Boveri and Valentin Haecker sought to microscopically identify the germ cells that carried the germ plasm during embryonic development. Boveri, under the guidance of Richard Hertwig, the latter of whom along with his brother is often referred to as the most brilliant students of Ernst Haeckel, noted that during embryonic development in *Ascaris megalocephala*, a parasitic nematode in juvenile horses, chromosomes were transmitted intact and unaltered in germline cells whereas being dissolved in somatic cells (Boveri, 1895; Satzinger, 2008). In accordance with Weismann's theory, he concluded that only in the cell lineage giving rise to future germ cells were the full set of chromosomes retained to transmit hereditary information to the next generation (Maderspacher, 2008). In deriving this conclusion, Boveri suggested calling those cells from which the above germ cell and somatic cell emerged as stem cells, explicitly noting that he had adopted the term from Ernst Haeckel (Maehle, 2011). Around the same time, Haecker, while working as Weismann's assistant at the time, made the keen observation that cell division during embryonic development in the crustacean *Cyclops* resulted in one cell remaining a germ cell and the other differentiating into a mesodermal lineage (Haecker, 1892). Considered as one of the earliest demonstrations of asymmetric cell division, Haecker also referred to the common precursor of the above daughter cells as stem cell (Lendahl, 2022). Thus, the term stem cell was expanded from its original denotation of the fertilized egg to refer to those cells that give rise to the germline.

By the late nineteenth century, it was becoming increasingly clear among embryologists that a more mechanistically oriented approach was needed to answer biological questions. Weismann's work further stimulated this urge and catalyzed the transition of the field from descriptive embryology to experimental embryology. Previous efforts by the German anatomist Wilhelm His had already laid the initial groundworks for establishing experimental embryology as a new discipline by providing an operational conceptual framework through the combination of the following principles: an experimental design, a precise intent (i.e. proximate causes, rather than phylogenetic), and potential integration with the cell theory (Dupont, 2017). Inspired by His, Wilhelm Roux, a former student of Ernst Haeckel, is often credited to have driven this major shift in approach and is considered to have solidified experimental embryology as a new legitimate branch of study.

In Roux's view, Weismann's theory could be experimentally tested and as he embarked on this task, he inaugurated a new research program more tailored to establishing the causal-analytical relationship of development which he called "Entwicklungsmechanik" or developmental mechanics (Roux, 1885). The major focus of his new program revolved around whether organisms and their parts develop in an autonomous manner independent of extraneous factors and neighboring parts or require interactions with their surroundings and additional stimuli (Schlosser, 2023). Thus, he introduced two opposing terms for each of the two modes of development, "self-differentiation" and "dependent differentiation". Roux himself believed the former to be the predominant mode of development, in line with Weismann's idea of allocation of determinants serving as the basis for cellular lineage specification. In 1888, he directly tested this hypothesis through his famous "pricking experiment" wherein he killed, but did not remove one of the two blastomeres in a developing 2-cell stage frog embryo with a heated needle (Roux, 1888). The result was what he interpreted as being a half-embryo. Hence, he concluded that embryos develop as a mosaic structure of self-differentiating parts with specific sets of determinants accounting for their developmental trajectories, thus apparently validating Weismann's theory (Gilbert, 2000).

Roux's experiment quickly garnered the attention of several other researchers of the time. Among the first was Hans Driesch. In 1891, Driesch performed a similar experiment using a 2-cell stage sea urchin embryo, but unlike Roux, he separated the developing blastomeres by agitation and left them to grow in separate dishes (Driesch, 1891). To his surprise, the result was not two free-floating half-embryos, but two complete, albeit halved in size, larvae. The implication of this seminal study was two-fold. First, the fact that each separated blastomere forms a whole embryo, rather than a half, suggests that an interaction (i.e. Roux's dependent differentiation) must be in place between the two cells to restrict their potential during normal development (Hamburger, 1989). Second, it also indicates that each blastomere retains all the biological information required to self-regulate and form a complete embryo. These findings were soon replicated in other systems, such as in the frog embryo by Thomas Hunt Morgan, who was also an embryologist prior to making his characteristic vocational "deviation" of becoming a geneticist (as described by Sydney Brenner) (Morgan, 1895).

While Roux used the concept of "mechanics" in his newly founded research program as an analogy, Driesch took this in a much more literal manner and dedicated his effort to delineating the mathematical and physical basis of development (Priven & Alfonso-Goldfarb, 2009). Towards the turn of the century, Driesch reached an impasse in formulating a mechanistic explanation for the self-regulative capacity of his sea urchin eggs and attributed this ability to the developing embryo's property as a "harmonious equipotential system" (Driesch, 1899). By this time, Driesch was convinced that the laws of natural sciences of his time could not explain his findings and had quit experimental research to resort to vitalism, to which his longtime friend Morgan lamented that "experimental embryology took a metaphysical turn" (Sander, 1993).

The conceptual framework set forth by these earlier studies was further experimentally substantiated and refined in subsequent work by the next generation of researchers. As early as the beginning of the 1900s, it had become apparent that embryonic development was a complex process regulated both temporally as well as spatially. In 1901, Hans Spemann, who was previously one of Boveri's most talented students, showed that lens (derived from lens placode in the surface ectoderm) formation in the European frog *Rana fusca* depended on contact with the developing optic cup (derived from the diencephalon of the forebrain) (Spemann, 1901). This work not only corroborated the importance of tissue interactions in lens development, but as one of the first experimental demonstrations of inductive events in embryonic development (known today as "induction"), went on to serve as a paradigmatic model for other tissue interactions (Saha, 1991). Subsequently, it came to be understood that self-differentiation and dependent differentiation were not mutually exclusive but were complementary and depended on the spatiotemporal context during embryonic development (Hamburger, 1989). It was also soon realized that while dependent differentiation played a critical role in early tissue development, at some point, normal development continues independent of such stimuli. This gradual transition over developmental time from dependent differentiation to self-differentiation has become to be known as "determination" (Schlosser, 2023). In his later works on studying determination in different regions of the frog gastrula, Spemann noted that determination proceeds in a progressive manner (Spemann, 1918). Accordingly, he introduced the terms "labile determination" to denote when cells of a transplanted tissue can still be redirected to a different fate in their new environment, and "stable differentiation" in which the graft continues in its intrinsic trajectory regardless of its surroundings, in other words, its fate was irreversibly fixed (Hamburger, 1989; Schlosser, 2023; Spemann & Geinitz, 1927). Today, labile and stable determination are known as "specification", and "commitment" or "determination", respectively (Schlosser, 2023; Slack, 1991). As such, several key operational terms used in modern stem cell research in the context of acquisition of cell fate have been established by a series of pioneering research throughout history, each serving as inspiration for the next breakthrough.

As for Haeckel's "Stammzelle", the term made its way into the English language at the end of the nineteenth century. This is commonly credited to Edmund B. Wilson, a longtime friend and colleague of Morgan, who introduced the term in English as "stem cell" through his book *The Cell in Development and Inheritance* after reviewing Boveri and Hacker's work (Ramalho-Santos & Willenbring, 2007; Wilson, 1900). Its meaning however was still restricted to refer to those cells that gave rise to the germline, as first intended by his German colleagues. In the following years, the concept was adopted in hematology in the context of a common hematopoietic precursor cell (Dantschakoff, 1908; Maximow, 2009; Pappenheim, 1896). After several decades of heated debate around the notion of one single cell capable of giving rise to all the different types of blood cells, conclusive evidence was provided by a group of Canadian scientists in the 1960s (Maehle, 2011; Ramalho-Santos & Willenbring, 2007).

In 1961, Ernest Armstrong McCulloch and James Edgar Till injected bone marrow cells into lethally irradiated mice and noted the formation of small colonies or nodules in the spleens of the hosts, the number of which linearly correlated with that of transplanted cells (Till & McCulloch, 1961). They also observed that these colonies contained cells that differentiated into erythrocyte, granulocyte, as well as megakaryocyte lineages. Shortly after this initial observation, their next study, with graduate student Andrew J. Becker as lead author, revealed that the colonies represented clones originating from a single transplanted cell as confirmed via abnormal chromosomal marker analysis (Becker et al., 1963). In the same year, collaborating with Louis Siminovitch, who was previously Till's post-doctoral mentor, they further demonstrated that these cells also possess the capacity for self-renewal as shown by the formation of new colonies following serial transplantation into a secondary recipient (Siminovitch et al., 1963). Collectively, these landmark studies are attributed to establishing the defining characteristics of stem cells—the ability to proliferate and self-renew, and give rise to other specialized cells through differentiation. The stem cells discovered by McCulloch, Till, and colleagues are known today as hematopoietic stem cells (defined as multipotent) and since this discovery, a hierarchical organization of stem cells has been established according to their differentiation potential (summarized in Jaenisch & Young, 2008).

3. The road to pluripotency

3.1 Lessons learned from somatic cell nuclear transfer

An important implication of Weismann's idea of qualitative division is that during development, cells progressively and irreversibly lose their intrinsic potency such that eventually each cell is only able to produce a progeny of one specific cell type (Sánchez Alvarado & Yamanaka, 2014). The notion of gradual and irreversible reduction in potency during development was later reiterated by the British embryologist and geneticist Conrad Hal Waddington in his famed book *The Strategy of the Genes* (Cerneckis et al., 2024; Waddington, 1957). In what is known today as Waddington's epigenetic landscape (Fig. 1, *left*), he depicted cellular differentiation as a ball (representing a cell in an embryo) successively rolling downhill a series of valleys with branching points (developmental trajectories) at those of which it will make decisions until reaching the bottom (terminally differentiated state). Of course, we now know that the ball can be rolled back up the hill

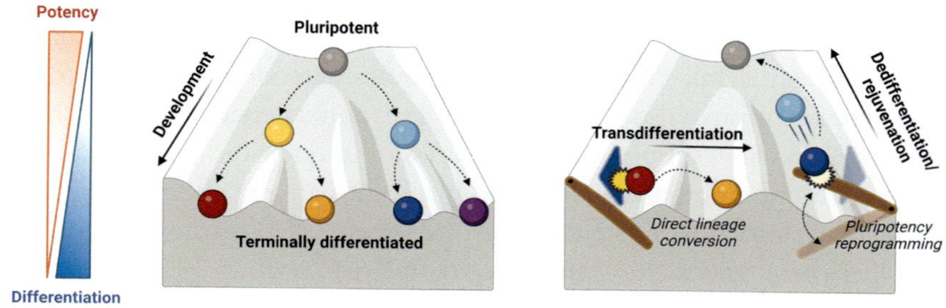

Fig. 1 *Left:* Schematic diagram of Waddington's epigenetic landscape. *Right:* Simplified illustration incorporating the notion of cell fate plasticity into the traditional description of Waddington's epigenetic landscape. *Adapted from Waddington, C.H., 1957. The Strategy of the Genes; A Discussion of Some Aspects of Theoretical Biology. Allen & Unwin, London. Image created via BioRender.com.*

(i.e. dedifferentiation, e.g. pluripotency reprogramming) or traverse ridges in horizontal movements (i.e. transdifferentiation, e.g. direct lineage conversion), but at the time it was unclear whether differentiation indeed progressed as a "one-way street" involving irreversible changes and/or loss of genetic information. Quite interestingly, this modified view of the plastic nature of cell fate is today often depicted as the pinball model of development (Fig. 1, *right*) (Chen et al., 2018; Goldberg et al., 2007; Sareen and Svendsen, 2010).

Intriguingly, the earliest evidence suggestive of cellular plasticity can also be traced back to the works of the very same person that devised the term "Stammzelle"—Ernst Haeckel (Sánchez Alvarado & Yamanaka, 2014). Predating Driesch's experiment by more than two decades, Haeckel observed in 1869 that a developing *Siphonophore* embryo split into two can grow to form complete larvae and this work is considered as the first demonstration of pluripotency of embryonic cells (Haeckel, 1869; Sánchez Alvarado & Yamanaka, 2014; Simsek & Özbudak, 2022). In a later study, Ethel Browne, as a graduate student under Morgan and Wilson, showed that transplanting the mouth tissue (hypostome) of a developing hydra embryo into the body wall of another hydra resulted in the development of an ectopic twin in the grafted region (Browne, 1909). Similar findings were also reported in salamander embryos by Spemann and Hilde Mangold through their organizer experiment discussed in subsequent sections (Spemann & Mangold, 1924). Collectively, these studies provided conclusive evidence that cell fate in embryonic cells is plastic and can be subjected to experimental manipulation (Sánchez Alvarado & Yamanaka, 2014).

A key conceptual prerequisite for somatic cell plasticity is that all cells possess the same full complement of genetic material required to form a complete organism (also known as nuclear equivalence) (Lensch, 2009). One conceptually simple way to address this question was to take the nucleus from a differentiated cell and put it into an egg and see whether it can develop into a complete organism (Maartens, 2017). Interestingly, such an experiment had already been contemplated in 1886 by Rauber who attempted to transfer the nucleus from a toad cell into a frog egg, although whether he actually conducted such an experiment is somewhat questionable as he merely reported that there was no development (Gurdon, 2017; Rauber, 1886). This idea was later revisited by Spemann in what he described would be a "fantastical experiment", wherein he proposed to transplant the nucleus of a differentiated cell into an egg to determine whether it would regain developmental potential (Spemann, 1938). Spemann was never able to perform this experiment however, due to technical limitations involving the transplant of free nuclei into other cells. This experimental challenge was overcome 14 years later by two American scientists, Robert William Briggs and Thomas Joseph King. In 1952, Briggs and King successfully transplanted blastula nuclei into enucleated oocytes in the North American leopard frog *Rana pipiens* that developed into swimming tadpoles (Briggs & King, 1952). In contrast to the blastula nuclei however, they subsequently showed that normal development was markedly decreased in later stage nuclei (e.g. gastrula or neurula) and only resulted in abnormal embryo development (Briggs & King, 1957). These results were interpreted as an indication that as development proceeds, the nuclei and genes undergo irreversible changes that repress their developmental potential (Gurdon & Byrne, 2003). Briggs and King's studies were soon followed up by a graduate student at Oxford, John Gurdon, under the advisership of Michail Fischberg (whose academic genealogy can also be traced back to Theodor Boveri) (De Robertis, 2014).

A rather remarkably inspiring story (with a touch of irony!) could be told of Gurdon's early academic years. Gurdon's first encounter with biology was not a kind one that can be summarized by a report written by his teacher: "...Gurdon has ideas about becoming a scientist... this is quite ridiculous. If he can't learn simple biological facts, he would have no chance of doing the work of a specialist, and it would be a sheer waste of time both on his part and of those who would have to teach him." (Gurdon, 2008). His school went on to conclude, "One thing that is absolutely clear is that you are not a scientist." ("A Conversation with John Gurdon & Shinya Yamanaka," 2020). Perhaps it was an important moment for stem cell research that went unnoticed when he joined Fischberg's laboratory after being declined admission into a Ph.D. program in entomology. The reader is referred to a personal historical account by Gurdon on these early years as well as his later career (Gurdon, 2006).

Building on Briggs and King's experiments, Gurdon conducted his somatic cell nuclear transfer (SCNT) studies in the African clawed frog *Xenopus laevis* that had been recently introduced to embryology by Fischberg and the Dutch experimental embryologist Pieter Nieuwkoop, the latter of whose work will be discussed later in this chapter (Fischberg et al., 1958; Gurdon & Hopwood, 2000). In the following years, he studied the developmental capacity of nuclei transplants derived from various stages ranging from blastula to intestinal cells of larvae, at first, confirming Briggs and King's observation that the ability of the transplanted nuclei to drive development declines with developmental time (Gurdon & Byrne, 2003). Importantly however, albeit in limited instances, he observed that nuclei from highly differentiated cells such as those from the intestinal epithelium in tadpoles, were able to promote normal development into adult frogs following transfer into enucleated eggs (Gurdon, 1962a, 1962b). In subsequent studies, Gurdon showed that the nuclear transplant tadpoles were able to develop into fertile adult frogs, conclusively demonstrating the generation of mature animals from somatic cell nuclei (Gurdon & Uehlinger, 1966). These pioneering studies provided support to the idea that differentiation may not necessarily involve irreversible changes/loss in genomic content (i.e. all cells contain the same genes), but rather it may depend on the change in expression (Gurdon & Byrne, 2003). According to Gurdon, at the time he was not entirely sure what implications such reprogramming in amphibia would have on the field and his findings were not always met with enthusiasm as evident by some criticisms: "a graduate student, working almost alone should not be able to repeat the results of well-established and highly respected workers Briggs and King" ("A Conversation with John Gurdon & Shinya Yamanaka," 2020; Gurdon, 2013). Nonetheless, this seminal work laid the foundation for subsequent "cloning" experiments and underscores key concepts of epigenetics today.

Equivalent studies in mammalian counterparts took longer owing to, in part, the markedly smaller size of mammalian eggs (120–150 μm) compared to those of amphibia (>1 mm) (Campbell, 2002). The first report of cloning in mammals came from Wilmut and colleagues in 1997 when they introduced the world's most famous sheep—"Dolly", as christened by one of its caretakers—by transplanting the nucleus derived from the mammary gland cell of a 6-year-old *Fin Dorset* ewe (Wilmut et al., 1997). Importantly, this monumental study served as proof-of-principle of the full developmental capacity of the nuclear genome (Daley, 2015). This was soon followed by cloning in mice by Wakayama et al., providing further validation of cloning in mammalian species through the transfer of adult somatic nuclei (Wakayama et al., 1998). Collectively, these studies demonstrated that the nuclei of terminally differentiated cells can be reprogrammed to a younger, less differentiated state, or even to totipotency such as the case with Dolly and the first cloned mice. Indeed, it was also shown that fertile adult mice can be generated using the nuclei of highly specialized cells such as mature B or T lymphocytes (Hochedlinger & Jaenisch, 2002). Importantly, by showing that the cells of the cloned animals carried the fully rearranged immunoglobulin or T cell receptor genes, this study provided additional confirmation that differentiated cells are indeed genetically totipotent and definitively proved that such reprogramming was not due to residual tissue stem cells (Hochedlinger & Jaenisch, 2006; Yamanaka & Blau, 2010).

3.2 From teratocarcinoma to embryonic stem cells

As early as the late 19th century, pathologists observed that certain tumors were comprised of a mixture of different types of cells reminiscent of both mature and fetal tissue (Sell, 2010). Reflective of its haphazard composition, Rudolf Virchow referred to such tumors as "teratoma", derived from the Greek words *teratos* and *onkoma* meaning "monster" and "swelling", respectively (Virchow, 1858). Later, the American pathologist James Ewing described teratomas as complex tumors consisting of tissues and organs of multiple germinal layers (Ewing, 1919, 1911).

Clinically speaking, teratomas are mostly found in gonads and are relatively rare in humans (Damjanov & Wewer-Albrechtsen, 2013). Despite having attracted the attention of pathologists for decades due to their unusual histological properties, they were rare in rodent experimental models and thus were difficult to study (Bustamante-Marin & Capel, 2023). An important breakthrough for studying this peculiar type of tumor was made by two researchers at the Jackson Laboratory in Bar Harbor Maine. In 1954, Leroy Stevens and Clarence Cook Little reported that the inbred strain 129 mice showed an increased incidence of spontaneous testicular teratoma (1%) (Stevens & Little, 1954). In this study, Stevens and Little also demonstrated that these tumors can be maintained through serial transplantations and attributed this capacity to the presence of pluripotent cells—"Pluripotent embryonic cells appear to give rise to both rapidly differentiating cells and others which, like themselves, remain undifferentiated" (Stevens & Little, 1954). This was the first description of the modern definition of embryonic stem cells (ESCs) (Evans, 2011). Following this discovery, another significant study by Larry Kleinsmith and Barry Pierce demonstrated that a single cell isolated from a tumor, when transplanted to another mouse via intraperitoneal injection, can produce another teratocarcinoma consisting of a wide range of differentiated cells (Kleinsmith & Pierce, 1964). This work conclusively identified the presence of a unique tumor stem cell (referred to as embryonal carcinoma cells) that can propagate indefinitely and give rise to multiple cell types. In addition, studies showing the derivation of teratomas and retransplantable teratocarcinomas from ectopic transplantation of mouse embryos further suggested a close relationship between embryonal carcinoma (EC) cells and developing early embryos (Solter, 2006; Solter et al., 1970; Stevens, 1970). This notion was further strengthened by the identification of marker proteins such as alkaline phosphatase (AP)—commonly expressed in mouse and human EC cells, cells of the inner cell mass of mouse blastocysts, and germ cells—as well as cell stage-specific embryonic antigen 1 (SSEA-1) (Benham et al., 1981; Bernstine et al., 1973; Gooi et al., 1981; Solter, 2006; Solter & Knowles, 1978). Soon after, Ralph Brinster substantiated the developmental capacity of EC cells by demonstrating that EC cells can participate in embryonic development to form chimeric offspring when injected into early mouse blastocysts (Andrews, 2024; Brinster, 1974). Similar findings on the formation of chimeric animals were subsequently reported by others to provide additional corroboration (Mintz & Illmensee, 1975; Papaioannou et al., 1975).

Building on these studies, two groups independently reported on the successful derivation of pluripotent cell cultures directly from the inner cell masses (ICMs) of mouse pre-implantation blastocysts (Evans & Kaufman, 1981; Martin, 1981). These ESCs, as coined by Gail Martin, shared the previously described mouse EC cell morphology of high nucleus-to-cytoplasm ratio, prominent nucleoli, and scant cytoplasm—a generic morphological description still used today for pluripotent stem cells (differences in colony morphology discussed below) (Martin & Evans, 1974). They were also shown to express the surface marker SSEA-1, formed embryoid bodies (EBs), generated teratomas upon ectopic transplantation, and produced chimeras when transferred to blastocysts. Shortly after these findings, Bradley and colleagues further demonstrated that the chimerism extended beyond the soma and included the germline (Bradley et al., 1984). In parallel with these discoveries, refinement in culture conditions, most notably the identification of leukemia inhibitory factor (LIF), enabled the long-term in vitro maintenance of mouse embryonic stem cell (mESCs) pluripotency (Smith et al., 1988). Collectively, these monumental studies signaled the beginning of modern stem cell research. The series of assays performed in the above studies, including the formation of EBs, teratomas, and chimera (for mouse cells) have long served as gold standards for validating the pluripotency and differentiation capacity of putative pluripotent stem cells (PSCs). Various other techniques have been developed to complement these conventional methods to date (Allison et al., 2018). The reader is also recommended to refer to a recent review highlighting additional surface markers that have been identified over the years (Andrews & Gokhale, 2024).

As a logical extension to the series of work discussed above, attention turned to whether corresponding cells could be generated from other species. In the 1990s, James Thomson, who received his VMD/Ph.D. under Davor Solter, was working at the Wisconsin Regional (now national) Primate Research Center to derive primate ESCs in the aims of establishing a model system that better recapitulates human development than mESCs. In 1995, Thomson reported on the first derivation of non-human primate (NHP) ESCs from rhesus monkey blastocysts and then from those of marmosets in the following year (Thomson et al., 1996, 1995). While somewhat overlooked at the time of publication, this was a significant breakthrough as it was not certain whether ESCs could be derived outside of rodents (Stojković & Daher, 2008). Remarkably, NHP ESCs were shown to differentiate in vitro and formed teratomas upon transplantation into immunodeficient mice (Thomson et al., 1996, 1995). Importantly, several aspects that differed from mouse EC cells/ESCs were unveiled including the ineffectiveness of LIF in maintaining pluripotency and surface marker expression patterns that bore closer resemblance to human EC cells. In addition, unlike mESC colonies that display a dome-like structure with indistinct borders, NHP ESCs exhibited a flatter morphology with clear borders that are more associated with human EC cells. To date, more than 80 NHP ESC lines (and over 100 induced pluripotent stem cell lines) from various species have been established and continue to serve as valuable tools for preclinical research as reviewed by Anwised and colleagues (Anwised et al., 2023).

Interestingly, according to Thomson, he initially did not consider extending his work to human counterparts as he believed that it would surely be done soon by others (Gitschier, 2008). To his surprise, months went by, and he decided to do the experiments himself. Aided by their previous experience with NHP ESCs, Thomson and his team published their work on the first derivation of human ESCs (hESCs) in 1998 (Thomson et al., 1998). In their study, Thomson and colleagues isolated the ICMs of 14 blastocysts obtained from donated in vitro fertilization embryos and derived five independent hESC lines, commonly known today as H1, H7, H9, H13, and H14. These cell lines expressed cell surface markers corresponding to undifferentiated NHP ESCs and human EC cells and formed teratomas consisting of derivatives of all three germinal layers upon injection into immunodeficient mice. Furthermore, in congruence with NHP ESCs, hESCs showed extensive differentiation when cultured without feeder layers, regardless of the presence or absence of LIF.

This seminal work was significant on multiple levels. First and foremost, it provided researchers with a platform to study human development in a hypothesis-driven manner starting from uncommitted precursors which had not been previously possible (Lensch, 2009). In addition, the advent of PSCs of human origin opened new possibilities for studying diseases and drug discovery as these cells could serve as the starting material for generating specific cell types and bypass the need for interspecies extrapolation. Finally, the derivation of hESCs also facilitated the advances in developing optimal culture conditions (e.g. addition of basic fibroblast growth factor as opposed to LIF in mESCs) for human pluripotent stem cells (hPSCs) (Xu et al., 2005). Indeed, several chemically defined culture media have been developed that are still widely used today. Some notable examples include the first reported TeSR1 (Tenneille Serum Replacer 1) and modified TeSR1 (mTeSR1) media formulated by Tenneille Ludwig and colleagues from James Thomson's group, as well as the Essential 8 (E8) medium from the same laboratory (Chen et al., 2011; Ludwig et al., 2006a, 2006b).

3.3 Induced pluripotent stem cells

Although the derivation of hESCs was undoubtedly a groundbreaking achievement, there were several limitations that dampened some of the initial excitement surrounding its discovery. One of the most often cited challenges hampering its translational potential are the ethical concerns and legal constraints associated with the use of human embryos. This complex landscape also informs deliberations on justifying the derivation of additional hESC lines. Indeed, from 2008 to 2016, only 21 hESC lines were actively used in public research wherein the H1, H7, and H9 lines accounted for nearly 80% of all publications involving the use of hESCs (Guhr et al., 2018). In addition, as most lines came from presumed otherwise healthy subjects, except for those derived from embryos confirmed to harbor disease-associated mutations via preimplantation genetic diagnostics, addressing specific questions in areas such as human disease modeling proved difficult. Further, concerns involving allogenic immune rejections presented another hindrance to realizing hESC therapeutic potential. One way to circumvent these obstacles was to generate hPSCs from autologous cells, the usage of which would have reduced ethical and legal entanglements and could be readily accessible for the derivation of multiple lines representing a wide spectrum of genetic backgrounds.

Conceptually, this could be achieved by restoring pluripotency in somatic cells and SCNT studies dating back to Gurdon's reprogramming in amphibia had already provided evidence for the reversibility of somatic cell fate. However, at this time the question arose of how could one reprogram an intact, terminally differentiated somatic cell back to pluripotency?

Critical pieces of information for answering the above question could be found from studies conducted in the late 1980s on a mammalian transcription factor, namely, myogenic determination gene number 1 (MyoD). In 1987, Davis and colleagues demonstrated that the over-expression of MyoD can convert murine fibroblasts into muscle-like cells (Davis et al., 1987). Shortly after this discovery, Weintraub et al. further showed that the ectopic expression of this transcription factor in various cell types, ranging from pigment, nerve, fat, to liver cells can drive gene expression towards a myogenic differentiation pathway (Weintraub et al., 1989). These studies demonstrated the importance of transcription factors in determining cell fate and how they can regulate downstream gene expression to direct developmental decisions. In addition, by the turn of the century, key transcription factors responsible for maintaining pluripotency and self-renewal in culture have been identified using ESCs including Oct4, Sox2, and Nanog (derived from the island "Tir na nOg" in Celtic myth meaning "Land of the Young") (Boyer et al., 2005; Chambers et al., 2003; Nichols et al., 1998). Collectively, these studies provided the operational framework for reprogramming, or inducing, pluripotency in intact somatic cells.

Although such a notion was conceived, the prevailing view was that this would be an extremely complicated task, likely to take 20–30 years, if not more, as Shinya Yamanaka had initially thought ("A Conversation with John Gurdon & Shinya Yamanaka," 2020). As the story goes, he also commented (jocosely), that when he first began recruiting graduate students upon setting up his laboratory at the Nara Institute of Science and Technology in 1999, he did not mention the predicted timeline of the study as he was afraid that it would scare them away. One of the three students that joined his lab then was Kazutoshi Takahashi (see below). James Thomson also shared a similar opinion on the prospects of reprogramming intact human somatic cells wherein upon embarking on this work he had told his postdoctoral fellow Junying Yu, "...we have to try this, even though it probably isn't going to work. And it's probably a 20-year problem" (Gitschier, 2008). Fortunately for science, there was a change in schedule and these four researchers would later become the lead authors and principal investigators of the first publications on human induced pluripotent stem cells as discussed below.

The first major breakthrough came in 2006 when Yamanaka and Takahashi reported on the first successful reprogramming of mouse fibroblasts to an embryonic-like state (Takahashi & Yamanaka, 2006). They started with a list of 24 candidate genes associated with maintaining pluripotency in ESCs and introduced them into mouse fibroblasts via retroviral delivery. Like the proverbial search for the needle in a haystack, they tested different combinations of these factors until they discovered that four transcription factors, namely, Oct3/4, Sox2, Klf4, and c-Myc (collectively known as OSKM or "Yamanaka factors") were sufficient to induce pluripotency. The resulting cells displayed remarkable resemblance to mESCs and were termed induced pluripotent stem cells (iPSCs), wherein Yamanaka whimsically borrowed the lower-case "i" from "iPod", in the hopes that the name would be easier to remember (Eguizabal et al., 2019).

Following this critical discovery, multiple labs across the world flocked to race towards the "Holy Grail" of generating human iPSCs (hiPSCs). The first labs to publish their work were those of Yamanaka and Thomson, who independently published their findings in Cell and Science, respectively, in November 2007 (Takahashi et al., 2007; Yu et al., 2007). Two other groups, those of George Daley and Kathrin Plath, also had their work submitted prior to Yamanaka and Thomson's publications, reflecting the robustness and reproducibility of this technique (Goldman, 2008; Lowry et al., 2008; Park et al., 2008b). Building on their success with mouse iPSCs, Yamanaka's group used the same cocktail of Yamanaka factors while Thomson's group used Oct4, Sox2, Nanog, and Lin28. In both cases however, hiPSCs derived from fibroblasts were karyotypically normal and expressed remarkable similarity to hESCs in terms of their morphology, gene expression, surface antigen profile, and capacity to generate cells of all three germ layers in vitro and in vivo. Two key drawbacks at the time were the low reprogramming efficiency and safety issues involved with the use of retroviral or lentiviral vector-mediated delivery of reprogramming factors. To address these concerns, several integration-free methods have been developed to date including the use of adenovirus or Sendai virus-mediated delivery, episomal plasmids, RNA, protein, and chemical reprogramming (Karami et al., 2023; Liu et al., 2020; Wang et al., 2023). In addition, the notion that the ectopic expression of a quartet of transcription factors can reprogram somatic cells to pluripotency also paved the way for transdifferentiation studies involving the direct conversion of fibroblasts to neurons, hepatocytes, and cardiomyocytes, to name a few (Ieda et al., 2010; Sekiya & Suzuki, 2011; Vierbuchen et al., 2010). The reader is referred to the following review that describes the current state-of-art of transdifferentiation or direct cell reprogramming (Wang et al., 2021).

The advent of iPSC technology revolutionized the fields of biological and biomedical sciences and ushered in a new era of stem cell-based research. The most immediate impact was that it liberated the field from most of the ethical and legal concerns regarding the use of human embryos in deriving hPSCs. Furthermore, hiPSCs unlocked new opportunities in drug discovery, toxicity screening (particularly relevant to this volume), regenerative medicine, and cell therapy by offering a readily accessible platform for obtaining patient-specific autologous PSCs. In principle, the ability to derive patient-/disease-specific hiPSCs also meant that researchers would be able study not only monogenic, but also polygenic diseases with incomplete penetrance as well as sporadic onsets involving environmental effects as they possess all the genetic elements that confer predisposition to disease development. Further, since hiPSCs can be generated from adults, this enables production of iPSC lines for which post-birth/adult-onset health conditions could be assessed and quantified and associated with the cell lines for study. Thus, hiPSCs may be utilized to study individual-level or group-level susceptibility to specific and characterized environmental chemical exposures (Kumar et al., 2012). Consequently, within the first two years following its discovery, multiple studies reported on the derivation of patient-specific iPSC lines. One of the earliest examples include the first large repository of disease-specific hiPSCs generated by In-Hyun Park and colleagues from George Daley's lab (former of whom now heads his own group at Yale) that encompassed lines from 1-month to 57-year-old patients covering 10 different disorders (Park et al., 2008a). Importantly, the iPSC lines, except for those from Parkinson disease and juvenile onset diabetes patients, were confirmed to carry disease-specific molecular features such as expanded CAG repeats in Huntington disease. Around the same time, Dimos et al. showed that iPSCs derived from fibroblasts of familial amyotrophic lateral sclerosis (ALS) patients can be differentiated into motor neurons, the major cell type degenerated in ALS (Dimos et al., 2008). Subsequent studies by others provided further proof-of-principle that patient-specific iPSCs can recapitulate disease-relevant phenotypical characteristics when differentiated into implicated cells in culture (Brennand et al., 2011; Ebert et al., 2009; Lee et al., 2009; Marchetto et al., 2010; Nguyen et al., 2011). A major challenge during this time however was that there was a lack of robust protocols that can faithfully differentiate iPSCs into a highly pure population of desired cell type(s) and thus, many of the early studies involved less characterized cultures containing variable cell populations (Passier et al., 2016). In the next section, we turn our attention to directed differentiation of hiPSCs to the neural lineage.

4. To have some nerve (Cells): Differentiation to neural lineages

Two distinct strategies have been developed to date for the derivation of neural cells, namely, directed differentiation and transdifferentiation. For the remainder of this discussion, we focus solely on the former, but the reader is referred to the following review and key papers regarding the topic of direct neuronal reprogramming (Bocchi et al., 2022; Pang et al., 2011; Vierbuchen et al., 2010).

4.1 Directed differentiation

Initial studies highlighting the potential of utilizing hPSCs for generating neural cells involved their isolation and expansion from spontaneously differentiated cell aggregates or EBs (Reubinoff et al., 2001; Zhang et al., 2001). These early studies provided important proof-of-concept and initial validation for the preconceptions of harnessing hPSC technology to study human brain disorders and promote translational research. However, early differentiation methods were prone to high variability resulting in heterogenous cultures of mixed cell types that differ in their relative abundance and developmental stage. Thus, an efficient method that can generate desired cell types in a more controlled and temporally regulated manner was warranted.

Despite the above-mentioned challenges, hESC-derived EBs provided valuable insight into identifying key molecular signaling processes and growth factors such as transforming growth factor β1, activin-A, and bone morphogenic protein 4 (BMP-4), involved in the differentiation of specific germ layers and their corresponding derivatives (Schuldiner et al., 2000). This led to the idea that it may be possible to control cellular differentiation to a particular lineage by manipulating associated signaling pathways through the treatment of specific growth factors. This notion formed the basis of "directed differentiation" that can be defined as: an in vitro differentiation strategy in which PSCs are differentiated towards a specific cell fate by the addition of exogenous growth factors and/or small molecules in a concentration- and time-dependent manner that recapitulates in vivo development (Cohen & Melton, 2011). The current gold standard for the directed differentiation of PSCs to neural lineage cells is a method called "dual SMAD inhibition (dSMADi)" previously developed by Lorenz Studer's group (Chambers et al., 2009). Indeed, this robust protocol resolved the difficulties associated with earlier studies and serves as the fundamental basis for deriving a wide array of neural cell types. To fully appreciate this remarkable method and the beautiful developmental biology embedded within though, one must revisit the works of Hans Spemann and Hilde Mangold.

4.2 The Spemann-Mangold organizer

In 1924, Spemann and his graduate student Hilde Mangold (previously Hilde Proescholdt) reported on one of the most famous studies in experimental embryology referred to today as the organizer experiment (Spemann & Mangold, 1924). As part of Mangold's Ph.D. thesis, this work involved transplantation experiments of salamander embryos in the gastrula stage, wherein they wisely chose to use differently pigmented gastrula to distinguish between donor and host tissue. During this time, Mangold performed 274 heteroplastic transplantation experiments wherein she grafted the dorsal blastopore lip (region where gastrulation begins) from an unpigmented donor gastrula to the ventral side of a pigmented host embryo (Fäßler, 2013). 29 of these operated embryos survived to reach stages compatible for histological analyses (Sander & Faessler, 2001). Findings from the longest surviving embryo (reaching the tail bud stage), "Um 132", wherein "Um" stands for blastopore, is epitomized in today's developmental biology textbooks as the organizer experiment (Hamburger, 1989).

In the roughly 15% of the surviving embryos, Mangold observed that the host embryo developed a secondary body axis, an ectopic twin on the ventral side composed of a second nervous system that was properly patterned along its anterior-posterior (A-P) and dorsal-ventral (D-V) axes. Importantly, Mangold described that the donor graft mostly contributed to the notochord while the surrounding host cells gave rise to the rest of the CNS. Spemann and Mangold concluded that the graft "induced" the surrounding host tissue to change fate and "organized" the proper A-P/D-V axes in the induced tissue (De Robertis, 2006). Owing to these properties, they named the dorsal blastopore lip the organizer. This landmark study formed the basis of the present-day notion that development proceeds through a series of cell-cell inductions (De Robertis, 2006). Today, the term organizer is used in a more comprehensive manner to denote a special group of cells that can determine the fate of surrounding tissue through the release of chemical signals (Kiecker & Lumsden, 2012). Tragically however, just a few months following this seminal discovery, Hilde Mangold died at the young age of 26 in a kitchen stove accident while warming up milk for her newborn son and did not live to witness the publication of her influential paper (De Robertis, 2009). As Nobel prizes are not posthumously awarded, only Spemann received the Nobel Prize in Physiology/Medicine in 1935 for this work. The reader is referred to a memoir by Viktor Hamburger on the firsthand recollection of his days as a graduate student in Spemann's lab alongside Mangold with detailed description on the scientific discoveries leading to the organizer experiment and thereafter (Hamburger, 1989).

The organizer experiment was later followed up by multiple studies including that of Johannes Holtfreter, a student of Spemann who later worked in Otto Mangold's lab (Hilde Mangold's husband, who himself was one of Spemann's students). Importantly, he made the striking discovery that organizer tissues devitalized by heat, cold, or desiccation still possess its inductive properties when transplanted into amphibian embryos (Holtfreter, 1933). The startling observation that even "dead" organizers retain their inductive capacity raised the idea that the neural inducing factors must be chemical signals. Over the following years, researchers around the world rushed to elucidate the identity of such inductive agents, but their efforts were hampered by, in large part, limitations in biochemical methods of the time. The field had hit a critical impasse as noted by Waddington, "...until this difficulty can be surmounted it appears impossible to discover the true nature of the natural evocator (or inducer)..." (Waddington, 1938).

It was not for another six decades in the early 1990s that the molecular identities of neural inductive signals from the organizer started to be unveiled through studies in Xenopus. As one of the first of such molecules to be identified, noggin (named as such by its ability to induce excessive head development) was shown to be able to induce neural tissue in Xenopus ectodermal explants (also called animal caps) when injected as RNA or purified soluble proteins (Lamb et al., 1993; Smith & Harland, 1992). Similar findings were reported in the identification of follistatin and chordin (Hemmati-Brivanlou et al., 1994; Sasai et al., 1995). In parallel with these discoveries, studies showed that disrupting activin (member of the TGF-β family) signaling in Xenopus ectodermal explants, via expression of dominant negative activin receptor, resulted in neuralization even in the absence of neural inducing signals (Hemmati-Brivanlou & Melton, 1994, 1992). In addition, previous studies had also shown that prolonged dissociation of Xenopus animal caps results in the conversion of cell fate from that of the epidermis to CNS after reaggregation, also in the absence of neural inducing signals (Godsave & Slack, 1989; Grunz & Tacke, 1989; Sato & Sargent, 1989). As cell dissociation and the expression of dominant negative mutants have the effect of interfering with cell-cell communication, these studies prompted two important notions—that (1) development proceeds, in the absence of additional signals, to a "default" neural fate, (2) by negative signaling (Hemmati-Brivanlou & Melton, 1997). Indeed, other studies further demonstrated that noggin and chordin blocks BMP signaling by directly binding to BMP-4 (Piccolo et al., 1996; Zimmerman et al., 1996). Collectively, as activin/nodal signaling leads to mesoendodermal specification and BMPs are potent epidermal inducers, these studies demonstrated that neural inductive agents function by, not so much as inducing neural fate per se, but by suppressing non-neural fate. In this sense, the term "neural induction" may even seem counterintuitive; after all, it is the epidermis that is induced (Hemmati-Brivanlou & Melton, 1997). Thus, in contrary to the initial idea that neural inducing molecules would act as positive signals, it came to be established that they exert their neural inducing activity through inhibitory mechanisms by blocking activin/nodal and BMP signaling, both of which relay their downstream signaling processes via a distinct class of SMAD proteins (receptor-regulated SMAD, R-SMAD) (Muñoz-Sanjuán & Brivanlou, 2002). Of note, R-SMADs consist of SMAD1, 2, 3, 5, and 8 and can be divided into two subfamilies based on their signaling specificity wherein SMAD2/3 respond to TGF-β and activin signaling, and SMAD1/5/8 primarily respond to BMPs (Shi & Massagué, 2003).

These findings provided the conceptual basis for the directed differentiation of hPSCs to neural cells via dSMADi that elegantly recapitulates the in vivo developmental principles unveiled by the aforementioned studies. In their influential paper, Chambers and colleagues used the small molecule SB431542 (TGF-β/activin/nodal inhibitor) and recombinant noggin (BMP inhibitor) to differentiate hPSCs to the neural lineage; hence the name "dual" SMAD inhibition (Chambers et al., 2009). They demonstrated that treating hESCs with SB431542 and noggin directed more than 80% of the cells to a neuroectodermal fate within seven days as confirmed by the development of $PAX6^+$ neuroepithelial cells. In remarkable accordance with in vivo development, co-treatment of just two inhibitors of SMAD signaling was sufficient to derive neural progeny. As such, the greatest advantage of this method lay in its efficiency, simplicity, and reproducibility in generating neural culture and as the authors predicted then, dSMADi has become the standard strategy in directing the differentiation of hPSCs to neural cells. Finally, while the initial method used recombinant noggin, several alternative BMP inhibitors have become available to date including LDN193189, dorsomorphin, as well as DMH1 previously tested by our group (Neely et al., 2012). In the next section, we briefly discuss how specific neuronal subtypes can be differentiated from hPSCs following the principles of regional patterning.

4.3 Regional patterning

Gastrulation is a milestone event during in vivo embryonic development that sets the stage for axial body plans and cellular diversification through the formation of the three germ layers—ectoderm, mesoderm, and endoderm—from pluripotent epiblast cells (Zhai et al., 2022). Perhaps not surprisingly, in line with its operational framework of mimicking in vivo embryonic development, most directed differentiation protocols begin by guiding hPSCs to one of the three germinal layers (Williams et al., 2012). Another central event during gastrulation is the formation of the notochord, a transient rod-shaped mesodermal structure that extends along the A-P axis of the embryo. The notochord is a defining feature of all chordates situated in the midline of the embryo and thus establishes the A-P, D-V, as well as the left-right axes (Darnell & Gilbert, 2017). In vertebrates, the notochord arises from the dorsal organizer (node in mice/humans, and Spemann-Mangold organizer in amphibia as discussed previously) and secretes chemical signals to neighboring tissues to provide fate and positional information (Stemple, 2005).

"Neural induction", the first step in vertebrate neurodevelopment, is initiated by the secretion of TGF-β and BMP antagonists from the node/notochord to the immediately overlying ectoderm (i.e. neuroectoderm). These antagonizing signaling renders the ectoderm to follow its "default" anterior neural fate and results in the formation of the neural plate or the neuroepithelial layer, the process of which is recapitulated in vitro by dSMADi as described previously (Chambers et al., 2009; Ozair et al., 2013). In addition, by carefully manipulating BMP, TGF-β, FGF, and WNT signaling, it is also possible to differentiate hPSCs into all four major ectodermal lineages including neuroectoderm, neural crest, cranial placode, and non-neural ectoderm (Tchieu et al., 2017).

During the next stage, "neurulation", the flat neural plate thickens and invaginates to sequentially form the neural groove and ultimately the neural tube. Consequently, the initial mediolateral patterns are converted to a D-V polarity wherein cells of the medial region become ventral (floorplate) and those of the lateral part become dorsal (roofplate) (Kiecker & Lumsden, 2012). Following neurulation, the neural tube undergoes an elaborate process of "regional patterning" by the concerted action of spatiotemporal morphogen gradients along the rostral-caudal (R-C) and D-V axes to give rise to the prosencephalon (forebrain), mesencephalon (midbrain), rhombencephalon (hindbrain), and the spinal cord. The prosencephalon is further divided into the telencephalon and the diencephalon. The former consists of the isocortex, hippocampal formation, olfactory areas, cortical subplate, and cerebral nuclei, whereas the latter gives rise to the thalamus, hypothalamus, and retina (Yao et al., 2023). Likewise, the rhombencephalon is divided into the metencephalon that becomes the cerebellum and pons, and the myelencephalon that forms the medulla. Each of these main brain structures contain multiple subregions that are occupied by specific neural cell types that contribute to overall brain function (Zeng, 2022).

As can be expected from above, the finely tuned choreography of morphogenic cues enables the generation of a myriad of specific neural cell populations along the R-C/D-V axes. How this remarkable cellular diversity can be achieved from a uniform layer of neuroepithelial cells has been a longstanding question in neurobiology since the days of Santiago Ramon y Cajal (Cadwell et al., 2019). For several decades, the prevailing view has been that regional patterning follows what is referred to as the "activation-transformation" model, proposed by Pieter Nieuwkoop in the 1950s based on his experiments in amphibia (Nieuwkoop & Nigtevecht, 1954). According to this classical model, neural tissues are first induced, or "activated", to differentiate into a "default" rostral (i.e. forebrain) identity and subsequently caudalized by "transforming" signals emanating from secondary organizers to develop more posterior structures such as the midbrain, hindbrain, and spinal cord. However, it is also worth noting that there is emerging evidence challenging this view. Of note, it has been suggested that the establishment of A-P identity may precede the acquisition of neural identity and that the mechanisms underlying spinal cord induction and patterning are distinct from those of the brain (Metzis et al., 2018; Polevoy et al., 2019).

Notwithstanding these challenges, Nieuwkoop's model has survived the passing of time remarkably well and has served as the basis for some of the earlier studies involving the directed differentiation of spinal cord motor neurons and midbrain dopaminergic neurons from hESCs (Li et al., 2005; Pankratz et al., 2007; Yan et al., 2005). Indeed, several aspects of this model have been shown to be recapitulated in PSC models. First, the differentiation of PSCs to a default dorsal telencephalic identity has been demonstrated in both human and mouse ESCs/iPSCs, validating the activation step (Chambers et al., 2009; Pankratz et al., 2007; Watanabe et al., 2005). Thus, in the absence of morphogens, neural induction of PSCs yields dorsal telencephalic cortical cell populations that are characterized by the expression of key forebrain markers such as FOXG1, OTX2, and the dorsal telencephalic specific PAX6 and EMX1/2.

Similarly, a wide array of studies using PSC models have provided evidence in support of the subsequent transformation, or caudalization step. The three main morphogens involved in the specification of the R-C axis are FGFs, WNTs, and retinoic acid (RA). In vivo, these morphogenic cues are produced by secondary local organizers along the R-C axis including the anterior neural ridge (ANR) at the most rostral end of the neural tube, zona limitans intrathalamica of the diencephalon, and the isthmic organizer (IsO) at the midbrain-hindbrain boundary (Kiecker and Lumsden, 2012). A previous study by Malin Parmar's group elaborately demonstrated that activation of the canonical WNT signaling pathway by the GSK3 inhibitor CHIR99021 (CHIR) can differentiate hESCs into neural progenitors of the telencephalic forebrain, midbrain, hindbrain/spinal cord in a dose-dependent manner (Kirkeby et al., 2012). In addition, this study also demonstrated that in the absence of morphogenic cues, progenitors differentiated into an anterior neural fate. Importantly, the identification of CHIR as a potent caudalizing agent sparked the development of PSC-based mesencephalic dopaminergic (mDAergic) neuronal differentiation protocols by multiple labs (Kirkeby et al., 2012; Kriks et al., 2011; Tao and Zhang, 2016). In recent years, several refinements have been made to enhance the efficiency and/or the ability to generate subtype-specific mDA neurons. Notable examples include biphasic/two-step WNT activation, the addition of FGF8b as well as dual activation of canonical and non-canonical WNT signaling (Kim et al., 2021; Nishimura et al., 2023; Piao et al., 2021). The reader is referred to the following reviews highlighting recent advances in this area (Garritsen et al., 2023; Kim et al., 2020).

In addition to the R-C axis, regional identity is also established according to D-V polarity. Patterning along the D-V region is mediated by SHH, BMP, and canonical WNT signaling pathways. First identified as a diffusible signaling molecule capable of inducing motor neuron and floorplate differentiation in neural plate explants, SHH is a critical ventralizing agent produced by the notochord and subsequently by the floorplate (Roelink et al., 1995; Yamada et al., 1993). As such, along with CHIR, recombinant SHH is a vital component used in the directed differentiation of hPSCs to $FOXA2^+/LMX1A^+/OTX2^+/EN1^+$ mDAergic neurons (Nolbrant et al., 2017). The ventralizing activity of SHH is also utilized to differentiate anterior neuroepithelial cells to progenitors of the medial ganglionic eminence, the most ventral region of the telencephalon from which basal forebrain cholinergic neurons and inhibitory GABAergic interneurons develop (Ananth et al., 2023; Li et al., 2009). These progenitor population can be distinguished by the expression of NKX2–1 that demarcates the D-V telencephalic border with PAX6 in vivo (Hébert & Fishell, 2008). Commonly used small molecule alternatives of recombinant SHH include purmorphamine and smoothened agonist (SAG). Conversely, it is also possible to enrich specific neuronal subtypes by utilizing morphogens or small molecules of opposing effects. For example, several studies employ the use of cyclopamine, the first identified small molecule inhibitor of hedgehog signaling, to skew the differentiation of PSCs to a dorsal cortical fate to obtain near-pure excitatory glutamatergic neuronal cultures with minimal inhibitory GABAergic neurons (Cao et al., 2017; Chen, 2016; Vazin et al., 2014).

As discussed thus far, the directed differentiation strategy of hPSCs is deeply rooted in relevance to principles of in vivo developmental biology. While the aforementioned studies mostly highlight examples regarding regionally patterned neurons and immature progenitors, there are ongoing efforts to extend the application to non-neuronal cells as well. For instance, there is a growing body of evidence indicating that the regional identity of astrocytes may confer functional diversity (Akdemir et al., 2020; Bradley et al., 2019; Clarke et al., 2020; Escartin et al., 2021; Khakh & Sofroniew, 2015). It will be exciting to see how the continuous refinement of this versatile platform will contribute to our efforts of developing bona fide in vitro models to study human CNS health and disease.

All the principles described thus far and the recapitulation of in vivo development via the directed differentiation of PSCs can be summarized and traced back to the works of Wichterle and colleagues from 2002 that pioneered this approach (Wichterle et al., 2002). In this landmark study, mESCs were first induced to differentiate and form EBs by removing LIF from culture. Subsequently, cells were directed towards a spinal cord motor neuron fate by RA and SHH as caudalizing and ventralizing signals, respectively. Remarkably, these cells were shown to acquire terminal motor neuron identity in a progressive manner by going through regionally patterned progenitor stages as they do in vivo. The proof-of-principle offered by this seminal study provided the critical initial insight that later served as the essence of directed differentiation of PSCs. Today, many protocols designed to differentiate hPSCs to regionally patterned neural cells follow these principles, with the addition of an initial dSMADi stage at the beginning.

5. Concluding remarks

Since its discovery in 2007, hiPSC technology has attained a commanding position in the study of human health and disease and is considered one of the most crowning achievements in the field of biological and biomedical sciences. hiPSCs were expected, rightfully so, to be a game-changing force in facilitating progress in drug discovery, disease modeling, regenerative medicine, cell-based therapy, and translational research, a prophecy that is playing out today (De Luca et al., 2019; Kleiman & Engle, 2021; McKinley et al., 2023; Okano & Morimoto, 2022; Rivetti di Val Cervo et al., 2021; Rowe & Daley, 2019; Sharma et al., 2020; Shi et al., 2017; Temple, 2023; Yamanaka, 2020).

We are continuously marveled by the versatility of this model in generating a wide array of neural cell types and the elegance with which it recapitulates in vivo development. Indeed, there is now a plethora of directed differentiation protocols for the derivation of not just cortical neurons, but also striatal medium spiny neurons, mesencephalic dopaminergic neurons, spinal cord motor neurons, peripheral motor/sensory neurons, astrocytes, oligodendrocytes, and microglia; many of which are rigorously validated for their relevance to in vivo developmental trajectories as well as functionality (Conforti et al., 2022; Giacomelli et al., 2022; Hasselmann & Blurton-Jones, 2020; Kim et al., 2020; Lanjewar & Sloan, 2021; Shim et al., 2024; Van Lent et al., 2024). Moreover, several exciting progresses are being made in understanding the mechanisms governing neuronal maturation, aging, and senescence including epigenetic, metabolic, and post-translational modifications, as well as the contributions from surrounding glial cells and extracellular matrices (Ciceri et al., 2024; De Luzy et al., 2024; Qi et al., 2017; Saurat et al., 2024; Wallace & Pollen, 2024). The ability to experimentally induce and/or expedite functional maturation and aging-related phenotypes will significantly enhance our capacity to construct superior disease modeling platforms and improve the preclinical validity of hiPSC-based neural models.

Additionally, akin to how discoveries from multiple disparate disciplines contributed to the development of hiPSCs, the field of stem cells has entered a vibrant era of science wherein advances in molecular genetics, single cell multi-omics technology, cellular neurobiology, engineering, and computational tools dynamically converge together, allowing us to unlock and explore previously inaccessible areas of research. Indeed, with the growing appreciation of interdisciplinary collaboration, various innovative platforms have been integrated into hiPSC-derived models, enabling us to address complex biological questions at multiple levels. Genome-editing tools such as the CRISPR-Cas system have enabled the production of genetically engineered isogenic hiPSC lines to interrogate the contribution of specific genes in disease pathology as well as their role in altering individual risk to disease onset in response to environmental risk factors (Hendriks et al., 2020; Kalamakis & Platt, 2023). Continued efforts to develop new hiPSC lines including the isogenic clonal derivatives of the KOLF 2.1J reference line are actively underway as well as endeavors to stratify these according to disease, sex, ethnicity, and age to expand the repertoire of biomedical health questions that can be addressed by these model systems (Pantazis et al., 2022). Furthermore, progress in single cell multi-omics technology aids us in discerning the molecular mechanisms underlying various pathologies as well as toxic exposures in cell-type specific resolution on multiple biological layers ranging from (epi)genomic, transcriptomic, proteomic, metabolomic, to phenomic levels, enabling us to better characterize genotype-phenotype relationships (Brooks et al., 2022). Advances in high-content imaging methods as well as electrophysiological platforms such as multielectrode array further enable us to implement multipronged experimental approaches and evaluate morphological, structural, and functional perturbations at cellular and circuit/network level of complexity (Lv et al., 2023; Menduti & Boido, 2023). In addition, continuous development in advanced culture systems such as 3D organoids, microfluidic-based organ or multi-organs-on-a-chip models will provide additional opportunity by adding spatiotemporal context and providing a platform that mimics the complex multicellular interactions and physiochemical properties of the in vivo brain microenvironment (Amartumur et al., 2024; Eichmüller & Knoblich, 2022).

The immense potential of hPSC-based models for neurotoxicity studies has garnered the attention of toxicologists starting in the late 2000s and early 2010s to complement traditional in vivo animal models, in vitro human cell lines, as well as to address the scarcity of human fetal-derived samples. While much of these early works involved one or just a few hESC line(s) and were mostly restricted to assessing developmental toxicity, these pioneering studies served as valuable proof-of-concept for the utilization of hPSCs for neurotoxicological research (Colleoni et al., 2011; Pal et al., 2011; Stummann et al., 2009; Taléns-Visconti et al., 2011). Building on these findings, subsequent studies further demonstrated the potential value of incorporating tissue engineering platforms to enhance cellular diversity and structural complexity, as well as utilizing bioinformatics and machine learning tools for large-scale toxicity screening (Schwartz et al., 2015). With the advent of hiPSC technology and refinement of directed differentiation protocols, various toxicity studies involving lineage-specific mature neurons have been conducted (Neely et al., 2017; Xie et al., 2023). Moreover, such advances have enabled us to ask important toxicological questions such as persistent effects following developmental exposures and dissecting underlying transcriptomic differences governing acute versus chronic exposures and how they relate to disease pathology (Neely et al., 2021; Saleh et al., 2024). Furthermore, the use of patient-derived iPSCs have allowed researchers to directly test the contributions of environmental neurotoxicants confirmed through epidemiological data to a specific neurodegenerative disease previously associated with those toxicants, thereby bridging the gap between cell autonomous mechanistic toxicology and environmental epidemiology (Paul et al., 2023). Finally, the depth of toxicological questions addressed in recent studies are significantly aided by the cutting-edge tools mentioned above and several novel approaches, such as population-based studies, as well as 3D chimeric brain models ("Chimeroids") that are actively being developed, to better extract meaningful biological data from potential background genetic noise (Antón-Bolaños et al., 2024; Huang et al., 2022; Jerber et al., 2021).

In this chapter, we described the scientific paths that led to the development of hiPSC technology by placing it into historical context. As we have seen, hiPSC neuronal models are the culmination of paradigm-shifting ideas and discoveries made across more than a century through the ingenious insights, remarkable technical acumen, extraordinary craftmanship spirit, and perseverance of pioneering researchers in science. It is truly remarkable to see how much progress has been achieved in less than 20 years of its discovery and it is an exciting era for science to see what lies ahead. We conclude this chapter with eager hope of seeing further advances to fulfill the promises of stem cells and the further integration of this versatile model in toxicology to enhance human health.

Funding

This work was supported in part by grants from the National Institutes of Health (USA), NIEHS R01 ES010563, R01 ES07331 and R01 ES031401 and NIA R01 AG080917.

Conflict of interest

The authors declare no conflict of interest.

References

A conversation with John Gurdon and Shinya Yamanaka, 2020. Stem Cell Reports 14, 351–356. https://doi.org/10.1016/j.stemcr.2020.02.007.

Akdemir, E.S., Huang, A.Y.-S., Deneen, B., 2020. Astrocytogenesis: where, when, and how. F1000Research 9, 233. https://doi.org/10.12688/f1000research.22405.1.

Akhurst, R.J., 2023. From shape-shifting embryonic cells to oncology: the fascinating history of epithelial mesenchymal transition. Semin. Cancer Biol. 96, 100–114. https://doi.org/10.1016/j.semcancer.2023.10.003.

Allison, T.F., Andrews, P.W., Avior, Y., Barbaric, I., Benvenisty, N., Bock, C., et al., 2018. Assessment of established techniques to determine developmental and malignant potential of human pluripotent stem cells. Nat. Commun. 9, 1925. https://doi.org/10.1038/s41467-018-04011-3.

Amartumur, S., Nguyen, H., Huynh, T., Kim, T.S., Woo, R.-S., Oh, E., et al., 2024. Neuropathogenesis-on-chips for neurodegenerative diseases. Nat. Commun. 15, 2219. https://doi.org/10.1038/s41467-024-46554-8.

Ananth, M.R., Rajebhosale, P., Kim, R., Talmage, D.A., Role, L.W., 2023. Basal forebrain cholinergic signalling: development, connectivity and roles in cognition. Nat. Rev. Neurosci. 24, 233–251. https://doi.org/10.1038/s41583-023-00677-x.

Andrews, P.W., 2024. The origins of human pluripotent stem cells: the road from a cancer to regenerative medicine. In Vitro Cell. Dev. Biol. Anim. 60, 514–520. https://doi.org/10.1007/s11626-024-00865-8.

Andrews, P.W., Gokhale, P.J., 2024. A short history of pluripotent stem cells markers. Stem Cell Reports 19, 1–10. https://doi.org/10.1016/j.stemcr.2023.11.012.

Antón-Bolaños, N., Faravelli, I., Faits, T., Andreadis, S., Kastli, R., Trattaro, S., et al., 2024. Brain Chimeroids reveal individual susceptibility to neurotoxic triggers. Nature 631, 142–149. https://doi.org/10.1038/s41586-024-07578-8.

Anwised, P., Moorawong, R., Samruan, W., Somredngan, S., Srisutush, J., Laowtammathron, C., et al., 2023. An expedition in the jungle of pluripotent stem cells of non-human primates. Stem Cell Reports 18, 2016–2037. https://doi.org/10.1016/j.stemcr.2023.09.013.

Becker, A.J., McCulloch, E.A., Till, J.E., 1963. Cytological demonstration of the clonal nature of spleen colonies derived from transplanted mouse marrow cells. Nature 197, 452–454. https://doi.org/10.1038/197452a0.

Benham, F.J., Andrews, P.W., Knowles, B.B., Bronson, D.L., Harris, H., 1981. Alkaline phosphatase isozymes as possible markers of differentiation in human testicular teratocarcinoma cell lines. Dev. Biol. 88, 279–287. https://doi.org/10.1016/0012-1606(81)90171-8.
Bernstine, E.G., Hooper, M.L., Grandchamp, S., Ephrussi, B., 1973. Alkaline phosphatase activity in mouse teratoma. Proc. Natl. Acad. Sci. U. S. A. 70, 3899–3903. https://doi.org/10.1073/pnas.70.12.3899.
Bline, A.P., Le Goff, A., Allard, P., 2020. What is lost in the Weismann barrier? J. Dev. Biol. 8. https://doi.org/10.3390/jdb8040035.
Bocchi, R., Masserdotti, G., Götz, M., 2022. Direct neuronal reprogramming: fast forward from new concepts toward therapeutic approaches. Neuron 110, 366–393. https://doi.org/10.1016/j.neuron.2021.11.023.
Boveri, T., 1895. Über die Befruchtungs-und Entwickelungsfähigkeit kernloser Seeigel-Eier und über die Möglichkeit ihrer Bastardirung. Arch. Entwicklmech Org. 2, 394–443.
Boyer, L.A., Lee, T.I., Cole, M.F., Johnstone, S.E., Levine, S.S., Zucker, J.P., et al., 2005. Core transcriptional regulatory circuitry in human embryonic stem cells. Cell 122, 947–956. https://doi.org/10.1016/j.cell.2005.08.020.
Bradley, A., Evans, M., Kaufman, M.H., Robertson, E., 1984. Formation of germ-line chimaeras from embryo-derived teratocarcinoma cell lines. Nature 309, 255–256. https://doi.org/10.1038/309255a0.
Bradley, R.A., Shireman, J., McFalls, C., Choi, J., Canfield, S.G., Dong, Y., et al., 2019. Regionally specified human pluripotent stem cell-derived astrocytes exhibit different molecular signatures and functional properties. Development 146. https://doi.org/10.1242/dev.170910.
Brennand, K.J., Simone, A., Jou, J., Gelboin-Burkhart, C., Tran, N., Sangar, S., et al., 2011. Modelling schizophrenia using human induced pluripotent stem cells. Nature 473, 221–225. https://doi.org/10.1038/nature09915.
Briggs, R., King, T.J., 1957. Changes in the nuclei of differentiating endoderm cells as revealed by nuclear transplantation. J. Morphol. 100, 269–311.
Briggs, R., King, T.J., 1952. Transplantation of living nuclei from blastula cells into enucleated frogs' eggs. Proc. Natl. Acad. Sci. U. S. A. 38, 455–463.
Brinster, R.L., 1974. The effect of cells transferred into the mouse blastocyst on subsequent development. J. Exp. Med. 140, 1049–1056. https://doi.org/10.1084/jem.140.4.1049.
Brooks, I.R., Garrone, C.M., Kerins, C., Kiar, C.S., Syntaka, S., Xu, J.Z., et al., 2022. Functional genomics and the future of iPSCs in disease modeling. Stem Cell Reports 17, 1033–1047. https://doi.org/10.1016/j.stemcr.2022.03.019.
Browne, E.N., 1909. The production of new hydranths in Hydra by the insertion of small grafts. J. Exp. Zool. 7, 1–23.
Bustamante-Marin, X.M., Capel, B., 2023. Oxygen availability influences the incidence of testicular teratoma in Dnd1Ter/+ mice. Front. Genet. 14, 1179256. https://doi.org/10.3389/fgene.2023.1179256.
Cadwell, C.R., Bhaduri, A., Mostajo-Radji, M.A., Keefe, M.G., Nowakowski, T.J., 2019. Development and arealization of the cerebral cortex. Neuron 103, 980–1004. https://doi.org/10.1016/j.neuron.2019.07.009.
Campbell, K.H.S., 2002. A background to nuclear transfer and its applications in agriculture and human therapeutic medicine. J. Anat. 200, 267–275. https://doi.org/10.1046/j.1469-7580.2002.00035.x.
Cao, S.-Y., Hu, Y., Chen, C., Yuan, F., Xu, M., Li, Q., et al., 2017. Enhanced derivation of human pluripotent stem cell-derived cortical glutamatergic neurons by a small molecule. Sci. Rep. 7, 3282. https://doi.org/10.1038/s41598-017-03519-w.

Cerneckis, J., Cai, H., Shi, Y., 2024. Induced pluripotent stem cells (iPSCs): molecular mechanisms of induction and applications. Signal Transduct. Target Ther. 9, 112. https://doi.org/10.1038/s41392-024-01809-0.

Chambers, I., Colby, D., Robertson, M., Nichols, J., Lee, S., Tweedie, S., et al., 2003. Functional expression cloning of Nanog, a pluripotency sustaining factor in embryonic stem cells. Cell 113, 643–655. https://doi.org/10.1016/S0092-8674(03)00392-1.

Chambers, S.M., Fasano, C.A., Papapetrou, E.P., Tomishima, M., Sadelain, M., Studer, L., 2009. Highly efficient neural conversion of human ES and iPS cells by dual inhibition of SMAD signaling. Nat. Biotechnol. 27, 275–280. https://doi.org/10.1038/nbt.1529.

Chen, G., Gulbranson, D.R., Hou, Z., Bolin, J.M., Ruotti, V., Probasco, M.D., et al., 2011. Chemically defined conditions for human iPSC derivation and culture. Nat. Methods 8, 424–429. https://doi.org/10.1038/nmeth.1593.

Chen, J.K., 2016. I only have eye for ewe: the discovery of cyclopamine and development of Hedgehog pathway-targeting drugs. Nat. Prod. Rep. 33, 595–601. https://doi.org/10.1039/C5NP00153F.

Chen, Q., Shi, J., Tao, Y., Zernicka-Goetz, M., 2018. Tracing the origin of heterogeneity and symmetry breaking in the early mammalian embryo. Nat. Commun. 9, 1819. https://doi.org/10.1038/s41467-018-04155-2.

Ciceri, G., Baggiolini, A., Cho, H.S., Kshirsagar, M., Benito-Kwiecinski, S., Walsh, R.M., et al., 2024. An epigenetic barrier sets the timing of human neuronal maturation. Nature 626, 881–890. https://doi.org/10.1038/s41586-023-06984-8.

Clarke, B.E., Taha, D.M., Ziff, O.J., Alam, A., Thelin, E.P., García, N.M., et al., 2020. Human stem cell-derived astrocytes exhibit region-specific heterogeneity in their secretory profiles. Brain 143, e85. https://doi.org/10.1093/brain/awaa258.

Cohen, D.E., Melton, D., 2011. Turning straw into gold: directing cell fate for regenerative medicine. Nat. Rev. Genet. 12, 243–252. https://doi.org/10.1038/nrg2938.

Colleoni, S., Galli, C., Gaspar, J.A., Meganathan, K., Jagtap, S., Hescheler, J., et al., 2011. Development of a neural teratogenicity test based on human embryonic stem cells: response to retinoic acid exposure. Toxicol. Sci. 124, 370–377. https://doi.org/10.1093/toxsci/kfr245.

Conforti, P., Bocchi, V.D., Campus, I., Scaramuzza, L., Galimberti, M., Lischetti, T., et al., 2022. In vitro-derived medium spiny neurons recapitulate human striatal development and complexity at single-cell resolution. Cell Rep. Methods 2, 100367. https://doi.org/10.1016/j.crmeth.2022.100367.

Daley, G.Q., 2015. Stem cells and the evolving notion of cellular identity. Philos. Trans. R Soc. Lond. B Biol. Sci. 370, 20140376. https://doi.org/10.1098/rstb.2014.0376.

Damjanov, I., Wewer-Albrechtsen, N., 2013. Testicular germ cell tumors and related research from a historical point of view. Int. J. Dev. Biol. 57, 197–200. https://doi.org/10.1387/ijdb.130143id.

Dantschakoff, W., 1908. Untersuchungen über die Entwickelung des Blutes und Bindegewebes bei den Vögeln. Anat. Embryol. (Berl) 37, 471–587.

Darnell, D., Gilbert, S.F., 2017. Neuroembryology. WIREs Dev. Biol. 6. https://doi.org/10.1002/wdev.215.

Davis, R.L., Weintraub, H., Lassar, A.B., 1987. Expression of a single transfected cDNA converts fibroblasts to myoblasts. Cell 51, 987–1000. https://doi.org/10.1016/0092-8674(87)90585-X.

Dayrat, B., 2003. The roots of phylogeny: how did Haeckel build his trees? Syst. Biol. 52, 515–527. https://doi.org/10.1080/10635150390218277.

De Luca, M., Aiuti, A., Cossu, G., Parmar, M., Pellegrini, G., Robey, P.G., 2019. Advances in stem cell research and therapeutic development. Nat. Cell Biol. 21, 801–811. https://doi.org/10.1038/s41556-019-0344-z.

De Luzy, I.R., Lee, M.K., Mobley, W.C., Studer, L., 2024. Lessons from inducible pluripotent stem cell models on neuronal senescence in aging and neurodegeneration. Nat. Aging 4, 309–318. https://doi.org/10.1038/s43587-024-00586-3.

De Robertis, E.M., 2014. Lessons from a great developmental biologist. Differentiation 88, 3–8. https://doi.org/10.1016/j.diff.2013.12.004.

De Robertis, E.M., 2009. Spemann's organizer and the self-regulation of embryonic fields. Mech. Dev. 126, 925–941. https://doi.org/10.1016/j.mod.2009.08.004.

De Robertis, E.M., 2006. Spemann's organizer and self-regulation in amphibian embryos. Nat. Rev. Mol. Cell. Biol. 7, 296–302. https://doi.org/10.1038/nrm1855.

Dimos, J.T., Rodolfa, K.T., Niakan, K.K., Weisenthal, L.M., Mitsumoto, H., Chung, W., et al., 2008. Induced pluripotent stem cells generated from patients with ALS can be differentiated into motor neurons. Science 321 (1979), 1218–1221. https://doi.org/10.1126/science.1158799.

Driesch, H., 1899. Die Lokalisation morphogenetischer Vorgänge: Ein Beweis vitalistischen Geschehens. Archiv für Entwicklungsmechanik 8, 35–111.

Driesch, H., 1891. Entwicklungsmechanisiche Studien, I. Der Werth der beiden ersten FurchungszeHen in der Echino-dermentwicklung. Experimentelle Erzeugen von Theil-und Doppelbildung. Zeitschr. wissenschaft. ZooL 53, 160–178.

Dröscher, A., 2014. Images of cell trees, cell lines, and cell fates: the legacy of Ernst Haeckel and August Weismann in stem cell research. Hist. Philos. Life Sci. 36, 157–186. https://doi.org/10.1007/s40656-014-0028-8.

Dupont, J.-C., 2017. Wilhelm His and mechanistic approaches to development at the time of Entwicklungsmechanik. Hist. Philos. Life Sci. 39, 21. https://doi.org/10.1007/s40656-017-0148-z.

Ebert, A.D., Yu, J., Rose, F.F., Mattis, V.B., Lorson, C.L., Thomson, J.A., et al., 2009. Induced pluripotent stem cells from a spinal muscular atrophy patient. Nature 457, 277–280. https://doi.org/10.1038/nature07677.

Eguizabal, C., Aran, B., Chuva de Sousa Lopes, S.M., Geens, M., Heindryckx, B., Panula, S., et al., 2019. Two decades of embryonic stem cells: a historical overview. hoy024. Hum. Reprod. Open 2019. https://doi.org/10.1093/hropen/hoy024.

Eichmüller, O.L., Knoblich, J.A., 2022. Human cerebral organoids—a new tool for clinical neurology research. Nat. Rev. Neurol. 18, 661–680. https://doi.org/10.1038/s41582-022-00723-9.

Escartin, C., Galea, E., Lakatos, A., O'Callaghan, J.P., Petzold, G.C., Serrano-Pozo, A., et al., 2021. Reactive astrocyte nomenclature, definitions, and future directions. Nat. Neurosci. 24, 312–325. https://doi.org/10.1038/s41593-020-00783-4.

Evans, M., 2011. Discovering pluripotency: 30 years of mouse embryonic stem cells. Nat. Rev. Mol. Cell. Biol. 12, 680–686. https://doi.org/10.1038/nrm3190.

Evans, M.J., Kaufman, M.H., 1981. Establishment in culture of pluripotential cells from mouse embryos. Nature 292, 154–156. https://doi.org/10.1038/292154a0.

Ewing, J., 1919. Neoplastic Diseases: A Treatise on Tumors. W.B. Saunders, Philadelphia.

Ewing, J., 1911. Teratoma testis and its derivatives. Surg. Gynecol. Obstet. 12, 230–261.

Fäßler, P.E., 2013. Hans Spemann 1869–1941 Experimentelle Forschung im Spannungsfeld von Empirie und Theorie: Ein Beitrag zur Geschichte der Entwicklungsphysiologie zu Beginn des 20. Jahrhunderts. Springer-Verlag.

Fischberg, M., Gurdon, J.B., Elsdale, T.R., 1958. Nuclear transplantation in *Xenopus laevis*. Nature 181, 424.

Garritsen, O., van Battum, E.Y., Grossouw, L.M., Pasterkamp, R.J., 2023. Development, wiring and function of dopamine neuron subtypes. Nat. Rev. Neurosci. 24, 134–152. https://doi.org/10.1038/s41583-022-00669-3.

Giacomelli, E., Vahsen, B.F., Calder, E.L., Xu, Y., Scaber, J., Gray, E., et al., 2022. Human stem cell models of neurodegeneration: from basic science of amyotrophic lateral sclerosis to clinical translation. Cell Stem Cell 29, 11–35. https://doi.org/10.1016/j.stem.2021.12.008.
Gilbert, S.F., 2000. Developmental Biology, sixth ed. Sinauer Associates, Sunderland.
Gitschier, J., 2008. Sweating the details: an interview with Jamie Thomson. PLoS Genet. 4, e1000182. https://doi.org/10.1371/journal.pgen.1000182.
Godsave, S.F., Slack, J.M.W., 1989. Clonal analysis of mesoderm induction in *Xenopus laevis*. Dev. Biol. 134, 486–490. https://doi.org/10.1016/0012-1606(89)90122-X.
Goldberg, A.D., Allis, C.D., Bernstein, E., 2007. Epigenetics: a landscape takes shape. Cell 128, 635–638. https://doi.org/10.1016/j.cell.2007.02.006.
Goldman, B., 2008. Embryonic stem cells 2.0. Nat. Rep. Stem Cells. https://doi.org/10.1038/stemcells.2008.67.
Gooi, H.C., Feizi, T., Kapadia, A., Knowles, B.B., Solter, D., Evans, M.J., 1981. Stage-specific embryonic antigen involves αl→ 3 fucosylated type 2 blood group chains. Nature 292, 156–158. https://doi.org/10.1038/292156a0.
Grunz, H., Tacke, L., 1989. Neural differentiation of *Xenopus laevis* ectoderm takes place after disaggregation and delayed reaggregation without inducer. Cell Differ. Dev. 28, 211–217. https://doi.org/10.1016/0922-3371(89)90006-3.
Guhr, A., Kobold, S., Seltmann, S., Seiler Wulczyn, A.E.M., Kurtz, A., Löser, P., 2018. Recent trends in research with human pluripotent stem cells: impact of research and use of cell lines in experimental research and clinical trials. Stem Cell Rep. 11, 485–496. https://doi.org/10.1016/j.stemcr.2018.06.012.
Gurdon, J.B., 2017. Nuclear transplantation, the conservation of the genome, and prospects for cell replacement. FEBS J. 284, 211–217.
Gurdon, J.B., 2013. The egg and the nucleus: a battle for supremacy. Development 140, 2449–2456. https://doi.org/10.1242/dev.097170.
Gurdon, J.B., 2006. From nuclear transfer to nuclear reprogramming: the reversal of cell differentiation. Annu. Rev. Cell Dev. Biol. 22, 1–22. https://doi.org/10.1146/annurev.cellbio.22.090805.140144.
Gurdon, J.B., 1962a. Adult frogs derived from the nuclei of single somatic cells. Dev. Biol. 4, 256–273.
Gurdon, J.B., 1962b. The developmental capacity of nuclei taken from intestinal epithelium cells of feeding tadpoles. Development 10, 622–640.
Gurdon, J.B., Byrne, J.A., 2003. The first half-century of nuclear transplantation. Proc. Natl. Acad. Sci. U. S. A. 100, 8048–8052. https://doi.org/10.1073/pnas.1337135100.
Gurdon, J.B., Hopwood, N., 2000. The introduction of Xenopus laevis into developmental biology: of empire, pregnancy testing and ribosomal genes. Int. J. Dev. Biol. 44, 43–50.
Gurdon, J.B., Uehlinger, V., 1966. Fertile" intestine nuclei. Nature 210, 1240–1241.
Gurdon, S.J., 2008. Sir John Gurdon: godfather of cloning. Interviewed by Ruth Williams. J. Cell Biol. 181, 178–179. https://doi.org/10.1083/jcb.1812pi.
Haeckel, E., 1869. Zur entwicklungsgeschichte der siphonophoren. In: C. van der Post, Jr. (Ed.), Utrecht.
Haeckel, E., 1868. Natürliche schöpfungsgeschichte. Verlag Georg Reimer, Berlin.
Haeckel, E., 1877. Anthropogenie, third ed. W. Engelmann, Leipzig.
Haecker, V., 1892. Die Kerntheilungsvorgänge bei der Mesodermund Entodermbildung von Cyclops. Archiv für mikroskopische Anatomie 39, 556–581.
Hamburger, V., 1989. The heritage of experimental embryology: Hans Spemann and the organizer. J. Hist. Biol. 22.
Hasselmann, J., Blurton-Jones, M., 2020. Human iPSC-derived microglia: a growing toolset to study the brain's innate immune cells. Glia 68, 721–739. https://doi.org/10.1002/glia.23781.

Hébert, J.M., Fishell, G., 2008. The genetics of early telencephalon patterning: some assembly required. Nat. Rev. Neurosci. 9, 678–685. https://doi.org/10.1038/nrn2463.
Hemmati-Brivanlou, A., Kelly, O.G., Melton, D.A., 1994. Follistatin, an antagonist of activin, is expressed in the Spemann organizer and displays direct neuralizing activity. Cell 77, 283–295. https://doi.org/10.1016/0092-8674(94)90320-4.
Hemmati-Brivanlou, A., Melton, D., 1997. Vertebrate embryonic cells will become nerve cells unless told otherwise. Cell 88, 13–17. https://doi.org/10.1016/S0092-8674(00)81853-X.
Hemmati-Brivanlou, A., Melton, D.A., 1994. Inhibition of activin receptor signaling promotes neuralization in Xenopus. Cell 77, 273–281. https://doi.org/10.1016/0092-8674(94)90319-0.
Hemmati-Brivanlou, A., Melton, D.A., 1992. A truncated activin receptor inhibits mesoderm induction and formation of axial structures in Xenopus embryos. Nature 359, 609–614. https://doi.org/10.1038/359609a0.
Hendriks, D., Clevers, H., Artegiani, B., 2020. CRISPR-Cas tools and their application in genetic engineering of human stem cells and organoids. Cell Stem Cell 27, 705–731. https://doi.org/10.1016/j.stem.2020.10.014.
Hochedlinger, K., Jaenisch, R., 2006. Nuclear reprogramming and pluripotency. Nature 441, 1061–1067. https://doi.org/10.1038/nature04955.
Hochedlinger, K., Jaenisch, R., 2002. Monoclonal mice generated by nuclear transfer from mature B and T donor cells. Nature 415, 1035–1038. https://doi.org/10.1038/nature718.
Holtfreter, J., 1933. Nachweis der Induktionsf?higkeit abget?teter Keimteile. Wilhelm Roux' Archiv fur Entwicklungsmechanik der Organismen 128, 584–633. https://doi.org/10.1007/BF00649865.
Huang, C.Y., Nicholson, M.W., Wang, J.Y., Ting, C.Y., Tsai, M.H., Cheng, Y.C., et al., 2022. Population-based high-throughput toxicity screen of human iPSC-derived cardiomyocytes and neurons. Cell Rep. 39, 110643. https://doi.org/10.1016/j.celrep.2022.110643.
Ieda, M., Fu, J.-D., Delgado-Olguin, P., Vedantham, V., Hayashi, Y., Bruneau, B.G., et al., 2010. Direct reprogramming of fibroblasts into functional cardiomyocytes by defined factors. Cell 142, 375–386. https://doi.org/10.1016/j.cell.2010.07.002.
Jaenisch, R., Young, R., 2008. Stem cells, the molecular circuitry of pluripotency and nuclear reprogramming. Cell 132, 567–582. https://doi.org/10.1016/j.cell.2008.01.015.
Jerber, J., Seaton, D.D., Cuomo, A.S.E., Kumasaka, N., Haldane, J., Steer, J., et al., 2021. Population-scale single-cell RNA-seq profiling across dopaminergic neuron differentiation. Nat. Genet. 53, 304–312. https://doi.org/10.1038/s41588-021-00801-6.
Kalamakis, G., Platt, R.J., 2023. CRISPR for neuroscientists. Neuron 111, 2282–2311. https://doi.org/10.1016/j.neuron.2023.04.021.
Karami, Z., Moradi, S., Eidi, A., Soleimani, M., Jafarian, A., 2023. Induced pluripotent stem cells: generation methods and a new perspective in COVID-19 research. Front. Cell Dev. Biol. 10. https://doi.org/10.3389/fcell.2022.1050856.
Khakh, B.S., Sofroniew, M.V., 2015. Diversity of astrocyte functions and phenotypes in neural circuits. Nat. Neurosci. 18, 942–952. https://doi.org/10.1038/nn.4043.
Kiecker, C., Lumsden, A., 2012. The role of organizers in patterning the nervous system. Annu. Rev. Neurosci. 35, 347–367. https://doi.org/10.1146/annurev-neuro-062111-150543.
Kim, T.W., Koo, S.Y., Studer, L., 2020. Pluripotent stem cell therapies for Parkinson disease: present challenges and future opportunities. Front. Cell Dev. Biol. 8. https://doi.org/10.3389/fcell.2020.00729.
Kim, T.W., Piao, J., Koo, S.Y., Kriks, S., Chung, S.Y., Betel, D., et al., 2021. Biphasic activation of WNT signaling facilitates the derivation of midbrain dopamine neurons from hESCs for translational use. Cell Stem Cell 28, 343–355.e5. https://doi.org/10.1016/j.stem.2021.01.005.

Kirkeby, A., Grealish, S., Wolf, D.A., Nelander, J., Wood, J., Lundblad, M., et al., 2012. Generation of regionally specified neural progenitors and functional neurons from human embryonic stem cells under defined conditions. Cell Rep. 1, 703–714. https://doi.org/10.1016/j.celrep.2012.04.009.

Kleiman, R.J., Engle, S.J., 2021. Human inducible pluripotent stem cells: realization of initial promise in drug discovery. Cell Stem Cell 28, 1507–1515. https://doi.org/10.1016/j.stem.2021.08.002.

Kleinsmith, L.J., Pierce Jr, G.B., 1964. Multipotentiality of single embryonal carcinoma cells. Cancer Res. 24, 1544–1551.

Kriks, S., Shim, J.-W., Piao, J., Ganat, Y.M., Wakeman, D.R., Xie, Z., et al., 2011. Dopamine neurons derived from human ES cells efficiently engraft in animal models of Parkinson's disease. Nature 480, 547–551. https://doi.org/10.1038/nature10648.

Kumar, K.K., Aboud, A.A., Bowman, A.B., 2012. The potential of induced pluripotent stem cells as a translational model for neurotoxicological risk. Neurotoxicology 33, 518–529. https://doi.org/10.1016/j.neuro.2012.02.005.

Lamb, T.M., Knecht, A.K., Smith, W.C., Stachel, S.E., Economides, A.N., Stahl, N., et al., 1993. Neural induction by the secreted polypeptide noggin. Science 262 (1979), 713–718. https://doi.org/10.1126/science.8235591.

Lanjewar, S.N., Sloan, S.A., 2021. Growing glia: cultivating human stem cell models of gliogenesis in health and disease. Front. Cell Dev. Biol. 9. https://doi.org/10.3389/fcell.2021.649538.

Lee, G., Papapetrou, E.P., Kim, H., Chambers, S.M., Tomishima, M.J., Fasano, C.A., et al., 2009. Modelling pathogenesis and treatment of familial dysautonomia using patient-specific iPSCs. Nature 461, 402–406. https://doi.org/10.1038/nature08320.

Lendahl, U., 2022. 100 plus years of stem cell research-20 years of ISSCR. Stem Cell Rep. 17, 1248–1267. https://doi.org/10.1016/j.stemcr.2022.04.004.

Lensch, M.W., 2009. Cellular reprogramming and pluripotency induction. Br. Med. Bull. 90, 19–35. https://doi.org/10.1093/bmb/ldp011.

Levit, G.S., Hossfeld, U., 2019. Ernst Haeckel in the history of biology. Curr. Biol. 29, R1276–R1284. https://doi.org/10.1016/j.cub.2019.10.064.

Li, X.-J., Du, Z.-W., Zarnowska, E.D., Pankratz, M., Hansen, L.O., Pearce, R.A., et al., 2005. Specification of motoneurons from human embryonic stem cells. Nat. Biotechnol. 23, 215–221. https://doi.org/10.1038/nbt1063.

Li, X.-J., Zhang, X., Johnson, M.A., Wang, Z.-B., LaVaute, T., Zhang, S.-C., 2009. Coordination of sonic hedgehog and Wnt signaling determines ventral and dorsal telencephalic neuron types from human embryonic stem cells. Development 136, 4055–4063. https://doi.org/10.1242/dev.036624.

Liu, G., David, B.T., Trawczynski, M., Fessler, R.G., 2020. Advances in pluripotent stem cells: history, mechanisms, technologies, and applications. Stem Cell Rev. Rep. 16, 3–32. https://doi.org/10.1007/s12015-019-09935-x.

Lowry, W.E., Richter, L., Yachechko, R., Pyle, A.D., Tchieu, J., Sridharan, R., et al., 2008. Generation of human induced pluripotent stem cells from dermal fibroblasts. Proc. Natl. Acad. Sci. U.S.A. 105, 2883–2888. https://doi.org/10.1073/pnas.0711983105.

Ludwig, T.E., Bergendahl, V., Levenstein, M.E., Yu, J., Probasco, M.D., Thomson, J.A., 2006a. Feeder-independent culture of human embryonic stem cells. Nat. Methods 3, 637–646. https://doi.org/10.1038/nmeth902.

Ludwig, T.E., Levenstein, M.E., Jones, J.M., Berggren, W.T., Mitchen, E.R., Frane, J.L., et al., 2006b. Derivation of human embryonic stem cells in defined conditions. Nat. Biotechnol. 24, 185–187. https://doi.org/10.1038/nbt1177.

Lv, S., He, E., Luo, J., Liu, Y., Liang, W., Xu, S., et al., 2023. Using human-induced pluripotent stem cell derived neurons on microelectrode arrays to model neurological disease: a review. Adv. Sci. 10. https://doi.org/10.1002/advs.202301828.
Maartens, A., 2017. An interview with John Gurdon. Development 144, 1581–1583. https://doi.org/10.1242/dev.152058.
Maderspacher, F., 2008. Theodor Boveri and the natural experiment. Curr. Biol. 18, R279–R286. https://doi.org/10.1016/j.cub.2008.02.061.
Maehle, A.-H., 2011. Ambiguous cells: the emergence of the stem cell concept in the nineteenth and twentieth centuries. Notes Rec. R Soc. Lond. 65, 359–378. https://doi.org/10.1098/rsnr.2011.0023.
Marchetto, M.C.N., Carromeu, C., Acab, A., Yu, D., Yeo, G.W., Mu, Y., et al., 2010. A model for neural development and treatment of rett syndrome using human induced pluripotent stem cells. Cell 143, 527–539. https://doi.org/10.1016/j.cell.2010.10.016.
Martin, G.R., 1981. Isolation of a pluripotent cell line from early mouse embryos cultured in medium conditioned by teratocarcinoma stem cells. Proc. Natl. Acad. Sci. U.S.A. 78, 7634–7638. https://doi.org/10.1073/pnas.78.12.7634.
Martin, G.R., Evans, M.J., 1974. The morphology and growth of a pluripotent teratocarcinoma cell line and its derivatives in tissue culture. Cell 2, 163–172. https://doi.org/10.1016/0092-8674(74)90090-7.
Masic, I., 2019. The most influential scientists in the development of public health (2): Rudolf Ludwig Virchow (1821–1902). Mater. Sociomed. 31, 151–152. https://doi.org/10.5455/msm.2019.31.151-152.
Maximow, A.A., 2009. Der Lymphozyt als gemeinsame Stammzelle der verschiedenen Blutelemente in der embryonalen Entwicklung und im postfetalen Leben der Säugetiere. Cell Ther. Transplant 1, 9–13.
McKinley, K.L., Longaker, M.T., Naik, S., 2023. Emerging frontiers in regenerative medicine. Science 380 (1979), 796–798. https://doi.org/10.1126/science.add6492.
Menduti, G., Boido, M., 2023. Recent advances in high-content imaging and analysis in iPSC-based modelling of neurodegenerative diseases. Int. J. Mol. Sci. 24, 14689. https://doi.org/10.3390/ijms241914689.
Metzis, V., Steinhauser, S., Pakanavicius, E., Gouti, M., Stamataki, D., Ivanovitch, K., et al., 2018. Nervous system regionalization entails axial allocation before neural differentiation. Cell 175, 1105–1118.e17. https://doi.org/10.1016/j.cell.2018.09.040.
Mintz, B., Illmensee, K., 1975. Normal genetically mosaic mice produced from malignant teratocarcinoma cells. Proc. Natl. Acad. Sci. U.S.A. 72, 3585–3589. https://doi.org/10.1073/pnas.72.9.3585.
Morgan, T.H., 1895. Half-embryos and whole-embryos from one of the first two blastomeres of the frog's egg. Anat. Anz. 10, 623–628.
Muñoz-Sanjuán, I., Brivanlou, A.H., 2002. Neural induction, the default model and embryonic stem cells. Nat. Rev. Neurosci. 3, 271–280. https://doi.org/10.1038/nrn786.
Neely, M.D., Davison, C.A., Aschner, M., Bowman, A.B., 2017. From the cover: manganese and rotenone-induced oxidative stress signatures differ in iPSC-derived human dopamine neurons. Toxicol. Sci. 159, 366–379. https://doi.org/10.1093/toxsci/kfx145.
Neely, M.D., Litt, M.J., Tidball, A.M., Li, G.G., Aboud, A.A., Hopkins, C.R., et al., 2012. DMH1, a highly selective small molecule BMP inhibitor promotes neurogenesis of hiPSCs: comparison of PAX6 and SOX1 expression during neural induction. ACS Chem. Neurosci. 3, 482–491. https://doi.org/10.1021/cn300029t.
Neely, M.D., Xie, S., Prince, L.M., Kim, H., Tukker, A.M., Aschner, M., et al., 2021. Single cell RNA sequencing detects persistent cell type- and methylmercury exposure

paradigm-specific effects in a human cortical neurodevelopmental model. Food Chem. Toxicol. 154, 112288. https://doi.org/10.1016/j.fct.2021.112288.
Nguyen, H.N., Byers, B., Cord, B., Shcheglovitov, A., Byrne, J., Gujar, P., et al., 2011. LRRK2 mutant iPSC-derived DA neurons demonstrate increased susceptibility to oxidative stress. Cell Stem Cell 8, 267–280. https://doi.org/10.1016/j.stem.2011.01.013.
Nichols, J., Zevnik, B., Anastassiadis, K., Niwa, H., Klewe-Nebenius, D., Chambers, I., et al., 1998. Formation of pluripotent stem cells in the mammalian embryo depends on the POU transcription factor Oct4. Cell 95, 379–391. https://doi.org/10.1016/S0092-8674(00)81769-9.
Nieuwkoop, P.D., Nigtevecht, G.V., 1954. Neural activation and transformation in explants of competent ectoderm under the influence of fragments of anterior notochord in urodeles. Development 2, 175–193. https://doi.org/10.1242/dev.2.3.175.
Nishimura, K., Yang, S., Lee, K.W., Ásgrímsdóttir, E.S., Nikouei, K., Paslawski, W., et al., 2023. Single-cell transcriptomics reveals correct developmental dynamics and high-quality midbrain cell types by improved hESC differentiation. Stem Cell Rep. 18, 337–353. https://doi.org/10.1016/j.stemcr.2022.10.016.
Nolbrant, S., Heuer, A., Parmar, M., Kirkeby, A., 2017. Generation of high-purity human ventral midbrain dopaminergic progenitors for in vitro maturation and intracerebral transplantation. Nat. Protoc. 12, 1962–1979. https://doi.org/10.1038/nprot.2017.078.
Okano, H., Morimoto, S., 2022. iPSC-based disease modeling and drug discovery in cardinal neurodegenerative disorders. Cell Stem Cell 29, 189–208. https://doi.org/10.1016/j.stem.2022.01.007.
Ozair, M.Z., Kintner, C., Brivanlou, A.H., 2013. Neural induction and early patterning in vertebrates. WIREs Dev. Biol. 2, 479–498. https://doi.org/10.1002/wdev.90.
Pal, R., Mamidi, M.K., Kumar Das, A., Bhonde, R., 2011. Human embryonic stem cell proliferation and differentiation as parameters to evaluate developmental toxicity. J. Cell Physiol. 226, 1583–1595. https://doi.org/10.1002/jcp.22484.
Pander, C.H., 1817. Dissertatio Inauguralis Sistens Historiam Metamorphoseos, Quam Ovum Incubatum Prioribus Quinque Diebus Subit. Typis Nitribitt.
Pang, Z.P., Yang, N., Vierbuchen, T., Ostermeier, A., Fuentes, D.R., Yang, T.Q., et al., 2011. Induction of human neuronal cells by defined transcription factors. Nature 476, 220–223. https://doi.org/10.1038/nature10202.
Pankratz, M.T., Li, X.-J., LaVaute, T.M., Lyons, E.A., Chen, X., Zhang, S.-C., 2007. Directed neural differentiation of human embryonic stem cells via an obligated primitive anterior stage. Stem Cells 25, 1511–1520. https://doi.org/10.1634/stemcells.2006-0707.
Pantazis, C.B., Yang, A., Lara, E., McDonough, J.A., Blauwendraat, C., Peng, L., et al., 2022. A reference human induced pluripotent stem cell line for large-scale collaborative studies. Cell Stem Cell 29, 1685–1702.e22. https://doi.org/10.1016/j.stem.2022.11.004.
Papaioannou, V.E., McBurney, M.W., Gardner, R.L., Evans, M.J., 1975. Fate of teratocarcinoma cells injected into early mouse embryos. Nature 258, 70–73. https://doi.org/10.1038/258070a0.
Pappenheim, A., 1896. Ueber entwickelung und ausbildung der erythroblasten. Arch. Pathol. Anat. Physiol. Klin. Med. 145, 587–643.
Park, I.-H., Arora, N., Huo, H., Maherali, N., Ahfeldt, T., Shimamura, A., et al., 2008a. Disease-specific induced pluripotent stem cells. Cell 134, 877–886. https://doi.org/10.1016/j.cell.2008.07.041.
Park, I.-H., Zhao, R., West, J.A., Yabuuchi, A., Huo, H., Ince, T.A., et al., 2008b. Reprogramming of human somatic cells to pluripotency with defined factors. Nature 451, 141–146. https://doi.org/10.1038/nature06534.

Passier, R., Orlova, V., Mummery, C., 2016. Complex tissue and disease modeling using hiPSCs. Cell Stem Cell 18, 309–321. https://doi.org/10.1016/j.stem.2016.02.011.
Paul, K.C., Krolewski, R.C., Lucumi Moreno, E., Blank, J., Holton, K.M., Ahfeldt, T., et al., 2023. A pesticide and iPSC dopaminergic neuron screen identifies and classifies Parkinson-relevant pesticides. Nat. Commun. 14, 2803. https://doi.org/10.1038/s41467-023-38215-z.
Piao, J., Zabierowski, S., Dubose, B.N., Hill, E.J., Navare, M., Claros, N., et al., 2021. Preclinical efficacy and safety of a human embryonic stem cell-derived midbrain dopamine progenitor product, MSK-DA01. Cell Stem Cell 28, 217–229.e7. https://doi.org/10.1016/j.stem.2021.01.004.
Piccolo, S., Sasai, Y., Lu, B., De Robertis, E.M., 1996. Dorsoventral patterning in Xenopus: inhibition of ventral signals by direct binding of chordin to BMP-4. Cell 86, 589–598. https://doi.org/10.1016/S0092-8674(00)80132-4.
Poczai, P., Santiago-Blay, J.A., 2022. Chip off the old block: generation, development, and ancestral concepts of heredity. Front. Genet. 13, 814436. https://doi.org/10.3389/fgene.2022.814436.
Polevoy, H., Gutkovich, Y.E., Michaelov, A., Volovik, Y., Elkouby, Y.M., Frank, D., 2019. New roles for Wnt and BMP signaling in neural anteroposterior patterning. EMBO Rep. 20. https://doi.org/10.15252/embr.201845842.
Priven, S.W., Alfonso-Goldfarb, A.M., 2009. Mathematics ab ovo: Hans Driesch and Entwicklungsmechanik. Hist. Philos. Life Sci. 31, 35–54.
Qi, Y., Zhang, X.-J., Renier, N., Wu, Z., Atkin, T., Sun, Z., et al., 2017. Combined small-molecule inhibition accelerates the derivation of functional cortical neurons from human pluripotent stem cells. Nat. Biotechnol. 35, 154–163. https://doi.org/10.1038/nbt.3777.
Ramalho-Santos, M., Willenbring, H., 2007. On the origin of the term "stem cell". Cell Stem Cell 1, 35–38. https://doi.org/10.1016/j.stem.2007.05.013.
Rauber, A., 1886. Personaltheil und germinaltheil des individuum. Zool. Anz. 9, 166–171.
Reubinoff, B.E., Itsykson, P., Turetsky, T., Pera, M.F., Reinhartz, E., Itzik, A., et al., 2001. Neural progenitors from human embryonic stem cells. Nat. Biotechnol. 19, 1134–1140. https://doi.org/10.1038/nbt1201-1134.
Rivetti di Val Cervo, P., Besusso, D., Conforti, P., Cattaneo, E., 2021. hiPSCs for predictive modelling of neurodegenerative diseases: dreaming the possible. Nat. Rev. Neurol. 17, 381–392. https://doi.org/10.1038/s41582-021-00465-0.
Roelink, H., Porter, J.A., Chiang, C., Tanabe, Y., Chang, D.T., Beachy, P.A., et al., 1995. Floor plate and motor neuron induction by different concentrations of the amino-terminal cleavage product of sonic hedgehog autoproteolysis. Cell 81, 445–455. https://doi.org/10.1016/0092-8674(95)90397-6.
Roux, W., 1888. Beiträge zur Entwickelungsmechanik des Embryo. V. Ueber die künstliche Hervorbringung "halber" Embryonen durch Zerstörung einer der beiden ersten Furchungszellen, sowie über die Nachentwickelung (Post generation) der fehlenden Körperhälfte. Virchows Arch. Bd. 114, 419–521.
Roux, W., 1885. Beiträge zur Entwickelungsmechanik des Embryos. I. Zur Orientirung über einige Probleme der organischen Entwickelung. Z. Biol. 21, 411–524.
Rowe, R.G., Daley, G.Q., 2019. Induced pluripotent stem cells in disease modelling and drug discovery. Nat. Rev. Genet. 20, 377–388. https://doi.org/10.1038/s41576-019-0100-z.
Saha, M., 1991. Spemann seen through a lens. A Conceptual History of Modern EmbryologySpringer US, Boston, MA, pp. 91–108. https://doi.org/10.1007/978-1-4615-6823-0_5.
Saleh, M.A., Amer-Sarsour, F., Berant, A., Pasmanik-Chor, M., Kobo, H., Sharabi, Y., et al., 2024. Chronic and acute exposure to rotenone reveals distinct Parkinson's

disease-related phenotypes in human iPSC-derived peripheral neurons. Free Radic. Biol. Med. 213, 164–173. https://doi.org/10.1016/j.freeradbiomed.2024.01.016.
Sánchez Alvarado, A., Yamanaka, S., 2014. Rethinking differentiation: stem cells, regeneration, and plasticity. Cell 157, 110–119. https://doi.org/10.1016/j.cell.2014.02.041.
Sander, K., 1993. Entelechy and the ontogenetic machine ? work and views of Hans Driesch from 1895 to 1910. Roux's Arch. Dev. Biol. 202, 67–69. https://doi.org/10.1007/BF00636530.
Sander, K., Faessler, P.E., 2001. Introducing the Spemann-Mangold organizer: experiments and insights that generated a key concept in developmental biology. Int. J. Dev. Biol. 45, 1–11.
Sareen, D., Svendsen, C.N., 2010. Stem cell biologists sure play a mean pinball. Nat. Biotechnol. 28, 333–335. https://doi.org/10.1038/nbt0410-333.
Sasai, Y., Lu, B., Steinbeisser, H., De Robertis, E.M., 1995. Regulation of neural induction by the Chd and Bmp-4 antagonistic patterning signals in Xenopus. Nature 376, 333–336. https://doi.org/10.1038/376333a0.
Sato, S.M., Sargent, T.D., 1989. Development of neural inducing capacity in dissociated Xenopus embryos. Dev. Biol. 134, 263–266. https://doi.org/10.1016/0012-1606(89)90096-1.
Satzinger, H., 2008. Theodor and Marcella Boveri: chromosomes and cytoplasm in heredity and development. Nat. Rev. Genet. 9, 231–238. https://doi.org/10.1038/nrg2311.
Saurat, N., Minotti, A.P., Rahman, M.T., Sikder, T., Zhang, C., Cornacchia, D., et al., 2024. Genome-wide CRISPR screen identifies neddylation as a regulator of neuronal aging and AD neurodegeneration. Cell Stem. Cell 31, 1162–1174. https://doi.org/10.1016/j.stem.2024.06.001.
Schlosser, G., 2023. From "self-differentiation" to organoids—the quest for the units of development. Dev. Genes. Evol. https://doi.org/10.1007/s00427-023-00711-z.
Schuldiner, M., Yanuka, O., Itskovitz-Eldor, J., Melton, D.A., Benvenisty, N., 2000. Effects of eight growth factors on the differentiation of cells derived from human embryonic stem cells. Proc. Natl. Acad. Sci. U.S.A. 97, 11307–11312. https://doi.org/10.1073/pnas.97.21.11307.
Schwartz, M.P., Hou, Z., Propson, N.E., Zhang, J., Engstrom, C.J., Costa, V.S., et al., 2015. Human pluripotent stem cell-derived neural constructs for predicting neural toxicity. Proc. Natl. Acad. Sci. U.S.A. 112, 12516–12521. https://doi.org/10.1073/pnas.1516645112.
Sekiya, S., Suzuki, A., 2011. Direct conversion of mouse fibroblasts to hepatocyte-like cells by defined factors. Nature 475, 390–393. https://doi.org/10.1038/nature10263.
Sell, S., 2010. On the stem cell origin of cancer. 2584–494. Am. J. Pathol. 176. https://doi.org/10.2353/ajpath.2010.091064.
Sharma, A., Sances, S., Workman, M.J., Svendsen, C.N., 2020. Multi-lineage human iPSC-derived platforms for disease modeling and drug discovery. Cell Stem Cell 26, 309–329. https://doi.org/10.1016/j.stem.2020.02.011.
Shi, Y., Inoue, H., Wu, J.C., Yamanaka, S., 2017. Induced pluripotent stem cell technology: a decade of progress. Nat. Rev. Drug Discov. 16, 115–130. https://doi.org/10.1038/nrd.2016.245.
Shi, Y., Massagué, J., 2003. Mechanisms of TGF-β signaling from cell membrane to the nucleus. Cell 113, 685–700. https://doi.org/10.1016/S0092-8674(03)00432-X.
Shim, G., Romero-Morales, A.I., Sripathy, S.R., Maher, B.J., 2024. Utilizing hiPSC-derived oligodendrocytes to study myelin pathophysiology in neuropsychiatric and neurodegenerative disorders. Front. Cell Neurosci. 17. https://doi.org/10.3389/fncel.2023.1322813.
Siminovitch, L., McCulloch, E.A., Till, J.E., 1963. The distribution of colony-forming cells among spleen colonies. J. Cell Comp. Physiol. 327–336. https://doi.org/10.1002/jcp.1030620313.

Simsek, M.F., Özbudak, E.M., 2022. Patterning principles of morphogen gradients. Open Biol. 12, 220224. https://doi.org/10.1098/rsob.220224.

Slack, J.M.W., 1991. From Egg to Embryo: Regional Specification in Early Development. Cambridge University Press.

Smith, A.G., Heath, J.K., Donaldson, D.D., Wong, G.G., Moreau, J., Stahl, M., et al., 1988. Inhibition of pluripotential embryonic stem cell differentiation by purified polypeptides. Nature 336, 688–690. https://doi.org/10.1038/336688a0.

Smith, W.C., Harland, R.M., 1992. Expression cloning of noggin, a new dorsalizing factor localized to the Spemann organizer in Xenopus embryos. Cell 70, 829–840. https://doi.org/10.1016/0092-8674(92)90316-5.

Solter, D., 2006. From teratocarcinomas to embryonic stem cells and beyond: a history of embryonic stem cell research. Nat. Rev. Genet. 7, 319–327. https://doi.org/10.1038/nrg1827.

Solter, D., Knowles, B.B., 1978. Monoclonal antibody defining a stage-specific mouse embryonic antigen (SSEA-1). Proc. Natl. Acad. Sci. U.S.A. 75, 5565–5569. https://doi.org/10.1073/pnas.75.11.5565.

Solter, D., Škreb, N., Damjanov, I., 1970. Extrauterine growth of mouse egg-cylinders results in malignant teratoma. Nature 227, 503–504. https://doi.org/10.1038/227503a0.

Spemann, H., 1938. Embryonic Development and Induction. Reprinted 1967. Hafner, New York.

Spemann, H., 1918. Über die Determination der ersten Organanlagen des Amphibienembryo I–VI. Arch. Entwicklmech Org. 43, 448–555.

Spemann, H., 1901. Ueber korrelationen in der entwicklung des auges. Verh. Anat. Ges. 15, 61–79.

Spemann, H., Geinitz, B., 1927. Über Weckung organisatorischer Fähigkeiten durch Verpflanzung in organisatorische Umgebung. Wilhelm Roux'. Arch. Entwicklmech Org. 109, 129–175.

Spemann, H., Mangold, H., 1924. über Induktion von Embryonalanlagen durch Implantation artfremder Organisatoren. Arch. Mikrosk. Anat. Enwicklmech. 100, 599–638.

Stemple, D.L., 2005. Structure and function of the notochord: an essential organ for chordate development. Development 132, 2503–2512. https://doi.org/10.1242/dev.01812.

Stevens, L.C., 1970. The development of transplantable teratocarcinomas from intratesticular grafts of pre- and postimplantation mouse embryos. Dev. Biol. 21, 364–382. https://doi.org/10.1016/0012-1606(70)90130-2.

Stevens, L.C., Little, C.C., 1954. Spontaneous testicular teratomas in an inbred strain of mice. Proc. Natl. Acad. Sci. U S A 40, 1080–1087. https://doi.org/10.1073/pnas.40.11.1080.

Stojković, M., Daher, S.R., 2008. Celebrating 10 years of hESC lines: an interview with James Thomson. Stem Cells 26, 2747–2748. https://doi.org/10.1634/stemcells.2008-0942.

Stummann, T.C., Hareng, L., Bremer, S., 2009. Hazard assessment of methylmercury toxicity to neuronal induction in embryogenesis using human embryonic stem cells. Toxicology 257, 117–126. https://doi.org/10.1016/j.tox.2008.12.018.

Surani, M.A., 2016. Breaking the germ line–soma barrier. Nat. Rev. Mol. Cell Biol. 17, 136. https://doi.org/10.1038/nrm.2016.12.

Takahashi, K., Tanabe, K., Ohnuki, M., Narita, M., Ichisaka, T., Tomoda, K., et al., 2007. Induction of pluripotent stem cells from adult human fibroblasts by defined factors. Cell 131, 861–872. https://doi.org/10.1016/j.cell.2007.11.019.

Takahashi, K., Yamanaka, S., 2006. Induction of pluripotent stem cells from mouse embryonic and adult fibroblast cultures by defined factors. Cell 126, 663–676. https://doi.org/10.1016/j.cell.2006.07.024.

Taléns-Visconti, R., Sanchez-Vera, I., Kostic, J., Perez-Arago, M.A., Erceg, S., Stojkovic, M., et al., 2011. Neural differentiation from human embryonic stem cells as a tool to study early brain development and the neuroteratogenic effects of ethanol. Stem Cells Dev. 20, 327–339. https://doi.org/10.1089/scd.2010.0037.

Tao, Y., Zhang, S.-C., 2016. Neural subtype specification from human pluripotent stem cells. Cell Stem Cell 19, 573–586. https://doi.org/10.1016/j.stem.2016.10.015.

Tchieu, J., Zimmer, B., Fattahi, F., Amin, S., Zeltner, N., Chen, S., et al., 2017. A modular platform for differentiation of human PSCs into all major ectodermal lineages. Cell Stem Cell 21, 399–410.e7. https://doi.org/10.1016/j.stem.2017.08.015.

Temple, S., 2023. Advancing cell therapy for neurodegenerative diseases. Cell Stem Cell 30, 512–529. https://doi.org/10.1016/j.stem.2023.03.017.

Thomson, J.A., Itskovitz-Eldor, J., Shapiro, S.S., Waknitz, M.A., Swiergiel, J.J., Marshall, V.S., et al., 1998. Embryonic stem cell lines derived from human blastocysts. Science 282 (1979), 1145–1147. https://doi.org/10.1126/science.282.5391.1145.

Thomson, J.A., Kalishman, J., Golos, T.G., Durning, M., Harris, C.P., Becker, R.A., et al., 1995. Isolation of a primate embryonic stem cell line. Proc. Natl. Acad. Sci. U S A 92, 7844–7848. https://doi.org/10.1073/pnas.92.17.7844.

Thomson, J.A., Kalishman, J., Golos, T.G., Durning, M., Harris, C.P., Hearn, J.P., 1996. Pluripotent cell lines derived from common marmoset (*Callithrix jacchus*) Blastocysts1. Biol. Reprod. 55, 254–259. https://doi.org/10.1095/biolreprod55.2.254.

Till, J.E., McCulloch, E.A., 1961. A direct measurement of the radiation sensitivity of normal mouse bone marrow cells. Radiat. Res. 14, 213–222.

Van Lent, J., Prior, R., Pérez Siles, G., Cutrupi, A.N., Kennerson, M.L., Vangansewinkel, T., et al., 2024. Advances and challenges in modeling inherited peripheral neuropathies using iPSCs. Exp. Mol. Med. 56, 1348–1364. https://doi.org/10.1038/s12276-024-01250-x.

Vazin, T., Ball, K.A., Lu, H., Park, H., Ataeijannati, Y., Head-Gordon, T., et al., 2014. Efficient derivation of cortical glutamatergic neurons from human pluripotent stem cells: a model system to study neurotoxicity in Alzheimer's disease. Neurobiol. Dis. 62, 62–72. https://doi.org/10.1016/j.nbd.2013.09.005.

Vierbuchen, T., Ostermeier, A., Pang, Z.P., Kokubu, Y., Südhof, T.C., Wernig, M., 2010. Direct conversion of fibroblasts to functional neurons by defined factors. Nature 463, 1035–1041. https://doi.org/10.1038/nature08797.

Virchow, R., 1858. Die Cellularpathologie in Ihrer Begründung auf Physiologische und Pathologische Gewebelehre. August Hirschwald, Berlin.

Virchow, R., 1861. Die Cellularpathologie in ihrer Begrundung auf physiologische und pathologische Gewebelehre.

Waddington, C.H., 1957. The Strategy of the Genes; A Discussion of Some Aspects of Theoretical Biology. Allen & Unwin, London.

Waddington, C.H., 1938. Studies on the nature of the amphibian organization centre—VII. Evocation by some further chemical compounds. Proc. R Soc. Lond. B Biol. Sci. 125, 365–372. https://doi.org/10.1098/rspb.1938.0032.

Wakayama, T., Perry, A.C.F., Zuccotti, M., Johnson, K.R., Yanagimachi, R., 1998. Full-term development of mice from enucleated oocytes injected with cumulus cell nuclei. Nature 394, 369–374. https://doi.org/10.1038/28615.

Wallace, J.L., Pollen, A.A., 2024. Human neuronal maturation comes of age: cellular mechanisms and species differences. Nat. Rev. Neurosci. 25, 7–29. https://doi.org/10.1038/s41583-023-00760-3.

Wang, H., Yang, Y., Liu, J., Qian, L., 2021. Direct cell reprogramming: approaches, mechanisms and progress. Nat. Rev. Mol. Cell Biol. 22, 410–424. https://doi.org/10.1038/s41580-021-00335-z.

Wang, J., Sun, S., Deng, H., 2023. Chemical reprogramming for cell fate manipulation: methods, applications, and perspectives. Cell Stem Cell 30, 1130–1147. https://doi.org/10.1016/j.stem.2023.08.001.

Watanabe, K., Kamiya, D., Nishiyama, A., Katayama, T., Nozaki, S., Kawasaki, H., et al., 2005. Directed differentiation of telencephalic precursors from embryonic stem cells. Nat. Neurosci. 8, 288–296. https://doi.org/10.1038/nn1402.

Weintraub, H., Tapscott, S.J., Davis, R.L., Thayer, M.J., Adam, M.A., Lassar, A.B., et al., 1989. Activation of muscle-specific genes in pigment, nerve, fat, liver, and fibroblast cell lines by forced expression of MyoD. Proc. Natl. Acad. Sci. U.S.A. 86, 5434–5438. https://doi.org/10.1073/pnas.86.14.5434.

Weismann, A., 1892. Das Keimplasma: Eine Theorie der Vererbung. Fischer,.

Weismann, A., 1885. Die Continuität des Keimplasma's als grundlage einer Theorie der Vererbung. Verlag von Gustav Fischer, Jena.

Wichterle, H., Lieberam, I., Porter, J.A., Jessell, T.M., 2002. Directed differentiation of embryonic stem cells into motor neurons. Cell 110, 385–397. https://doi.org/10.1016/S0092-8674(02)00835-8.

Williams, L.A., Davis-Dusenbery, B.N., Eggan, K.C., 2012. SnapShot: directed differentiation of pluripotent stem cells. Cell 149, 1174.e1. https://doi.org/10.1016/j.cell.2012.05.015.

Wilmut, I., Schnieke, A.E., McWhir, J., Kind, A.J., Campbell, K.H.S., 1997. Viable offspring derived from fetal and adult mammalian cells. Nature 385, 810–813. https://doi.org/10.1038/385810a0.

Wilson, E.B., 1900. The Cell in Development and Inheritance. Macmillan.

Wolff, C.F., 1759. Theoria Generationis. Typis et sumtu Io. Christ. Hendel.

Xie, J., Wu, S., Szadowski, H., Min, S., Yang, Y., Bowman, A.B., et al., 2023. Developmental Pb exposure increases AD risk via altered intracellular Ca2+ homeostasis in hiPSC-derived cortical neurons. J. Biol. Chem. 299, 105023. https://doi.org/10.1016/j.jbc.2023.105023.

Xu, R.-H., Peck, R.M., Li, D.S., Feng, X., Ludwig, T., Thomson, J.A., 2005. Basic FGF and suppression of BMP signaling sustain undifferentiated proliferation of human ES cells. Nat. Methods 2, 185–190. https://doi.org/10.1038/nmeth744.

Yamada, T., Pfaff, S.L., Edlund, T., Jessell, T.M., 1993. Control of cell pattern in the neural tube: motor neuron induction by diffusible factors from notochord and floor plate. Cell 73, 673–686. https://doi.org/10.1016/0092-8674(93)90248-O.

Yamanaka, S., 2020. Pluripotent stem cell-based cell therapy—promise and challenges. Cell Stem Cell 27, 523–531. https://doi.org/10.1016/j.stem.2020.09.014.

Yamanaka, S., Blau, H.M., 2010. Nuclear reprogramming to a pluripotent state by three approaches. Nature 465, 704–712. https://doi.org/10.1038/nature09229.

Yan, Y., Yang, D., Zarnowska, E.D., Du, Z., Werbel, B., Valliere, C., et al., 2005. Directed differentiation of dopaminergic neuronal subtypes from human embryonic stem cells. Stem Cells 23, 781–790. https://doi.org/10.1634/stemcells.2004-0365.

Yao, Z., van Velthoven, C.T.J., Kunst, M., Zhang, M., McMillen, D., Lee, C., et al., 2023. A high-resolution transcriptomic and spatial atlas of cell types in the whole mouse brain. Nature 624, 317–332. https://doi.org/10.1038/s41586-023-06812-z.

Yu, J., Vodyanik, M.A., Smuga-Otto, K., Antosiewicz-Bourget, J., Frane, J.L., Tian, S., et al., 2007. Induced pluripotent stem cell lines derived from human somatic cells. Science 318 (1979), 1917–1920. https://doi.org/10.1126/science.1151526.

Zeng, H., 2022. What is a cell type and how to define it? Cell 185, 2739–2755. https://doi.org/10.1016/j.cell.2022.06.031.

Zhai, J., Xiao, Z., Wang, Y., Wang, H., 2022. Human embryonic development: from peri-implantation to gastrulation. Trends Cell Biol. 32, 18–29. https://doi.org/10.1016/j.tcb.2021.07.008.

Zhang, S.-C., Wernig, M., Duncan, I.D., Brüstle, O., Thomson, J.A., 2001. In vitro differentiation of transplantable neural precursors from human embryonic stem cells. Nat. Biotechnol. 19, 1129–1133. https://doi.org/10.1038/nbt1201-1129.

Zimmerman, L.B., De Jesús-Escobar, J.M., Harland, R.M., 1996. The Spemann organizer signal noggin binds and inactivates bone morphogenetic protein 4. Cell 86, 599–606. https://doi.org/10.1016/S0092-8674(00)80133-6.

CHAPTER TWO

Immortalized neuronal lines versus human induced pluripotent stem cell-derived neurons as in vitro toxicology models

Xueqi Tang and Aaron B. Bowman*
School of Health Sciences, Purdue University, West Lafayette, IN, United States
*Corresponding author. e-mail address: bowma117@purdue.edu

Contents

Abstract

Immortalized neuronal lines are widely applied in toxicology studies for their easy accessibility, low costs, minimized ethical issues, and potentialities for high-throughput screenings. Advancements in differentiation methods have revealed the capacity of select cell lines to undergo manipulation, resulting in a low-proliferative state while concurrently expressing biomarkers of mature neurons. Despite this potentiality, cellular biology of immortalized lines, especially proliferation-related

Advances in Neurotoxicology, Volume 12
ISSN 2468-7480, https://doi.org/10.1016/bs.ant.2024.07.004

signaling, can be distinct from post-mitotic human brain neurons. Therefore, it raised the concern for applying immortalized lines in neurotoxicology studies due to the not yet defined impact on cellular responsiveness. Besides, while the homogeneity of cell lines on one side simplified the experimental design, on the other side it ceased to represent the complexity of cell types present in human central nervous system (CNS). Therefore, neuronal models derived from human induced pluripotent stem cells (hiPSCs) are now seeing expanded use to enable advanced modeling of the CNS. This chapter aims to comprehensively review the utilization of both models in neurotoxicology, and to offer insights into the strengths and limitations inherent in each system. Through the compilation of available literature, the ultimate objective of this chapter is to foster informed experimental designs in the field.

1. Characterization and utilization of cell lines

Based on the source from which the lines were originally derived from, immortalized neuronal lines can be divided into two main categories, (1) lines that were derived from cancerous cells, and (2) lines that were developed from healthy tissues. Continuous culture and sub-cloning of bone marrow or tumor tissues obtained from biopsy is one of the original routes to establish cancerous cell lines (Andrews, 1984; Biedler et al., 1973). Transplanted normal brain tissue can be immortalized by retrovirus transduction of myc pro-oncogenes (Lotharius et al., 2002; Pollock et al., 2006). Therefore, most cell lines exhibit high expression and activity levels of myc. Besides activated oncogenes, some cancerous lines possess abnormal chromosome arrangements including higher modal chromosome numbers, trisomic or higher ploidy chromosomes, or aneuploidy. The myc overexpression and karyotypic abnormalities on one side enabled or permit rapid and continuous proliferation of the cells, yet on the other hand impact the translational power of neuronal models developed with these lines. For instance, it has been observed in rodents that myc can mediate p53 independent apoptosis, and therefore may result in decreased sensitivity to chemicals toxicity acting through p53 pathways, and failures of detecting p53 activation in adverse outcome pathway studies (Culbreth et al., 2012).

1.1 Cell line models derived from carcinoma – the SH-SY5Y line as an example

1.1.1 Development and differentiation methods of SH-SY5Y cell line

One of the most widely used and well-developed neuronal lines from cancerous cells is SH-SY5Y, commonly abbreviated as SY5Y. SH-SY5Y is a human neuroblastoma line derived from its parental line SK-N-SH, which was developed by Biedler et al. from a four-year-old female neuroblastoma patient and first reported in 1973 (Biedler et al., 1973). Five years after the establishment of SK-N-SH line, SH-SY5Y was reported as a sub-cloned proportion of the SK-N-SH line, characterized by homogenous neuroblast-like morphology and significantly higher expression of dopamine-β-hydroxylase comparing to other clones cultured in parallel (Biedler et al., 1978). Considering its potential of representing dopaminergic neuronal-like features, SH-SY5Y line is applied as an in vitro model for studying Parkinson's disease (PD) etiology and environmentally relevant risks for PD and other neurotoxicity. In a prior screening study, SH-SY5Y showed the highest sensitivity to toxic compounds compared to 8 different lines derived from different species and organs and displayed distinct response patterns compared to the rat N2a neuroblastoma line, thus validated its potential as a sensitive and human-relevant model (Xia et al., 2008). The rapid spread of SH-SY5Y utilization soon brought up the discussion how to optimize the cell line for a better and more standardized modeling. One of the approaches was through retinoic acid (RA) differentiation reported by Påhlman et al. in 1984, describing cell cycle arrest and morphological promotion towards neuronal-like phenotype (Påhlman et al., 1984). RA differentiation, normally accompanied by serum deprivation, has been confirmed to significantly decrease the proliferation rate, elongate neurites, promote expression of neuronal biomarkers (including SMI31, MAP2, GAP43, and β-III tubulin), and potentiated dopaminergic phenotype (represented by increased gene expression of dopaminergic markers *SLC18A1* and *GCH1*) (Attoff et al., 2020; Dravid et al., 2021; Lopes et al., 2017; Shipley et al., 2016). Following the RA differentiation protocol, further manipulation of culture media and maintenance conditions were reported, achieving improvements in neuron-like morphology, expression of neuronal and synaptic biomarkers, synthesis of neurotransmitters, and response to electrophysiological stimulations. Optimized differentiation protocol also introduced extracellular matrix, B27 neuronal survival support supplement, and neurotrophic factors (eg, brain derived neurotrophic factor, BDNF), which enhanced and stabilized their mature neuron-like morphology (Dravid et al., 2021; Shipley et al., 2016). Besides RA and neurotrophic differentiation, other methods to induce a post-mitotic like state include 5-fluorodeoxyuridine (5-fdu), phorbol-12-myristate-13-acetate (PMA), and staurosporine (ST) have all been evidenced to decrease SH-SY5Y proliferation and promote neuronal-like morphology and biomarker expression (Ducray et al., 2020; Påhlman et al., 1991;

Strother et al., 2021). Furthermore, recent studies have reported cultivating SH-SY5Y cells into neuronal types other than dopaminergic, including promoting glutamatergic-like phenotype with B27 supplement, and advancing cholinergic-like features via differentiate with NGF, neuregulin β1, and vitamin D_3 (D'Aloia et al., 2024; Martin et al., 2022).

1.1.2 Differentiation of SH-SY5Y cells introduce variations to their toxicological responses

As described above, variations exist in SH-SY5Y differentiation protocols, which lead to reasonable and often expressed concerns regarding the stability and reproducibility of the model (Xicoy et al., 2017). Although the differentiation can yield a more homogenized neuron-like system compared to the undifferentiated culture, the introduction of differentiation factors can cause variability in cellular responses to toxicants. An extensively studied example is 6-hydroxydopamine (6-OHDA). As a neurotoxicant chiefly used in Parkinson's disease modeling, alterations of cellular sensitivity to 6-OHDA by differentiation interferes with data interpretation in Parkinson's-related studies using SH-SY5Y models. The reported impact of RA differentiation on SH-SY5Y sensitivity to 6-OHDA cytotoxicity is both differentiation method and 6-OHDA treatment condition dependent. Shorter term of RA application showed less impact on SH-SY5Y sensitivity to 6-OHDA cytotoxicity. 4-day application of 10 μM RA did not significantly affect cell viability under 24-h 6-OHDA treatment compared to undifferentiated cells, while 7–10 days of differentiation nearly doubled the sensitivity to 6-OHDA-induced cell death, with 50% viability loss 6-OHDA concentration (GI_{50}) decreased from 35 μM (for 4-day RA) to 15 μM (for 7-day RA) (Lopes et al., 2010). Differentiated cells display inconsistent yet potentially higher tolerance to low concentrations of 6-OHDA (under 25 μM), while are significantly more sensitive to high-concentration treatments compared to undifferentiated cultures (Cheung et al., 2009; Simões et al., 2021). Similarly, other Parkinson-induction neurotoxicants, including rotenone, paraquat, 1-methyl-4-phenyl-1,2,3,6-tetrahydropyridine (MPTP), and 1-methyl-4-phenyl-pyridinium ion (MPP^+) have been confirmed to cause more cell viability loss in differentiated SH-SY5Y cells compared to the undifferentiated line under the same treatment condition (Elmorsy et al., 2023; Khwanraj et al., 2015). Besides affecting sensitivity to exogenously applied neurotoxicants, RA application may also alter cellular response to genetic modification in Parkinson's disease modeling. 10 μM RA for 6 days in media deprived of pyruvate and uridine was reported to decrease survival of Parkinson's disease cybrids generated from ρ^0 cells derived from SH-SY5Ys (Appukuttan et al., 2013).

Beyond affecting sensitivity to toxicant-induced cytotoxicity, signaling pathway activities that are altered by continuous RA treatment can be part of the concerns for utilizing RA-differentiated SH-SY5Y as a neurotoxicity model. Mechanistic studies reported changes in apoptosis-related gene and protein baseline expression, overall capacity of antioxidant defenses, dopamine transporter expression, and baseline mitochondrial membrane potentials being consequences of the RA treatment and being mechanisms of altered sensitivity to Parkinson-inducing toxicants (Forster et al., 2016; Khwanraj et al., 2015; Lopes et al., 2017).

Implication on sensitivity by differentiation were also reported in other neurotoxicants. The Forsby group reported significant downregulated *BNDF* gene expression when SH-SY5Y were simultaneously treated with 1 μM RA (supplied with N2) and exposed to developmental toxicant acrylamide for 9 days (Attoff et al., 2020). However, the expression levels of *BDNF* were not statistically significant when the treatment and exposure only last for 6 days (Hinojosa et al., 2023). The authors attributed this to the length of the treatment and number of replicates available. These described variations in SH-SY5Y-involved studies brought up concerns about the reliability of detecting unknown mode of action of toxicants, despite the popularity of differentiated SH-SY5Ys in neurotoxicology. To enable stable data generation with this cell line, and promote future high-throughput and automated pipeline screening potentials, establishment of standardized differentiation methods is of vital importance.

1.1.3 Other cell lines derived from human carcinoma

Human pluripotent stem cell line developed from carcinoma also possesses the potential to be differentiated towards neural fate. A stem cell line NTera-2 (NT2) derived from human male teratocarcinoma can derive postmitotic neuronal and glial populations with RA treatment. 4–5 weeks of 10 μM RA differentiation followed by 3–4 weeks of mitotic inhibitor cocktail treatment (1 μM cytosine arabinoside, 10 μM fluorodexoyuridine, and 10 μM uridine) in NT2 cells could provide a mixed population of MAP2 and β-III tubulin (TUBB3) positive neurons and GFAP positive glial cells (Laurenza et al., 2013; Stern et al., 2014). NT2 cells also have the potential to form 3D structures, and display cell migration features that can be evaluated for developmental toxicity (Fabbri et al., 2023; Hill et al., 2012).

1.2 Cell line models derived from healthy brain tissues – LUHMES cells as an example

1.2.1 Development and differentiation methods of LUHMES

Cell lines derived from non-cancerous tissues are another major type of models for in vitro neurotoxicology studies, with the most widely used line in this category being the Lund human mesencephalic (LUHMES) neuronal line. The LUHMES cells were derived from a subclone of the MESC2.10 line developed from 8-week-old female human embryonic ventral mesencephalon and immortalized by introducing a tetracycline-responsive v-myc gene (Lotharius et al., 2005, 2002). Proliferating MESC2.10 express NR4A2 mRNA, which is an orphan nuclear receptor involved in dopaminergic differentiation, thus suggesting its dopaminergic lineage profile (Luo et al., 2008). With tetracycline treatment, MESC2.10 cells exit cell cycle; they exhibit neurites, express dopaminergic markers, and release dopamine. However, differentiation with tetracycline only was insufficient to stabilize the dopaminergic phenotype of MESC2.10 nor to generate tyrosine hydroxylase (TH) expressing neurons when transplanted in vivo. Therefore, differentiation with dibutyryl cyclic AMP (db-cAMP, 1 mM) and glial cell line-derived neurotrophic factor (GDNF, 2 ng/mL) in addition to tetracycline were tested in N2 medium and proved to increase in vitro TH expression and stabilize TH immunoreactivity when grafted into rat striatum (Paul et al., 2007). Similarly, LUHMES cells differentiated with the tetracycline, cAMP, and GDNF protocol exhibit intense signal of TUBB3, TH, and VMAT2 in immunofluorescence, as well as stimulated dopaime release. This differentiation method of LUHMES has been adopted by the majority of studies using this line with minor modifications. Despite the robustness of intensifying dopaminergic phenotype of the population, only part of the key features of dopaminergic neurons were recapitulated during the differentiation. Dopaminergic neuron-specific transcription factor pituitary homeobox 3 (PITX3) expression level was reported to have no significant change over the differentiation process. A recent approach aiming to optimize the differentiation cocktail investigated the effects of increasing GDNF supplementation to 20 ng/mL, with the addition of 20 ng/mL BDNF (brain derived neurotrophic factor), 20 ng/mL recombinant human TGFβ III (transforming growth factor beta III), 10 ng/mL recombinant human LIF (leukemia inhibitory factor) and 0.2 mM ascorbic acid. With the addition of the listed neurogenic compounds, neuralization and dopaminergic neuronal purity was improved, represented by higher percentage of TH positive cells and elevated dopamine uptake activity (Loser et al., 2021).

Beyond targeted evaluation of dopaminergic specific markers, the enhanced neuronal phenotype in differentiated LUHMES was also validated at omics levels. 6-day differentiation with tetracycline, cAMP, and GDNF resulted in proteomic and transcriptomic changes that were significant enough to distinguish the two populations. Enrichment analysis suggested that amongst the differentially expressed genes and abundant proteins, differentiated LUHMES had down-regulated cell proliferation and up-regulated neuronal phenotype, which aligned with reported single endpoint measurements (Delp et al., 2018a; Tüshaus et al., 2021). Furthermore, proteomics level analysis revealed that differentiated LUHMES displayed increased abundances of Parkinson's disease-risk related proteins, including the exclusively expressed microtubule-associated protein tau (MAPT) and synapsin 1 (SYN1), paralleled by moderate increase in Parkinson's disease-risk gene products PARK1 and PARK5 (Tüshaus et al., 2021). It was also reported that differentiated LUHMES had a significantly higher capacity to upregulate mitochondrial function and glycolysis upon stimuli compared to undifferentiated cells. Therefore, the differentiation decreased their susceptibility to mitochondrial inhibitors when measured by resazurin reduction and lactate dehydrogenase (LDH) release (Delp et al., 2018a). In agreement with the elevated capacity for metabolism upregulation, differentiated LUHMES cells were more responsive to ATP rescue compounds, with a EC_{50} of 12.3 nmol/L GW8510 in LUHMES and compared to 110 nmol/L in primary rat dopaminergic neurons (Zhang et al., 2014). Yet in contrast, differentiated LUHMES cells were also reported to be more vulnerable than undifferentiated cells to other known neurotoxicants, including colchicine, methylmercury, and vincristine (Tong et al., 2017). A list of available comparisons made between differentiated versus undifferentiated LUHMES and SH-SY5Y cells investigating their sensitivity to neurotoxicants highly relevant to PD pathogenesis or used in PD modeling are provided in Table 1.

1.2.2 LUHMES versus SH-SY5Y as a dopaminergic neuronal line

Compared to the SH-SY5Y system, LUHMES may provide more information about network connectivity considering its higher capability of generating electrophysiological activities and developing 3D structures. Detection of functional voltage-gated Na^+ and K^+ channels and spontaneous firing with bursts has been reported across various differentiation methods in LUHMES and the firing can arise as early as differentiation day 5 (Harischandra et al., 2020). The development of a functionally coupled

Table 1 Implications of neuronal differentiation on cell line sensitivity to Parkinson's disease-inducing neurotoxicants.

Line	Toxicant	Diff. IC_{50} (μM)	Undiff. IC_{50} (μM)	Exposure time (h)	Measurement	Diff. method		Diff. length	References
						Compounds	Base		
SH-SY5Y	Rotenone	**7.2**	16.3	24	CellTitreGlo intracellular ATP	10 μM all-trans RA	EMEM/F12, 3 % FBS	7 d	Tong et al. (2017)
	6-OHDA	20.44	**16.21**						
	FCCP	**6.25**	50	6	Resazurin reduction	10 μM all-trans RA	Low glucose DMEM, 1 % FBS	3 d	Simões et al. (2021)
	Paraquat	**1.4**	6.7	48	MTT assay	10 μM all-trans RA for 6 d, 80 nM TPA for 6 d	Not specified	6 d	Elmorsy et al. (2023)
	Rotenone	**23.4**	38						
	MPP^+	1000	**500**	24	MTT assay	10 μM all-trans RA	MEM/F12, 10 % FBS	3 d	Khwanraj et al. (2015)
	6-OHDA	35	**15**	24	MTT assay	10 μM all-trans RA	DMEM, 2 mM L-glutamine, 1 % FBS	7 d	Lopes et al. (2010)
	6-OHDA	35	32					4 d	

LUHMES	Rotenone	1.5	**0.17**	24	CellTitreGlo intracellular ATP	1 μg/mL tetracycline, 1 mM cAMP, 2 ng/mL GDNF	Advanced DMEM/ F12, 2 mM L-glutamine, 1 × N2	7 d	Tong et al. (2017)
	6-OHDA	0.94	1.18						
	Rotenone	48.3	**2.1**	24	LDH release	2.25 μM tetracycline, 1 mM dibutyryl cAMP, 2 ng/mL GDNF	Advanced DMEM/ F12, 2 mM L-glutamine, 1 × N2	6 d	Delp et al. (2018a)

6-OHDA, 6-hydroxydopamine; Diff., differentiation; FBS, fetal bovine serum; FCCP, carbonyl cyanide-4-(trifluoromethoxy) phenylhydrazone; GDNF, glial cell line-derived neurotrophic factor; LDH, Lactate dehydrogenase; MPP^+, 1-methyl-4-phenyl-pyridinium ion; RA, retinoic acid; TPA, 12-O-tetradecanoyl-phorbol-13-acetate. The differentiation status that displayed significantly lower IC_{50} in each study was highlighted in bold.

network in differentiated LUHMES cells were also confirmed by the detection of long-lasting oscillations of Ca^{2+} signal. This network activity could be impacted by Na_v channel and DAT modulators, suggesting its potential of evaluating neurotoxins that may affect circulatory functions of dopaminergic neurons (Loser et al., 2021). Neuronal networks were established in LUHMES 3D differentiation as well via mechanical shaking triggered aggregation. 21-day 3D differentiation in LUHMES cells resulted in neurospheroids with validated sufficient oxygen and nutrient supply and significant dopaminergic neuron marker expression throughout the structure. Cell viability assessments further confirmed the sensitivity of 3D cultures to known neurotoxicants, with robust potential for application in 1536 well high throughput screening design (Smirnova et al., 2016; Tong et al., 2024).

Despite the advantages of LUHMES cells over SH-SY5Ys, including more standardized differentiation method, higher sensitivity to toxicants, and capability in displaying electrophysiological features, there are still variables in LUHMES that may require investigators to take caution in their interpretation of data from this cell model with rigorous consideration of limitations of the system to provide reproducible and meaningful extrapolation to the in vivo situation. One example is a promising drug target for Alzheimer's disease, β-secretase BACE1, was not detectable in LUHMES independent of differentiation status, yet highly expressed in human brain extracts (Tüshaus et al., 2021; Willem et al., 2009). Similarly, the expression level of BACE1 was lower in differentiated SH-SY5Ys compared to primary cultured mice cortical neurons. Furthermore, pharmacological inhibition of BACE1 in SH-SY5Ys triggered uncoupling of the protease from its target amyloid precursor protein, which was not observed in primary cortical neurons (Colombo et al., 2013). Besides variation of gene expression level across different culture systems, long-term culture of LUHMES themselves can introduce genetic drift. The investigation on genomic changes in LUHMES by Gutbier et al. was inspired by the drastic increase of their tolerance to MPP^+. LUHMES cells from the original provider laboratory (University of Konstanz, UKN) died upon a 72-h treatment of MPP^+ at 5 μM, yet the same line originated from American Type Culture Collection (ATCC) showed an IC_{50} of 65.06 μM in 48 h MPP^+ treatment (Gutbier et al., 2018; Krug et al., 2014; Zhang et al., 2014). Phenotypic characterization suggested that after 6 d differentiation, there was no difference between the two subpopulations in the gene expression levels of neuronal markers including TUBB3, SYP, and

SNAP25, nor any difference in the immunoreactivity to β-III tubulin and postsynaptic density protein 95 (PSD95). However, the ATCC sub-population of this line displayed a significantly higher neurite area, and a downregulation of DAT and TH by the end of the differentiation (day 6) compared to the UKN sub-population. The uptake rate of MPP^+ is dramatically higher in the UKN subpopulation compared to ATCC, which explained its higher sensitivity to MPP^+ induced cell death. Furthermore, whole genome sequencing identified structural variants, as well as mutations and consequent protein expression alterations in stress-response related genes (Gutbier et al., 2018). This variation suggested that subtle differences may present and affect cellular sensitivity to toxicants despite similarity under standard characterization methods, and again emphasizes the importance of choosing the best match between modeling system and scientific questions.

1.2.3 Other cell lines derived from normal brain tissues

Like the development of LUHMES, transplantation of fetal brain tissue followed by myc family gene overexpression yielded several other neural stem cell (NSC) and neural progenitor cell (NPC) lines that have normal human karyotype and possess multipluripotency. ReNcell CX and VM are a set of NSC lines that were derived from cortex from a 14-week gestation (male) and midbrain from a 10-week gestation (male) respectively (Donato et al., 2007; Pollock et al., 2006). ReNcell CX has been reported to be analogous to neuroectodermal sphere differentiated from H9 human embryonic stem cells at transcriptomic level (Oh et al., 2017). Differentiation of ReNcell with growth factor depletion (removal of epidermal growth factor, EGF, and fibroblast growth factor 2, FGF2) plus optional db-cAMP and GDNF addition enhance their neuronal morphology and expression levels of TUBB3. In terms of sensitivity to toxicants, ReNcell CX NPCs displayed a higher resistance to p53 and caspase 3/7 related apoptosis, which may be attributed to the overexpression status of myc oncogenes (Culbreth et al., 2012; Druwe et al., 2015). Similar to the LUHMES line, differentiation in ReNcell VM cells boosted their resistance to proliferation deficits and viability loss triggered by mitochondria-targeting compounds (Nierode et al., 2016). This increase of tolerance along with maturation was observed in NPCs differentiated from hiPSCs as well, which may justify the use of immortalized lines developed from normal brain tissue over carcinoma lines in toxicology evaluations (Joshi et al., 2019; Slavin et al., 2021).

2. Characterization and utilization of hiPSC-derived neurons

Cultured hiPSC-derived neurons have been accepted as a powerful tool for in vitro alternative investigations in neurotoxicology, with a rapid expansion in application. The average number of publications per year with keywords "hiPSC" and "neurotoxicity" has doubled since 2017 and there were over 2000 manuscripts published in the year 2023 (Dimensions, 2018). Beyond its popularity, the effectiveness of hiPSC-derived system is also supported by evidence. In a responsiveness study with botulinum neurotoxins (BoNT), results from hiPSC-derived motor neurons aligned with the first-in-human study by displaying equivalent or greater responsiveness to recombinant BoNT serotype E compared to serotype A (De Lamotte et al., 2021; Pons et al., 2019). This contrasted findings from an earlier study conducted with rat primary spinal cord neurons which concluded that BoNT serotype A was more effective than serotype E (Donald et al., 2018). These data strongly suggested the translational potential of hiPSC-derived systems to recapitulate human responses.

2.1 hiPSCs preserve disease-related genetic variants

With dual SMAD inhibition method (small molecule and peptide modulator inhibition on bone morphogenic protein, TGF-β, and canonical Wnt pathways) that is discussed in further detail elsewhere in this volume, hiPSCs can be differentiated into neuroectodermal linages via simulating temporal fate-defining cues in human brain development (Chambers et al., 2009; Yan et al., 2023). This process can be performed in house, which would therefore enable developmental exposures during the induction phase; otherwise, commercialized iCell® neurons from Cellular Dynamics International (FUJIFILM Cellular Dynamics, Inc., Madison, WI) and other vendors provide access to purified postmitotic neurons across different lineages. The transcriptomic stability of commercially available hiPSC-derived glutamatergic neurons has been confirmed with a consistent increase of mature neuronal markers triggered by the differentiation; plus the expression were maintained at compatible levels amongst multiple lines originating from different vendors (Snyder et al., 2018). Neuronal culture from either source, in comparison to immortalized cell lines, possess normal karyotypes, are transgene free, and have exited cell cycle. Furthermore, although whole-genome sequencing with SH-SY5Y indicated its genetic integrity for studying amyotrophic lateral sclerosis, Alzheimer's disease, and Parkinson's disease, unknown genetic variants that may be corelated with heterogenous diseases etiology would solely be inherited in patient-derived hiPSCs (Krishna et al., 2014).

The transmission of disease-related genetic variants from patients to patient-derived hiPSCs has been justified in comparative studies. Pantothenate kinase 2 (*PANK2*) mutation is responsible for a neurodegenerative disorder characterized by disrupted mitochondrial iron homeostasis and abnormal iron accumulation in patients' globus pallidus (Kruer et al., 2011). Induced *PANK2* mutation in either *Drosophila* or mice were unable to recapitulate the iron accumulation feature, while fibroblasts derived from patients displayed deficits in iron handling (Campanella et al., 2012; Santambrogio et al., 2015). Transplantation of astrocytes differentiated from schizophrenia patient-derived hiPSCs into mouse forebrain caused transcriptomic alterations affecting synaptic dysfunction and inflammation pathways and resulted in behavioral changes in cognitive functions (Koskuvi et al., 2022). These examples validatED hiPSCs and neurons differentiated from hiPSCs being an effective model to study etiology and pathogenesis mechanisms especially for heterogenous neurodegenerative and mental disorders.

Beyond carrying diverse disease-related genetic information, hiPSC-derived neurons can also be a more responsive model. In the comparative study by Verheijen et al., the authors reported more DEGs detected in sAD patient -derived neurons compared to brain samples, with more AD-related proteins mapped. Furthermore, copper exposure ($CuCl_2$, 20 μM, 10 days) further exaggerated the overrepresentation of AD-related AKT and p53 pathways in sAD-cortical neurons in sAD-cortical neurons (Verheijen et al., 2022). Involvements of AKT and p53 in response to excess copper were also detected in SH-SY5Y cells, but not until copper concentrations reached millimolar levels (Sadžak et al., 2021). In the *PANK2* mutation study mentioned earlier, neurons differentiated from reprogrammed patient fibroblasts showed an at least 2-fold increase in iron exporter ferroportin mRNA expression, while small interfering RNA knockdown of *PANK2* only led to an approximately 1.2-fold increase in SH-SY5Y cells (Arber et al., 2017; Poli et al., 2010). These findings suggested that hiPSC-derived neurons can be a more sensitive model for evaluating disease-related susceptibility to toxicants. Genetic mutations inherited in patient-derived iPSCs may also display stronger phenotypes compared to mutations in cell lines by exogenous modifications.

2.2 hiPSCs preserve inter-individual variabilities

In addition to preserve mutations corelated with neurodegenerative diseases, hiPSCs also carried along inherited genetic variants representing sex, geographical, and ancestry diversity. Sex-sensitive variations in exposure patterns, biomarker levels, and neurobehavioral performances has been well recognized with accumulative epidemiology studies (Mergler, 2012). An investigation in adolescent exposure to pesticides suggested that boys were exposed directly when working in the fields while girls tended to be indirectly exposed via household aerosol, drinking water, and contact with contaminated materials (Rohlman et al., 2022). Developmental exposures to neurotoxicants, especially lead, were associated with sex specific decrement in verbal and full-scale intelligence quotient (IQ) scores (Goodman et al., 2023; Hamadani et al., 2011). The compelling evidence encouraged a trend of taking sex differences into consideration in in vitro neurotoxicological study design. In immortalized cell line models, although there are lines derived from both male (ReNcell CX, ReNcell VM, NTERA-2) and female (SH-SY5Y and LUHMES) available, the development and differentiation procedures introduced too many variations to allow direct comparison for investigating sex differences. This concern can be overcome by parallel culture and differentiation of hiPSCs from various donors. Mature neurons differentiated from male hiPSCs were more affected by delta-9-tetrahydrocannabinol (Δ^9-THC) exposure compared to neurons derived from a female donor, which agreed with findings in rodents where males were particularly affected with pronounced hippocampal neuronal damage (de Salas-Quiroga et al., 2020; Miranda et al., 2020). Male-derived NPCs were also reported to be more sensitive to subapoptotic concentrations of methylmercury exposure represented by impaired neurite extension, cell migration, and downregulation of neurite outgrowth modulating genes (Edoff et al., 2017). These findings again reflected the reported susceptibility of males in epidemiology and animal studies (Björklund et al., 2007). The differential sensitivity between male and female-derived hiPSCs and differentiated neurons can be attributed to different baseline expression and activity level of toxicant responsive pathways. Assessments with patient-derived hiPSC neurons revealed sex specific transcriptomic variations in the field of amyotrophic lateral sclerosis (ALS) and schizophrenia, altered mitochondrial dynamics that are related to Alzheimer's disease, as well as electrophysiological property variations in sensory neurons which may contributing to anesthetic drug discovery

(Flannagan et al., 2023; Tiihonen et al., 2019; Workman et al., 2023; Zurek et al., 2024). Furthermore, isogenic iPSCs with different sex chromosome complements is under development aiming to resemble sex differences with minimized genetic variabilities (Waldhorn et al., 2022).

With the recent recognition of the under-representation of non-European ancestry population in genomic-wide association studies (GWAS), multi-geographical and ancestry investigations is emerging as another aspect of taking advantage of the genetic variants embedded in hiPSCs derived from diverse donors. High throughput screening approaches, which will be discussed in the following sections, have identified sub-populational and inter-individual differences in toxicants-induced viability loss and neuronal functional impairments (Ford et al., 2022; Huang et al., 2022). Although there is not yet definitive conclusion of how genetic underpinnings affect in vitro response to neurotoxicant exposures, estimation of inter-individual variation is a necessary step for inclusive characterization of next-generation in vitro neurotoxicity assessment models. Empowered by the latest large-scale differentiation methods and multi-omics approaches, this knowledge gap can be bridged by hiPSC-derived systems in the near future.

3. Incorporating immortalized cell line and hiPSC-derived models in toxicological study design

3.1 To address the complexity of human central nervous system

Immortalized cell line models discussed in the earlier sections, especially the two most widely used lines SH-SY5Y and LUHMES, are predominantly characterized as dopaminergic neuron-like systems. These models provided accessible approaches for investigating the toxicology related to the dopaminergic system. Yet, they fail to address the interconnectivities between dopaminergic and other types of neurons and glial cells in the human brain, nor to mention the susceptibility of other neuronal lineages to environmental insults. Compiling evidence suggested that neuronal subtype plays a significant role in their sensitivity to toxicant-induced cell death or degeneration. Striatal NPCs showed the highest percentage of cell death compared to cortical and midbrain NPCs when exposed to 200 μM Mn for 24 h in their corresponding differentiation media (Joshi et al., 2019). Doxorubicin and rotenone, two compounds that were known to have dopaminergic-specific neurotoxicity, were reported to differentially trigger morphological degeneration in glutamatergic versus GABAergic neurons (Cohen and Tanaka, 2018). Moreover, there are developmental-stage related phenotypes that can only be captured by stem cell-derived progenitors but not differentiating cell lines. Arsenic exposures in motor neuron progenitors differentiated from P19 stem cells have been reported to increase pluripotency markers including Nestin and Nanog (Hong and Bain, 2012; Perego et al., 2023). However, RA differentiation in cell lines targets strong inhibition or even depletion of Nestin, therefore potentially masking this upregulation effect of arsenic in embryonic or pluripotent cells (Das and Bhattacharyya, 2014; Lopes et al., 2010). These findings indicate that neuronal subtype and developmental stage specific toxicology investigations can be achieved more efficiently if not exclusively with hiPSC models.

During the investigations of neuronal-type specific toxicity, the influence of medium composition on the outcome has been underestimated and insufficiently acknowledged. To derive specific cell fate, neuronal induction media, including supplied small molecule combinations and base medium formulations, can vary across protocols and affect the toxicokinetic of specific compounds thus implicating cellular sensitivity. As mentioned earlier, Joshi et al. reported that striatal NPCs displayed the most severe viability loss when exposed to acute Mn compared to cortical and midbrain NPCs. However, when the exposure media was controlled, differential percentage of survival was not detectable across NPC lineages anymore. Furthermore, viability and Mn uptake measurement of an immortalized striatal neuronal line, ST*Hdh*, exposed to Mn in different hiPSC culture media confirmed that a substantial degree of the observed lineage specific Mn cytotoxicity can be explained by the influence of media type on Mn transportation (Joshi et al., 2019). This conflict between supplementation requirement of specific neuronal lineages and their implication on toxicokinetics again highlighted the necessity of comprehensive understanding of the in vitro neuronal system to enable reliable translation from in vitro findings to toxicity identifications. It also suggests that intra-system validation could augment the confidence of the observations and therefore to integrate the use of immortalized cell line and hiPSC-derived neurons is a promising avenue for further exploration.

Screening in immortalized cell lines with follow-up study in hiPSC-derived neurons provided an example of incorporating the two systems. Four compounds were identified from the 1851 ToxCast™ (the U.S. Environmental

Protection Agency's Toxicity Forecaster) chemical pool to significantly alter cellular cholesterol biosynthesis in SK-N-SH, A549, and Hep-G2 cell lines despite the differences in their inherent profiles of cholesterol metabolize and biosynthesis related enzymes. Validation with cortical glutamatergic NPCs confirmed the disruption on cholesterol metabolism of two out of four highlighted chemicals, fenpropimorph and spiroxamine (Wages et al., 2020). These two highlighted compounds were known to disrupt fungal sterol biosynthesis yet did not have sufficient evidence to be considered to have the same effects in humans. Several studies conducted around the same time of the screening suggested the potential metabolic disruption in the central nervous and reproductive system (Bonvallot et al., 2018; Massei et al., 2019). These findings resulted in an appendage of these compounds and their implications on cholesterol metabolism into the adverse outcome pathway network to further address their potential neurotoxicity in Tau-driven memory loss (Tsamou and Roggen, 2022).

Another approach to integrate both models into neurotoxicology evaluations is to co-culture iPSC-derived neurons with cell lines. Successful development of iPSC-differentiated motor neurons and human immortalized myoblasts co-culture was reported by De Lamotte et al. with functionally active neuromuscular junctions. The feasibility of this system being applied to toxicological evaluations was confirmed with the inhibition of myotube contractions by botulinum neurotoxins (BoNT) (De Lamotte et al., 2021). Co-culture of astrocytes differentiated from hiPSCs and LUHMES cells eliminated viability loss induced by RSL-3 ferroptosis activator compared to monoculture of LUHMES via significantly boosting intracellular glutathione content (Renner et al., 2024). These studies exemplified how intra-cell type interactions can be embraced into in vitro modeling of the nervous system by taking advantage of both immortalized cell lines and hiPSCs for better extrapolation of in vitro findings into in vivo toxicity.

3.2 To achieve high throughput and predictive toxicity modeling

High throughput screening (HTS) with in vitro models is an increasing tendency since the initiation of the Tox21 program since 2008. The Tox21 program aims to develop fast and efficient assessments for systematical profiling of environmental chemicals and approved drugs toxicity. In combination with the strategic focus on the 3Rs (Replacement, Reduction, and Refinement), both immortalized cell lines and hiPSC-derived culture systems have been intensively characterized and validated for cell-based HTS assay development. Assays for HTS can be grouped into viability, morphological, and functional endpoints (Schmidt et al., 2017). Examples of HTS toxicity assessments using these read-outs with immortalized cell lines and hiPSC-derived neurons will be discussed in the following section.

3.2.1 Via viability screenings

Viability is the initial outcome measurement applied in toxicity identification screenings. Colormetric (represented by the tetrazole reduction assay), luminescent (represented by intracellular ATP measurement by Promega CellTiter-Glo assay), or fluorescent (represented by resazurin conversion measurement by Promega CellTiter-Blue assay) assessments of metabolic enzyme activity or ATP quantity can reflect irreversible cell damage and be easily measured with array design. The reliability of viability assays has been confirmed by consistent cytotoxicity detection in the majority tested compounds across independent studies. One of the first large-scale toxicity screenings conducted by NIH included 1353 unique compounds with available traditional toxicity evaluation data and tested their cytotoxicity with CellTiter-Glo across 10 cell lines from different species and tissues. 6% (428 out of 1353) of the total screened compounds were classified as high confidence cytotoxic, represented by significant dose-dependent viability loss (Xia et al., 2008). This study set up an example of how HTS can be performed with cell-based assays and how the data should be interpreted. It was pointed out that exposure time frames, concentration ranges, and source of cell lines need to be taken into consideration before concluding the toxicity of a compound, especially for those that were tested negative. For example, in mixed culture systems with multiple cell types, single read-out of viability can be misleading. Hogberg et al. reported a study where the resazurin assay failed to detect neuronal-specific cell loss induced by developmental toxicants (methylmercury chloride, lead chloride, and valproic acid, exposed for 24 h) in a mixed culture of rat primary neurons, microglia, and astrocytes. This absence of response was attributed to the non-specificity of resazurin mitochondrial activity measurement in distinguishing cell types, and the increase of glial proliferation triggered by neuronal death (Hogberg et al., 2010). Another evaluation of cell type-specific responses in neurons, astrocytes, and mixed cultures with 72 h exposure also suggested that cell types played a significant role affecting the vulnerability to neurotoxicants-induced mitochondrial dysfunction (Woehrling et al., 2010). Therefore, other aspects that are closely related with cellular survival were brought into the assessment battery, including cell type-specific markers, proliferation measurements (BrdU and Ki67 staining), apoptosis markers (caspase 3/7 expression), and mRNA expression levels has been incorporated into the testing battery, which greatly improved the sensitivity for neurotoxicant identification. More recent studies yielded a 40–50% detection rate through evaluating multiple aspects of cytotoxicity with literature-reported neurotoxicants (Breier et al., 2008; Go et al., 2023; Tong et al., 2017). The development of these readouts also facilitated further in-depth mode of action and adverse outcome pathway identification.

The HTS examples mentioned above were all conducted with immortalized cell lines, taking advantage of the homogeneity of the culture to allow more efficient interpretation of the results. In the meantime, validation of screening platforms with hiPSC-derived neurons is rapidly progressing aiming to enable advanced toxicity identification in complex systems. Viability measurements were applied as a straightforward method for initial responsiveness evaluation. MTT (3-(4,5-dimethylthiazol-2-yl)-2,5-diphenyltetrazolium bromide) assay indicated that neural stem cells derived from 8 different hiPSC lines, including both male and female, displayed compatible viability loss when exposed to rotenone for 24 h. Following differentiation, neurons, and astrocytes differentiated from these hiPSC lines exhibited cytotoxicity to 23 out of 37 neurotoxicants in the NTP80 chemical bank (Pei et al., 2016). Another study measuring mitochondrial enzyme activity by MTS and intracellular ATP obtained nearly identical dose-response to developmental neurotoxicants in hiPSCs between the two independent assays (Kamata et al., 2020). Moreover, mature neurons differentiated with both 2D and 3D methods showed robust responses to known neurotoxicants in acute exposures, with significant dose-dependent patterns (Kobolak et al., 2020; Slavin et al., 2021). These findings suggested promising positive detection rate and reproducibility across various hiPSC sources and differentiation methods for neurotoxicity identifications.

3.2.2 Via morphology screenings

As previously discussed, while cytotoxicity screenings demonstrated favorable positive detection rate and reproducibility in primary evaluations during HTS, it is necessary to supplement viability readouts with more in-depth analyses to ensure reliable identification of adverse outcomes. The development of high density and throughput imaging methods, as well as computational aid imaging analysis enabled high throughput morphological evaluations. In a comparative study between 2D and 3D cultured mature hiPSC neurons, significant difference was not detected by intracellular ATP measurement but via cellular calcium imaging (Slavin et al., 2021). This illustrated the effectiveness of imaging approaches in complex culture systems which would enable detection of sub-cytotoxic adverse effects, especially in degenerative deficits.

Inhibition of neurite outgrowth is a solid marker of developmental neurotoxicity which replicates in vivo observations with exposures applied to NPCs or during the differentiation phase of immortalized cell lines. Decreased neurite outgrowth as an adverse outcome has been tested positive for known developmental neurotoxicants and displayed consistency across different cell lines independent of differentiation induction methods. For instance, exposures to arsenic in differentiating cell lines, including P19 (mouse embryonal carcinoma, differentiated by 1% dimethyl sulfoxide), Neuro-2a (mouse neuroblasts, differentiated by 20 μM RA with 2% FBS), PC12 (rat pheochromocytoma, differentiated by 50 ng/mL NGF), and SH-SY5Y (human neuroblastoma, differentiated by 10 μM RA with 1% FBS), have been reported to induce deficits in neural-like morphology development when treated with a range of 0.5–5 μM arsenic over 24–120 h (Frankel et al., 2009; Hong and Bain, 2012; Niyomchan et al., 2015; Wang et al., 2010; Zhou et al., 2018). The observed morphological impairments aligned with the recognized developmental neurotoxicity of arsenic, with reported epidemiological and *in vivo* neurobehavioral damage including reduced intelligence quotient and repressed neuronal processes development (Chattopadhyay et al., 2002; Smeester and Fry, 2018; Tolins et al., 2014). Similarly, exposure to methylmercury, atrazine, and lead has been reported to impact neurite outgrowth in differentiating cell lines and hiPSC-derived NPCs (Chan et al., 2017; Xie et al., 2021; Zhou et al., 2018). Besides observed in single chemical tests, differentiation deficits were also detected in a screening study where SH-SY5Ys were exposed to rotenone, valproic acid, acrylamide, and methylmercury chloride during RA treatment. All chemicals inhibited neurite outgrowth and differentiating SH-SY5Ys displayed highest sensitivity to rotenone, which may be attributed to its dopaminergic phenotype. Furthermore, two negative control compounds, tolbutamide and clofibrate, were reported to have minimal impact on either the neurite morphology or neurodevelopmental genes expression, which further validated the potential of this model in identifying developmental neurotoxicity in unknown compounds (Hinojosa et al., 2023).

Amongst these studies, decreased length of outgrowing neurites was the most representative parameter, accompanied by one or more related degenerative features including increased soma size, decreased number of neurites per cell body, and decreased complexity of developed dendrites. In-depth analysis can be performed via manual evaluations of high-resolution images, yet algorithmic multi-parameter assessments are under development for morphological HTS. A screening study by Stiegler et al. investigating LUHMES morphological changes when exposed to chemicals from the NTP80 collection described detailed predictive parameter development (Stiegler et al., 2011). The authors provided justification of applying a half maximum effective concentration (EC_{50}) ratio $EC_{50(\mathrm{Viability})}/EC_{50(\mathrm{Neurite\ Area})}$ threshold. The ratio in this study was set at 4, implying that all compounds with a $EC_{50(\mathrm{Viability})}/EC_{50(\mathrm{Neurite\ Area})} \geq 4$ would be marked as a neurite specific toxicity hit. By re-visiting the hit compounds and comparing their no-observed-adverse-effect-levels (NOAELs) in the neurite outgrowth test to onset of toxicity concentrations in the Tox21 database, the LUHMES neurite outgrowth method showed high reproducibility, high sensitivity, and no bias for molecular weight. Meanwhile, in comparison with another available toxicant identification method, the benchmark concentrations, EC_{50} ratio identification yielded compatible hits. The benchmark concentration identification, which evaluates the distance from the background noise of a given assay, could be applied as a complementation of the neurite outgrowth for better classification with borderline compounds (Delp et al., 2018b). The combination of EC_{50} ratio and benchmark concentration identification methods was applied in a toxicity screening with mature peripheral neurons differentiated from H9 human embryonic stem cells. 21 out of 36 tested compounds were defined as positive hits, including a new drug class epothilones which has known peripheral neurotoxicity with no effect on cell viability (Argyriou et al., 2011; Hoelting et al., 2016). This further validated the reliability of utilizing neurite outgrowth as a neurotoxicity screening parameter, and meanwhile confirmed its feasibility to predict adverse outcomes in unknown chemicals.

3.3 To identify mode of actions and adverse outcome pathways

Single hypothesis-driven studies and HTS utilizing immortalized line or hiPSC-derived neuronal models discussed previously exemplified decades of efforts aiming to recapitulate traditional in vivo toxicity outcomes under in vitro conditions, and towards the eventual goal to develop adverse outcome pathway (AOP) network. AOP describes a cascade of measurable key events (KEs) that are triggered by the biological interactions between a chemical and its targets, and sequentially lead to an adverse outcome (Bal-Price et al., 2018).

The AOP network integrates existing mechanistic knowledge of neurotoxicity into the development of regulatory and predictive models, and in return sheds light on the gap for potentially unknown mode of actions (MOAs). Identification of MOA can be achieved by a fit-for-purpose combination of viability and morphological assays discussed above, functional assessments driven by electrophysiology and multielectrode array, and biomarker expression level evaluations with evolving multi-omics approaches. Detailed mechanistic insights acquired during MOA explorations would then contribute to a holistic investigation of KE cascade and linking the adverse outcomes of toxicants to their AOPs.

Electrophysiological assessment is an essential quality control parameter for neuronal cultures and a key endpoint measurement to provide functional insights in MOA studies. Multielectrode array (MEA) assay, a representative assessment for electrophysiological connectivity, is more sensitive in hiPSC-derived mature cortical neurons compared to MTT viability test (Kang et al., 2022). It has been applied in evaluating neurotransmission deficits triggered by different classes of chemicals, including not only environmental pollutants, but also chemotherapeutic drugs, seizurogenic compounds, illicit drugs, and new psychoactive substances (Abdo et al., 2015; Kang et al., 2022; Slavin et al., 2021; Tukker et al., 2020). Simultaneous administration of the testing compound and pharmacological modulator of neurotransmitter receptors or ion channels facilitates target and MOA characterization. Furthermore, it has been reported that high-density MEA is capable of detecting disease specific (Parkinson's disease and amyotrophic lateral sclerosis) spatiotemporal phenotypes in hiPSC neurons, indicating potentials for delivering disease-specific insights of toxicity risks (Ronchi et al., 2021).

Gene expression, especially comprehensive assessments with RNA sequencing, is a powerful tool for identifying potentially shared MOAs across multiple toxicants and for pinning down predictive genetic markers. This is achieved through RNA sequencing after paralleled exposures to compounds of interest mining for shared differentially expressed genes, and follow-up validation with quantitative reverse transcription polymerase chain reaction (qRT-PCR). Via this approach, axon guidance factor semaphorins 5A (*SEMA5A*), cholinergic receptor nicotinic alpha 7 subunit (*CHRNA7*), and metallothionein 1G (*MT1G*) were highlighted from exposures in SH-SY5Y and LUHMES as potential biomarkers for predicting developmental neurotoxicity (Hinojosa et al., 2023; Tong et al., 2020). Beyond direct comparisons between exposed versus non-exposed transcripts, exploring commonalities between shared differentially expressed genes in experimental data and disease-associated toxicogenomics databases (such as INSIdE nano) may provide *in silico* prediction of genetic biomarkers and neurodegenerative risks (Gupta et al., 2020).

Epigenomics, which is another route through which developmental exposures lead to adverse outcomes, displayed potentials of revealing AOPs as well. Wan et al. reported validation of DNA methylation detected in a lead-exposed children population with SH-SY5Ys. Undifferentiated SH-SY5Ys exposed to a serial concentration of lead (50, 100, and 200 μg/L) for 30 days captured the majority of differentially methylated fragments that were identified to be positively correlated with blood lead level and children's intelligence quotients scores (Wan et al., 2024). Exposure to lead or pesticide atrazine prior to differentiation induced disruptions in SH-SY5Y epigenome, highlighted by significant alterations in the levels and depositions of histone trimethylation marker H3K9me3 and H3K27me3 (Lin et al., 2021; Xie et al., 2021). These markers are targets of nuclear factor erythroid 2-related factor 2 (NRF2) and provide feedback regulation on the expression level of NRF2, which is involved in lead, arsenic, and metal nanoparticle developmental toxicity detected in human NPC cell line ReNcell CX (cortex) and keratinocytes (Choi and Lee, 2024; Choi et al., 2021; Park et al., 2024).

4. Discussion and future outlook

4.1 Current limitations that cannot yet be addressed with in vitro models

Despite the discussed potentialities of capturing the "human context" within culture dishes, there are still challenges that are not yet fully addressed in vitro, especially features related to the complexity of human central nervous system and real-world exposure conditions.

First, comparisons between hiPSC-derived neurons and human fetal samples provided some directions for further understanding of the gap between cultured systems versus in vivo brain. Compared to fetal neurons, hiPSC-derived neurons have been reported to display low level of DNA damage repair, and altered gene expression level of ion channels, as well as higher expression levels of immature progenitor markers (Snyder et al., 2018). Moreover, a prior transcriptomic study indicated that immune system and protein ubiquitination related Reactome pathways were overrepresented in fetal AD patient brain samples but not in neurons differentiated from Alzheimer's disease (AD) patients (Verheijen et al., 2022). These differences suggest that hiPSC-derived neuronal cultures are a mixed culture of mature neurons and progenitors, with glial cells were not sufficiently represented (Tydlacka et al., 2008). To address this disadvantage, co-culture protocol of multiple cell types in both 2D and 3D systems has been developed. The application of complex 3D culture with hiPSCs is discussed elsewhere in this volume. Accurate analyses of neurotransmitters were enabled by mixed cultures, and can be potentially incorporated into the neurotoxicity testing battery (Cervetto et al., 2023).

Second, the transportation and distribution of toxicants in living human organisms, especially in the nervous system, exhibit a considerably greater level of complexity in contrast with laboratory settings. In the case of simulating *PANK2* mutation-associated iron handling deficit with hiPSCs that was discussed in Section 2.1, it was observed that despite recapitulating the dysfunction in fibroblasts, there was no detectable iron accumulation once the fibroblasts were reprogrammed and differentiated into mature neurons, suggesting that human brain metal regulation cannot yet be fully modeled by in vitro neuronal cultures (Arber et al., 2017). Furthermore, currently availably literature is mainly focused on acute exposures (less than 7 days) with a single exposure compound. However, it is commonly reported in epidemiology studies that vulnerable populations are exposed to a mixture of contaminants, and the neurotoxicity profile of mixtures can be different from single chemical exposure. Individual exposure to lead, cadmium, arsenic, or mercury at their NOAELs did not affect intracellular calcium levels in differentiated PC12 cells, while mixed exposure resulted in significant increase (Zhou et al., 2018). Moreover, exposure-induced neurodegenerations are normally developed over long-term exposure, represented by functional loss, instead of neuronal death within days. However, due to the active proliferation of immortalized cell lines, the translational power of chronic studies with cell lines is yet under debate. Long-term culture of hiPSC-derived neurons provided a platform to evaluate chronic toxicity across extended time frame, therefore application of hiPSCs may bring new knowledge to the field.

4.2 Future directions and possibilities for cell based neurotoxicological research

The expanded use of hiPSCs in neurotoxicological studies, paralleled by the rapid advancements of high throughput and comprehensive assessment methodologies, led to the rise of collaborative efforts to further characterize hiPSC models in large-scale design. Multiple projects, for instance the Horizon 2020 and the ENDpoiNTs project, supported by the US National Institute of Health and the European Union aim to coordinate laboratories that work with hiPSCs to comprise cell banks to establish proof-of-concept testing methods and *in silico* physiologically-based toxicokinetic (PBTK) modeling, as well as generate new knowledge for understanding basis mechanisms of neurotoxicology (Lupu et al., 2020). Furthermore, individual variabilities are being incorporated, as briefly discussed in Section 3.2, via -omics approaches to improve the confidence of in vitro toxicity predictions. Besides evaluating variations via individual assessments within each hiPSC line, the emerging concept of "cell village" brings the possibility of understanding more complicated interactions by culturing and differentiating multiple lines from various donors in the same dish to create a stem cell community. Utility of village models followed by single-cell resolution evaluations may illustrate genetic, epigenetic, molecular, and phenotypic heterogeneity across individuals (Neavin et al., 2023; Wells et al., 2023). Integration of scaled hiPSC models and gene editing approaches would shed new light on explaining inter-individual variations in susceptibility to environmental insults.

5. Concluding remarks

This chapter reviewed the use of immortalized cell lines and hiPSC-derived neurons in neurotoxicological research by comparing essential

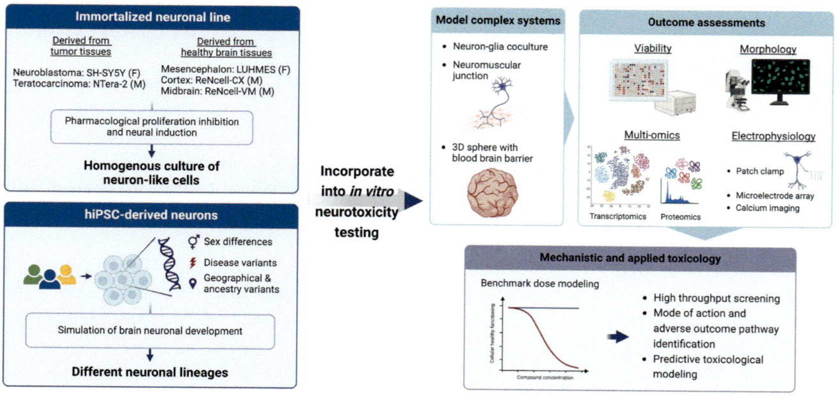

Fig. 1 Summary of the utilization of immortalized cell line and hiPCS-derived neurons as in vitro models for next generation neurotoxicological studies. *Created with biorender.com.*

features and differential responsiveness to toxicants of these two models as illustrated and summarized in Fig. 1. While immortalized cell lines offer highly accessible and homogeneous culture systems, the inherited pro-oncogenes and proliferative status impedes their translational power in modeling post-mitotic human brain neurons. Human iPSC-derived neurons provide a more physiologically relevant platform, though variabilities introduced by laboratory processing, such as donors and differentiation conditions, are not yet fully characterized and warrant continuous attention. Moving forward, informed decisions on modeling system choice need to be made to further advance cell-based alternative approaches for assessing neurotoxicity.

Acknowledgment

This study was supported by the National Institute of Health/National Institute of Environmental Health Sciences (R01 NIH/NIEHS ES010563 and ES07331; and R01 NIH/NIEHS AG080917).

References

Abdo, N., Xia, M., Brown, C.C., Kosyk, O., Huang, R., Sakamuru, S., et al., 2015. Population-based in vitro hazard and concentration–response assessment of chemicals: the 1000 genomes high-throughput screening study. Env. Health Perspect. 123, 458–466. https://doi.org/10.1289/ehp.1408775.

Andrews, P.W., 1984. Retinoic acid induces neuronal differentiation of a cloned human embryonal carcinoma cell line in vitro. Dev. Biol. 103, 285–293. https://doi.org/10.1016/0012-1606(84)90316-6.

Appukuttan, T.A., Ali, N., Varghese, M., Singh, A., Tripathy, D., Padmakumar, M., et al., 2013. Parkinson's disease cybrids, differentiated or undifferentiated, maintain morphological and biochemical phenotypes different from those of control cybrids. J. Neurosci. Res. 91, 963–970. https://doi.org/10.1002/jnr.23241.

Arber, C., Angelova, P.R., Wiethoff, S., Tsuchiya, Y., Mazzacuva, F., Preza, E., et al., 2017. iPSC-derived neuronal models of PANK2-associated neurodegeneration reveal mitochondrial dysfunction contributing to early disease. PLoS One 12, e0184104. https://doi.org/10.1371/journal.pone.0184104.

Argyriou, A.A., Marmiroli, P., Cavaletti, G., Kalofonos, H.P., 2011. Epothilone-induced peripheral neuropathy: a review of current knowledge. J. Pain. Symptom Manage 42, 931–940. https://doi.org/10.1016/j.jpainsymman.2011.02.022.

Attoff, K., Johansson, Y., Cediel-Ulloa, A., Lundqvist, J., Gupta, R., Caiment, F., et al., 2020. Acrylamide alters CREB and retinoic acid signalling pathways during differentiation of the human neuroblastoma SH-SY5Y cell line. Sci. Rep. 10, 16714. https://doi.org/10.1038/s41598-020-73698-6.

Bal-Price, A., Pistollato, F., Sachana, M., Bopp, S.K., Munn, S., Worth, A., 2018. Strategies to improve the regulatory assessment of developmental neurotoxicity (DNT) using in vitro methods. Toxicol. Appl. Pharmacol. 354, 7–18. https://doi.org/10.1016/j.taap.2018.02.008.

Biedler, J.L., Helson, L., Spengler, B.A., 1973. Morphology and growth, tumorigenicity, and cytogenetics of human neuroblastoma cells in continuous culture. Cancer Res. 33, 2643–2652.

Biedler, J.L., Roffler-Tarlov, S., Schachner, M., Freedman, L.S., 1978. Multiple neurotransmitter synthesis by human neuroblastoma cell lines and clones. Cancer Res. 38, 3751–3757.
Björklund, O., Kahlström, J., Salmi, P., Ögren, S.O., Vahter, M., Chen, J.-F., et al., 2007. The effects of methylmercury on motor activity are sex- and age-dependent, and modulated by genetic deletion of adenosine receptors and caffeine administration. Toxicology 241, 119–133. https://doi.org/10.1016/j.tox.2007.08.092.
Bonvallot, N., Canlet, C., Blas-Y-Estrada, F., Gautier, R., Tremblay-Franco, M., Chevolleau, S., et al., 2018. Metabolome disruption of pregnant rats and their offspring resulting from repeated exposure to a pesticide mixture representative of environmental contamination in Brittany. PLoS One 13, e0198448. https://doi.org/10.1371/journal.pone.0198448.
Breier, J.M., Radio, N.M., Mundy, W.R., Shafer, T.J., 2008. Development of a high-throughput screening assay for chemical effects on proliferation and viability of immortalized human neural progenitor cells. Toxicol. Sci. 105, 119–133. https://doi.org/10.1093/toxsci/kfn115.
Campanella, A., Privitera, D., Guaraldo, M., Rovelli, E., Barzaghi, C., Garavaglia, B., et al., 2012. Skin fibroblasts from pantothenate kinase-associated neurodegeneration patients show altered cellular oxidative status and have defective iron-handling properties. Hum. Mol. Genet. 21, 4049–4059. https://doi.org/10.1093/hmg/dds229.
Cervetto, C., Pistollato, F., Amato, S., Mendoza-de Gyves, E., Bal-Price, A., Maura, G., et al., 2023. Assessment of neurotransmitter release in human iPSC-derived neuronal/glial cells: a missing in vitro assay for regulatory developmental neurotoxicity testing. Reprod. Toxicol. 117, 108358. https://doi.org/10.1016/j.reprotox.2023.108358.
Chambers, S.M., Fasano, C.A., Papapetrou, E.P., Tomishima, M., Sadelain, M., Studer, L., 2009. Highly efficient neural conversion of human ES and iPS cells by dual inhibition of SMAD signaling. Nat. Biotechnol. 27, 275–280. https://doi.org/10.1038/nbt.1529.
Chan, M.C., Bautista, E., Alvarado-Cruz, I., Quintanilla-Vega, B., Segovia, J., 2017. Inorganic mercury prevents the differentiation of SH-SY5Y cells: amyloid precursor protein, microtubule associated proteins and ROS as potential targets. J. Trace Elem. Med. Biol. 41, 119–128. https://doi.org/10.1016/j.jtemb.2017.02.002.
Chattopadhyay, S., Bhaumik, S., Nag Chaudhury, A., Das Gupta, S., 2002. Arsenic induced changes in growth development and apoptosis in neonatal and adult brain cells in vivo and in tissue culture. Toxicol. Lett. 128, 73–84. https://doi.org/10.1016/S0378-4274(01)00535-5.
Cheung, Y.-T., Lau, W.K.-W., Yu, M.-S., Lai, C.S.-W., Yeung, S.-C., So, K.-F., et al., 2009. Effects of all-trans-retinoic acid on human SH-SY5Y neuroblastoma as in vitro model in neurotoxicity research. Neurotoxicology 30, 127–135. https://doi.org/10.1016/j.neuro.2008.11.001.
Choi, J., Lee, H., 2024. MLL1 histone methyltransferase and UTX histone demethylase functionally cooperate to regulate the expression of NRF2 in response to ROS-induced oxidative stress. Free. Radic. Biol. Med. 217, 48–59. https://doi.org/10.1016/j.freeradbiomed.2024.03.018.
Choi, J.H., Lee, H., Lee, H., Lee, H., 2021. Dopant-dependent toxicity of CeO2 nanoparticles is associated with dynamic changes in H3K4me3 and H3K27me3 and transcriptional activation of NRF2 gene in HaCaT human keratinocytes. Int. J. Mol. Sci. 22, 3087. https://doi.org/10.3390/ijms22063087.
Cohen, J.D., Tanaka, Y., 2018. Comparative sensitivity of human-induced pluripotent stem cell-derived neuronal subtypes to chemically induced neurodegeneration. Appl. Vitro Toxicol. 4, 347–364. https://doi.org/10.1089/aivt.2017.0028.
Colombo, A., Wang, H., Kuhn, P.-H., Page, R., Kremmer, E., Dempsey, P.J., et al., 2013. Constitutive α- and β-secretase cleavages of the amyloid precursor protein are partially coupled in neurons, but not in frequently used cell lines. Neurobiol. Dis. 49, 137–147. https://doi.org/10.1016/j.nbd.2012.08.011.

Culbreth, M.E., Harrill, J.A., Freudenrich, T.M., Mundy, W.R., Shafer, T.J., 2012. Comparison of chemical-induced changes in proliferation and apoptosis in human and mouse neuroprogenitor cells. Neurotoxicology 33, 1499–1510. https://doi.org/10.1016/j.neuro.2012.05.012.

D'Aloia, A., Pastori, V., Blasa, S., Campioni, G., Peri, F., Sacco, E., et al., 2024. A new advanced cellular model of functional cholinergic-like neurons developed by reprogramming the human SH-SY5Y neuroblastoma cell line. Cell Death Discov. 10, 24. https://doi.org/10.1038/s41420-023-01790-7.

Das, E., Bhattacharyya, N.P., 2014. MicroRNA-432 contributes to dopamine cocktail and retinoic acid induced differentiation of human neuroblastoma cells by targeting NESTIN and RCOR1 genes. FEBS Lett. 588, 1706–1714. https://doi.org/10.1016/j.febslet.2014.03.015.

De Lamotte, J.D., Roqueviere, S., Gautier, H., Raban, E., Bouré, C., Fonfria, E., et al., 2021. hiPSC-derived neurons provide a robust and physiologically relevant in vitro platform to test botulinum neurotoxins. Front. Pharmacol. 11. https://doi.org/10.3389/fphar.2020.617867.

de Salas-Quiroga, A., García-Rincón, D., Gómez-Domínguez, D., Valero, M., Simón-Sánchez, S., Paraíso-Luna, J., et al., 2020. Long-term hippocampal interneuronopathy drives sex-dimorphic spatial memory impairment induced by prenatal THC exposure. Neuropsychopharmacology 45, 877–886. https://doi.org/10.1038/s41386-020-0621-3.

Delp, J., Gutbier, S., Cerff, M., Zasada, C., Niedenführ, S., Zhao, L., et al., 2018a. Stage-specific metabolic features of differentiating neurons: implications for toxicant sensitivity. Toxicol. Appl. Pharmacol. 354, 64–80. https://doi.org/10.1016/j.taap.2017.12.013.

Delp, J., Gutbier, S., Klima, S., Hoelting, L., Pinto-Gil, K., Hsieh, J.-H., et al., 2018b. A high-throughput approach to identify specific neurotoxicants/ developmental toxicants in human neuronal cell function assays. ALTEX 35 235–253. https://doi.org/10.14573/altex.1712182.

Dimensions, 2018. Digital science. Dimensions [Software] [WWW Document]. Digital Science. URL ⟨https://app.dimensions.ai⟩ (accessed 5.20.24).

Donald, S., Elliott, M., Gray, B., Hornby, F., Lewandowska, A., Marlin, S., et al., 2018. A comparison of biological activity of commercially available purified native botulinum neurotoxin serotypes A1 to F1 in vitro, ex vivo, and in vivo. Pharmacol. Res. Perspect. 6. https://doi.org/10.1002/prp2.446.

Donato, R., Miljan, E.A., Hines, S.J., Aouabdi, S., Pollock, K., Patel, S., et al., 2007. Differential development of neuronal physiological responsiveness in two human neural stem cell lines. BMC Neurosci. 8, 36. https://doi.org/10.1186/1471-2202-8-36.

Dravid, A., Raos, B., Svirskis, D., O'Carroll, S.J., 2021. Optimised techniques for high-throughput screening of differentiated SH-SY5Y cells and application for neurite outgrowth assays. Sci. Rep. 11, 23935. https://doi.org/10.1038/s41598-021-03442-1.

Druwe, I., Freudenrich, T.M., Wallace, K., Shafer, T.J., Mundy, W.R., 2015. Sensitivity of neuroprogenitor cells to chemical-induced apoptosis using a multiplexed assay suitable for high-throughput screening. Toxicology 333, 14–24. https://doi.org/10.1016/j.tox.2015.03.011.

Ducray, A.D., Wiedmer, L., Herren, F., Widmer, H.R., Mevissen, M., 2020. Quantitative characterization of phenotypical markers after differentiation of SH-SY5Y cells. CNS Neurol. Disord. Drug. Targets 19, 618–629. https://doi.org/10.2174/1871527319666200708132716.

Edoff, K., Raciti, M., Moors, M., Sundström, E., Ceccatelli, S., 2017. Gestational age and sex influence the susceptibility of human neural progenitor cells to low levels of MeHg. Neurotox. Res. 32, 683–693. https://doi.org/10.1007/s12640-017-9786-x.

Elmorsy, E., Al-Ghafari, A., Al Doghaither, H., Hashish, S., Salama, M., Mudyanselage, A.W., et al., 2023. Differential effects of paraquat, rotenone, and MPTP on cellular bioenergetics of undifferentiated and differentiated human neuroblastoma cells. Brain Sci. 13, 1717. https://doi.org/10.3390/brainsci13121717.

Fabbri, R., Cacopardo, L., Ahluwalia, A., Magliaro, C., 2023. Advanced 3D models of human brain tissue using neural cell lines: state-of-the-art and future prospects. Cells 12, 1181. https://doi.org/10.3390/cells12081181.

Flannagan, K., Stopperan, J.A., Hauger, B.M., Troutwine, B.R., Lysaker, C.R., Strope, T.A., et al., 2023. Cell type and sex specific mitochondrial phenotypes in iPSC derived models of Alzheimer's disease. Front. Mol. Neurosci. 16. https://doi.org/10.3389/fnmol.2023.1201015.

Ford, L.C., Jang, S., Chen, Z., Zhou, Y.-H., Gallins, P.J., Wright, F.A., et al., 2022. A population-based human in vitro approach to quantify inter-individual variability in responses to chemical mixtures. Toxics 10, 441. https://doi.org/10.3390/toxics10080441.

Forster, J.I., Köglsberger, S., Trefois, C., Boyd, O., Baumuratov, A.S., Buck, L., et al., 2016. Characterization of differentiated SH-SY5Y as neuronal screening model reveals increased oxidative vulnerability. SLAS Discovery 21, 496–509. https://doi.org/10.1177/1087057115625190.

Frankel, S., Concannon, J., Brusky, K., Pietrowicz, E., Giorgianni, S., Thompson, W.D., et al., 2009. Arsenic exposure disrupts neurite growth and complexity in vitro. Neurotoxicology 30, 529–537. https://doi.org/10.1016/j.neuro.2009.02.015.

Go, S.M., Lee, B., Ahn, C., Jeong, S.H., Jo, N.R., Park, S.M., et al., 2023. Initial phase establishment of an in vitro method for developmental neurotoxicity test using Ki-67 in human neural progenitor cells. J. Physiol. Pharmacol. 74. https://doi.org/10.26402/jpp.2023.2.07.

Goodman, C.V., Green, R., DaCosta, A., Flora, D., Lanphear, B., Till, C., 2023. Sex difference of pre- and post-natal exposure to six developmental neurotoxicants on intellectual abilities: a systematic review and meta-analysis of human studies. Environ. Health 22, 80. https://doi.org/10.1186/s12940-023-01029-z.

Gupta, G., Gliga, A., Hedberg, J., Serra, A., Greco, D., Odnevall Wallinder, I., et al., 2020. Cobalt nanoparticles trigger ferroptosis-like cell death (oxytosis) in neuronal cells: potential implications for neurodegenerative disease. FASEB J. 34, 5262–5281. https://doi.org/10.1096/fj.201902191RR.

Gutbier, S., May, P., Berthelot, S., Krishna, A., Trefzer, T., Behbehani, M., et al., 2018. Major changes of cell function and toxicant sensitivity in cultured cells undergoing mild, quasi-natural genetic drift. Arch. Toxicol. 92, 3487–3503. https://doi.org/10.1007/s00204-018-2326-5.

Hamadani, J., Tofail, F., Nermell, B., Gardner, R., Shiraji, S., Bottai, M., et al., 2011. Critical windows of exposure for arsenic-associated impairment of cognitive function in pre-school girls and boys: a population-based cohort study. Int. J. Epidemiol. 40, 1593–1604. https://doi.org/10.1093/ije/dyr176.

Harischandra, D.S., Rokad, D., Ghaisas, S., Verma, S., Robertson, A., Jin, H., et al., 2020. Enhanced differentiation of human dopaminergic neuronal cell model for preclinical translational research in Parkinson's disease. Biochimica et. Biophysica Acta (BBA) - Mol. Basis Dis. 1866, 165533. https://doi.org/10.1016/j.bbadis.2019.165533.

Hill, E.J., Jiménez-González, C., Tarczyluk, M., Nagel, D.A., Coleman, M.D., Parri, H.R., 2012. NT2 derived neuronal and astrocytic network signalling. PLoS One 7, e36098. https://doi.org/10.1371/journal.pone.0036098.

Hinojosa, M.G., Johansson, Y., Cediel-Ulloa, A., Ivanova, E., Gabring, N., Gliga, A., et al., 2023. Evaluation of mRNA markers in differentiating human SH-SY5Y cells for estimation of developmental neurotoxicity. Neurotoxicology 97, 65–77. https://doi.org/10.1016/j.neuro.2023.05.011.

Hoelting, L., Klima, S., Karreman, C., Grinberg, M., Meisig, J., Henry, M., et al., 2016. Stem cell-derived immature human dorsal root ganglia neurons to identify peripheral neurotoxicants. Stem Cell Transl. Med. 5, 476–487. https://doi.org/10.5966/sctm.2015-0108.

Hogberg, H.T., Kinsner-Ovaskainen, A., Coecke, S., Hartung, T., Bal-Price, A.K., 2010. mRNA expression is a relevant tool to identify developmental neurotoxicants using an in vitro approach. Toxicol. Sci. 113, 95–115. https://doi.org/10.1093/toxsci/kfp175.

Hong, G.-M., Bain, L.J., 2012. Arsenic exposure inhibits myogenesis and neurogenesis in P19 stem cells through repression of the β-catenin signaling pathway. Toxicol. Sci. 129, 146–156. https://doi.org/10.1093/toxsci/kfs186.

Huang, C.Y., Nicholson, M.W., Wang, J.Y., Ting, C.Y., Tsai, M.H., Cheng, Y.C., et al., 2022. Population-based high-throughput toxicity screen of human iPSC-derived cardiomyocytes and neurons. Cell Rep. 39, 110643. https://doi.org/10.1016/j.celrep.2022.110643.

Joshi, P., Bodnya, C., Ilieva, I., Neely, M.D., Aschner, M., Bowman, A.B., 2019. Huntington's disease associated resistance to Mn neurotoxicity is neurodevelopmental stage and neuronal lineage dependent. Neurotoxicology 75, 148–157. https://doi.org/10.1016/j.neuro.2019.09.007.

Kamata, S., Hashiyama, R., Hana-ika, H., Ohkubo, I., Saito, R., Honda, A., et al., 2020. Cytotoxicity comparison of 35 developmental neurotoxicants in human induced pluripotent stem cells (iPSC), iPSC-derived neural progenitor cells, and transformed cell lines. Toxicol. Vitro 69, 104999. https://doi.org/10.1016/j.tiv.2020.104999.

Kang, K., Kim, C.-Y., Kim, J., Ryu, B., Lee, S.-G., Baek, J., et al., 2022. Establishment of neurotoxicity assessment using microelectrode array (MEA) with hiPSC-derived neurons and evaluation of new psychoactive substances (NPS). Int. J. Stem Cell 15, 258–269. https://doi.org/10.15283/ijsc21217.

Khwanraj, K., Phruksaniyom, C., Madlah, S., Dharmasaroja, P., 2015. Differential expression of tyrosine hydroxylase protein and apoptosis-related genes in differentiated and undifferentiated SH-SY5Y neuroblastoma cells treated with MPP $^{+}$. Neurol. Res. Int. 2015, 1–11. https://doi.org/10.1155/2015/734703.

Kobolak, J., Teglasi, A., Bellak, T., Janstova, Z., Molnar, K., Zana, M., et al., 2020. Human induced pluripotent stem cell-derived 3D-neurospheres are suitable for neurotoxicity screening. Cells 9, 1122. https://doi.org/10.3390/cells9051122.

Koskuvi, M., Lehtonen, Š., Trontti, K., Keuters, M., Wu, Y., Koivisto, H., et al., 2022. Contribution of astrocytes to familial risk and clinical manifestation of schizophrenia. Glia 70, 650–660. https://doi.org/10.1002/glia.24131.

Krishna, A., Biryukov, M., Trefois, C., Antony, P.M., Hussong, R., Lin, J., et al., 2014. Systems genomics evaluation of the SH-SY5Y neuroblastoma cell line as a model for Parkinson's disease. BMC Genomics 15, 1154. https://doi.org/10.1186/1471-2164-15-1154.

Kruer, M.C., Hiken, M., Gregory, A., Malandrini, A., Clark, D., Hogarth, P., et al., 2011. Novel histopathologic findings in molecularly-confirmed pantothenate kinase-associated neurodegeneration. Brain 134, 947–958. https://doi.org/10.1093/brain/awr042.

Krug, A.K., Gutbier, S., Zhao, L., Pöltl, D., Kullmann, C., Ivanova, V., et al., 2014. Transcriptional and metabolic adaptation of human neurons to the mitochondrial toxicant MPP+. Cell Death Dis. 5, e1222. https://doi.org/10.1038/cddis.2014.166.

Laurenza, I., Pallocca, G., Mennecozzi, M., Scelfo, B., Pamies, D., Bal-Price, A., 2013. A human pluripotent carcinoma stem cell-based model for in vitro developmental neurotoxicity testing: effects of methylmercury, lead and aluminum evaluated by gene expression studies. Int. J. Developmental Neurosci. 31, 679–691. https://doi.org/10.1016/j.ijdevneu.2013.03.002.

Lin, L.F., Xie, J., Sánchez, O.F., Bryan, C., Freeman, J.L., Yuan, C., 2021. Low dose lead exposure induces alterations on heterochromatin hallmarks persisting through SH-SY5Y cell differentiation. Chemosphere 264, 128486. https://doi.org/10.1016/j.chemosphere.2020.128486.

Lopes, F.M., da Motta, L.L., De Bastiani, M.A., Pfaffenseller, B., Aguiar, B.W., de Souza, L.F., et al., 2017. RA differentiation enhances dopaminergic features, changes redox parameters, and increases dopamine transporter dependency in 6-hydroxydopamine-induced neurotoxicity in SH-SY5Y cells. Neurotox. Res. 31, 545–559. https://doi.org/10.1007/s12640-016-9699-0.

Lopes, F.M., Schröder, R., Júnior, M.L.C. da F., Zanotto-Filho, A., Müller, C.B., Pires, A.S., et al., 2010. Comparison between proliferative and neuron-like SH-SY5Y cells as an in vitro model for Parkinson disease studies. Brain Res. 1337, 85–94. https://doi.org/10.1016/j.brainres.2010.03.102.

Loser, D., Schaefer, J., Danker, T., Möller, C., Brüll, M., Suciu, I., et al., 2021. Human neuronal signaling and communication assays to assess functional neurotoxicity. Arch. Toxicol. 95, 229–252. https://doi.org/10.1007/s00204-020-02956-3.

Lotharius, J., Barg, S., Wiekop, P., Lundberg, C., Raymon, H.K., Brundin, P., 2002. Effect of mutant α-synuclein on dopamine homeostasis in a new human mesencephalic cell line. J. Biol. Chem. 277, 38884–38894. https://doi.org/10.1074/jbc.M205518200.

Lotharius, J., Falsig, J., van Beek, J., Payne, S., Dringen, R., Brundin, P., et al., 2005. Progressive degeneration of human mesencephalic neuron-derived cells triggered by dopamine-dependent oxidative stress is dependent on the mixed-lineage kinase pathway. J. Neurosci. 25, 6329–6342. https://doi.org/10.1523/JNEUROSCI.1746-05.2005.

Luo, G.R., Chen, Y., Li, X.P., Liu, T.X., Le, W.D., 2008. Nr4a2 is essential for the differentiation of dopaminergic neurons during zebrafish embryogenesis. Mol. Cell. Neurosci. 39, 202–210. https://doi.org/10.1016/j.mcn.2008.06.010.

Lupu, D., Andersson, P., Bornehag, C.-G., Demeneix, B., Fritsche, E., Gennings, C., et al., 2020. The ENDpoiNTs project: novel testing strategies for endocrine disruptors linked to developmental neurotoxicity. Int. J. Mol. Sci. 21, 3978. https://doi.org/10.3390/ijms21113978.

Martin, E.-R., Gandawijaya, J., Oguro-Ando, A., 2022. A novel method for generating glutamatergic SH-SY5Y neuron-like cells utilizing B-27 supplement. Front. Pharmacol. 13, 943627. https://doi.org/10.3389/fphar.2022.943627.

Massei, R., Hollert, H., Krauss, M., von Tümpling, W., Weidauer, C., Haglund, P., et al., 2019. Toxicity and neurotoxicity profiling of contaminated sediments from Gulf of Bothnia (Sweden): a multi-endpoint assay with Zebrafish embryos. Env. Sci. Eur. 31, 8. https://doi.org/10.1186/s12302-019-0188-y.

Mergler, D., 2012. Neurotoxic exposures and effects: gender and sex matter! Hänninen Lecture 2011. Neurotoxicology 33, 644–651. https://doi.org/10.1016/j.neuro.2012.05.009.

Miranda, C.C., Barata, T., Vaz, S.H., Ferreira, C., Quintas, A., Bekman, E.P., 2020. hiPSC-based model of prenatal exposure to cannabinoids: effect on neuronal differentiation. Front. Mol. Neurosci. 13. https://doi.org/10.3389/fnmol.2020.00119.

Neavin, D.R., Steinmann, A.M., Farbehi, N., Chiu, H.S., Daniszewski, M.S., Arora, H., et al., 2023. A village in a dish model system for population-scale hiPSC studies. Nat. Commun. 14, 3240. https://doi.org/10.1038/s41467-023-38704-1.

Nierode, G.J., Perea, B.C., McFarland, S.K., Pascoal, J.F., Clark, D.S., Schaffer, D.V., et al., 2016. High-throughput toxicity and phenotypic screening of 3D human neural progenitor cell cultures on a microarray chip platform. Stem Cell Rep. 7, 970–982. https://doi.org/10.1016/j.stemcr.2016.10.001.

Niyomchan, A., Watcharasit, P., Visitnonthachai, D., Homkajorn, B., Thiantanawat, A., Satayavivad, J., 2015. Insulin attenuates arsenic-induced neurite outgrowth impairments by activating the PI3K/Akt/SIRT1 signaling pathway. Toxicol. Lett. 236, 138–144. https://doi.org/10.1016/j.toxlet.2015.05.008.

Oh, J., Jung, C., Lee, M., Kim, J., Son, M., 2017. Comparative analysis of human embryonic stem cell-derived neural stem cells as an in vitro human model. Int. J. Mol. Med. https://doi.org/10.3892/ijmm.2017.3298.

Påhlman, S., Meyerson, G., Lindgren, E., Schalling, M., Johansson, I., 1991. Insulin-like growth factor I shifts from promoting cell division to potentiating maturation during neuronal differentiation. Proc. Natl Acad. Sci. U S A 88, 9994–9998. https://doi.org/10.1073/pnas.88.22.9994.

Påhlman, S., Ruusala, A.-I., Abrahamsson, L., Mattsson, M.E.K., Esscher, T., 1984. Retinoic acid-induced differentiation of cultured human neuroblastoma cells: a comparison with phorbolester-induced differentiation. Cell Differ. 14, 135–144. https://doi.org/10.1016/0045-6039(84)90038-1.

Park, H.-R., Azzara, D., Cohen, E.D., Boomhower, S.R., Diwadkar, A.R., Himes, B.E., et al., 2024. Identification of novel NRF2-dependent genes as regulators of lead and arsenic toxicity in neural progenitor cells. J. Hazard. Mater. 463, 132906. https://doi.org/10.1016/j.jhazmat.2023.132906.

Paul, G., Christophersen, N.S., Raymon, H., Kiaer, C., Smith, R., Brundin, P., 2007. Tyrosine hydroxylase expression is unstable in a human immortalized mesencephalic cell line—studies in vitro and after intracerebral grafting in vivo. Mol. Cell. Neurosci. 34, 390–399. https://doi.org/10.1016/j.mcn.2006.11.010.

Pei, Y., Peng, J., Behl, M., Sipes, N.S., Shockley, K.R., Rao, M.S., et al., 2016. Comparative neurotoxicity screening in human iPSC-derived neural stem cells, neurons and astrocytes. Brain Res. 1638, 57–73. https://doi.org/10.1016/j.brainres.2015.07.048.

Perego, M.C., McMichael, B.D., McMurry, N.R., Ventrello, S.W., Bain, L.J., 2023. Arsenic impairs differentiation of human induced pluripotent stem cells into cholinergic motor neurons. Toxics 11, 644. https://doi.org/10.3390/toxics11080644.

Poli, M., Derosas, M., Luscieti, S., Cavadini, P., Campanella, A., Verardi, R., et al., 2010. Pantothenate kinase-2 (Pank2) silencing causes cell growth reduction, cell-specific ferroportin upregulation and iron deregulation. Neurobiol. Dis. 39, 204–210. https://doi.org/10.1016/j.nbd.2010.04.009.

Pollock, K., Stroemer, P., Patel, S., Stevanato, L., Hope, A., Miljan, E., et al., 2006. A conditionally immortal clonal stem cell line from human cortical neuroepithelium for the treatment of ischemic stroke. Exp. Neurol. 199, 143–155. https://doi.org/10.1016/j.expneurol.2005.12.011.

Pons, L., Vilain, C., Volteau, M., Picaut, P., 2019. Safety and pharmacodynamics of a novel recombinant botulinum toxin E (rBoNT-E): results of a phase 1 study in healthy male subjects compared with abobotulinumtoxinA (Dysport®). J. Neurol. Sci. 407, 116516. https://doi.org/10.1016/j.jns.2019.116516.

Renner, N., Schöb, F., Pape, R., Suciu, I., Spreng, A.-S., Ückert, A.-K., et al., 2024. Modeling ferroptosis in human dopaminergic neurons: pitfalls and opportunities for neurodegeneration research. Redox Biol. 73, 103165. https://doi.org/10.1016/j.redox.2024.103165.

Rohlman, D., Davis, J.W., Ismail, A., Abdel-Rasoul, G.M., Henry, O., Olson, J.R., et al., 2022. Pesticide risk perception and safety behavior among adolescent pesticide applicators in Egypt. Saf. Health Work. 13, S32. https://doi.org/10.1016/j.shaw.2021.12.814.

Ronchi, S., Buccino, A.P., Prack, G., Kumar, S.S., Schröter, M., Fiscella, M., et al., 2021. Electrophysiological phenotype characterization of human iPSC-derived neuronal cell lines by means of high-density microelectrode arrays. Adv. Biol. 5. https://doi.org/10.1002/adbi.202000223.

Sadžak, A., Vlašić, I., Kiralj, Z., Batarelo, M., Oršolić, N., Jazvinšćak Jembrek, M., et al., 2021. Neurotoxic effect of flavonol myricetin in the presence of excess copper. Molecules 26, 845. https://doi.org/10.3390/molecules26040845.

Santambrogio, P., Dusi, S., Guaraldo, M., Rotundo, L.I., Broccoli, V., Garavaglia, B., et al., 2015. Mitochondrial iron and energetic dysfunction distinguish fibroblasts and induced neurons from pantothenate kinase-associated neurodegeneration patients. Neurobiol. Dis. 81, 144–153. https://doi.org/10.1016/j.nbd.2015.02.030.

Schmidt, B.Z., Lehmann, M., Gutbier, S., Nembo, E., Noel, S., Smirnova, L., et al., 2017. In vitro acute and developmental neurotoxicity screening: an overview of cellular platforms and high-throughput technical possibilities. Arch. Toxicol. 91, 1–33. https://doi.org/10.1007/s00204-016-1805-9.

Shipley, M.M., Mangold, C.A., Szpara, M.L., 2016. Differentiation of the SH-SY5Y human neuroblastoma cell line. J. Vis. Exp. 53193. https://doi.org/10.3791/53193.

Simões, R.F., Ferrão, R., Silva, M.R., Pinho, S.L.C., Ferreira, L., Oliveira, P.J., et al., 2021. Refinement of a differentiation protocol using neuroblastoma SH-SY5Y cells for use in neurotoxicology research. Food Chem. Toxicol. 149, 111967. https://doi.org/10.1016/j.fct.2021.111967.

Slavin, I., Dea, S., Arunkumar, P., Sodhi, N., Montefusco, S., Siqueira-Neto, J., et al., 2021. Human iPSC-derived 2D and 3D platforms for rapidly assessing developmental, functional, and terminal toxicities in neural cells. Int. J. Mol. Sci. 22, 1908. https://doi.org/10.3390/ijms22041908.

Smeester, L., Fry, R.C., 2018. Long-term health effects and underlying biological mechanisms of developmental exposure to arsenic. Curr. Env. Health Rep. 5, 134–144. https://doi.org/10.1007/s40572-018-0184-1.

Smirnova, L., Harris, G., Delp, J., Valadares, M., Pamies, D., Hogberg, H.T., et al., 2016. A LUHMES 3D dopaminergic neuronal model for neurotoxicity testing allowing long-term exposure and cellular resilience analysis. Arch. Toxicol. 90, 2725–2743. https://doi.org/10.1007/s00204-015-1637-z.

Snyder, C., Yu, L., Ngo, T., Sheinson, D., Zhu, Y., Tseng, M., et al., 2018. In vitro assessment of chemotherapy-induced neuronal toxicity. Toxicol. Vitro 50, 109–123. https://doi.org/10.1016/j.tiv.2018.02.004.

Stern, M., Gierse, A., Tan, S., Bicker, G., 2014. Human Ntera2 cells as a predictive in vitro test system for developmental neurotoxicity. Arch. Toxicol. 88, 127–136. https://doi.org/10.1007/s00204-013-1098-1.

Stiegler, N.V., Krug, A.K., Matt, F., Leist, M., 2011. Assessment of chemical-induced impairment of human neurite outgrowth by multiparametric live cell imaging in high-density cultures. Toxicol. Sci. 121, 73–87. https://doi.org/10.1093/toxsci/kfr034.

Strother, L., Miles, G.B., Holiday, A.R., Cheng, Y., Doherty, G.H., 2021. Long-term culture of SH-SY5Y neuroblastoma cells in the absence of neurotrophins: a novel model of neuronal ageing. J. Neurosci. Methods 362, 109301. https://doi.org/10.1016/j.jneumeth.2021.109301.

Tiihonen, J., Koskuvi, M., Storvik, M., Hyötyläinen, I., Gao, Y., Puttonen, K.A., et al., 2019. Sex-specific transcriptional and proteomic signatures in schizophrenia. Nat. Commun. 10, 3933. https://doi.org/10.1038/s41467-019-11797-3.

Tolins, M., Ruchirawat, M., Landrigan, P., 2014. The developmental neurotoxicity of arsenic: cognitive and behavioral consequences of early life exposure. Ann. Glob. Health 80, 303. https://doi.org/10.1016/j.aogh.2014.09.005.

Tong, Z., Hogberg, H., Kuo, D., Sakamuru, S., Xia, M., Smirnova, L., et al., 2017. Characterization of three human cell line models for high-throughput neuronal cytotoxicity screening. J. Appl. Toxicol. 37, 167–180. https://doi.org/10.1002/jat.3334.

Tong, Z.-B., Braisted, J., Chu, P.-H., Gerhold, D., 2020. The MT1G gene in LUHMES neurons is a sensitive biomarker of neurotoxicity. Neurotox. Res. 38, 967–978. https://doi.org/10.1007/s12640-020-00272-3.

Tong, Z.-B., Huang, R., Braisted, J., Chu, P.-H., Simeonov, A., Gerhold, D.L., 2024. 3D-Suspension culture platform for high throughput screening of neurotoxic chemicals using LUHMES dopaminergic neurons. SLAS Discov. 29, 100143. https://doi.org/10.1016/j.slasd.2024.01.004.

Tsamou, M., Roggen, E.L., 2022. Building a network of adverse outcome pathways (AOPs) incorporating the Tau-driven AOP toward memory loss (AOP429). J. Alzheimers Dis. Rep. 6, 271–296. https://doi.org/10.3233/ADR-220015.

Tukker, A.M., Wijnolts, F.M.J., de Groot, A., Westerink, R.H.S., 2020. Applicability of hiPSC-derived neuronal cocultures and rodent primary cortical cultures for in vitro seizure liability assessment. Toxicol. Sci. 178, 71–87. https://doi.org/10.1093/toxsci/kfaa136.

Tüshaus, J., Kataka, E.S., Zaucha, J., Frishman, D., Müller, S.A., Lichtenthaler, S.F., 2021. Neuronal differentiation of LUHMES cells induces substantial changes of the proteome. Proteomics 21. https://doi.org/10.1002/pmic.202000174.

Tydlacka, S., Wang, C.-E., Wang, X., Li, S., Li, X.-J., 2008. Differential activities of the ubiquitin–proteasome system in neurons versus glia may account for the preferential accumulation of misfolded proteins in neurons. J. Neurosci. 28, 13285–13295. https://doi.org/10.1523/JNEUROSCI.4393-08.2008.

Verheijen, M.C.T., Krauskopf, J., Caiment, F., Nazaruk, M., Wen, Q.F., van Herwijnen, M.H.M., et al., 2022. iPSC-derived cortical neurons to study sporadic Alzheimer disease: a transcriptome comparison with post-mortem brain samples. Toxicol. Lett. 356, 89–99. https://doi.org/10.1016/j.toxlet.2021.12.009.

Wages, P.A., Joshi, P., Tallman, K.A., Kim, H.-Y.H., Bowman, A.B., Porter, N.A., 2020. Screening ToxCast™ for chemicals that affect cholesterol biosynthesis: studies in cell culture and human induced pluripotent stem cell–derived neuroprogenitors. Env. Health Perspect. 128. https://doi.org/10.1289/EHP5053.

Waldhorn, I., Turetsky, T., Steiner, D., Gil, Y., Benyamini, H., Gropp, M., et al., 2022. Modeling sex differences in humans using isogenic induced pluripotent stem cells. Stem Cell Rep. 17, 2732–2744. https://doi.org/10.1016/j.stemcr.2022.10.017.

Wan, C., Ma, H., Liu, J., Liu, F., Liu, J., Dong, G., et al., 2024. Quantitative relationships of FAM50B and PTCHD3 methylation with reduced intelligence quotients in school aged children exposed to lead: evidence from epidemiological and in vitro studies. Sci. Total. Environ. 907, 167976. https://doi.org/10.1016/j.scitotenv.2023.167976.

Wang, X., Meng, D., Chang, Q., Pan, J., Zhang, Z., Chen, G., et al., 2010. Arsenic inhibits neurite outgrowth by inhibiting the LKB1–AMPK signaling pathway. Env. Health Perspect. 118, 627–634. https://doi.org/10.1289/ehp.0901510.

Wells, M.F., Nemesh, J., Ghosh, S., Mitchell, J.M., Salick, M.R., Mello, C.J., et al., 2023. Natural variation in gene expression and viral susceptibility revealed by neural progenitor cell villages. Cell Stem Cell 30, 312–332.e13. https://doi.org/10.1016/j.stem.2023.01.010.

Willem, M., Lammich, S., Haass, C., 2009. Function, regulation and therapeutic properties of β-secretase (BACE1). Semin. Cell Dev. Biol. 20, 175–182. https://doi.org/10.1016/j.semcdb.2009.01.003.

Woehrling, E.K., Hill, E.J., Coleman, M.D., 2010. Evaluation of the importance of astrocytes when screening for acute toxicity in neuronal cell systems. Neurotox. Res. 17, 103–113. https://doi.org/10.1007/s12640-009-9084-3.

Workman, M.J., Lim, R.G., Wu, J., Frank, A., Ornelas, L., Panther, L., et al., 2023. Large-scale differentiation of iPSC-derived motor neurons from ALS and control subjects. Neuron 111, 1191–1204.e5. https://doi.org/10.1016/j.neuron.2023.01.010.

Xia, M., Huang, R., Witt, K.L., Southall, N., Fostel, J., Cho, M.-H., et al., 2008. Compound cytotoxicity profiling using quantitative high-throughput screening. Env. Health Perspect. 116, 284–291. https://doi.org/10.1289/ehp.10727.

Xicoy, H., Wieringa, B., Martens, G.J.M., 2017. The SH-SY5Y cell line in Parkinson's disease research: a systematic review. Mol. Neurodegener. 12, 10. https://doi.org/10.1186/s13024-017-0149-0.

Xie, J., Lin, L., Sánchez, O.F., Bryan, C., Freeman, J.L., Yuan, C., 2021. Pre-differentiation exposure to low-dose of atrazine results in persistent phenotypic changes in human neuronal cell lines. Environ. Pollut. 271, 116379. https://doi.org/10.1016/j.envpol.2020.116379.

Yan, Y.-W., Qian, E.S., Woodard, L.E., Bejoy, J., 2023. Neural lineage differentiation of human pluripotent stem cells: advances in disease modeling. World J. Stem Cell 15, 530–547. https://doi.org/10.4252/wjsc.v15.i6.530.

Zhang, X., Yin, M., Zhang, M., 2014. Cell-based assays for Parkinson's disease using differentiated human LUHMES cells. Acta Pharmacol. Sin. 35, 945–956. https://doi.org/10.1038/aps.2014.36.

Zhou, F., Xie, J., Zhang, S., Yin, G., Gao, Y., Zhang, Y., et al., 2018. Lead, cadmium, arsenic, and mercury combined exposure disrupted synaptic homeostasis through activating the Snk-SPAR pathway. Ecotoxicol. Env. Saf. 163, 674–684. https://doi.org/10.1016/j.ecoenv.2018.07.116.

Zurek, N.A., Ehsanian, R., Goins, A.E., Adams, I.M., Petersen, T., Goyal, S., et al., 2024. Electrophysiological analyses of human dorsal root ganglia and human induced pluripotent stem cell-derived sensory neurons from male and female donors. J. Pain. 25, 104451. https://doi.org/10.1016/j.jpain.2023.12.008.

CHAPTER THREE

Brain organoids as a translational model of human developmental neurotoxicity

Thomas Hartung[a,b,c,*], Maren Schenke[a], and Lena Smirnova[a]
[a]Center for Alternatives to Animal Testing (CAAT), Bloomberg School of Public Health and Whiting School of Engineering, Johns Hopkins University, Baltimore, MD, United States
[b]Doerenkamp-Zbinden Chair for Evidence-based Toxicology, Baltimore, MD, United States
[c]CAAT Europe, University of Konstanz, Konstanz, Baden-Württemberg, Germany
*Corresponding author. e-mail address: thartun1@jhu.edu

Contents

Abstract

Human brain organoids, derived from pluripotent stem cells, have emerged as a promising alternative to traditional animal-based methods for developmental neurotoxicity (DNT) testing. These 3D cell culture systems recapitulate key features of the developing human brain, offering advantages such as assessing toxicant effects on neurodevelopmental processes in a human-relevant context and the potential for personalized toxicology. However, challenges remain in improving reproducibility, incorporating supporting cell types, and establishing regulatory acceptance. Continued development and collaboration among researchers, industry, and regulatory agencies hold great promise for advancing DNT testing using brain organoids while considering ethical implications surrounding their use.

Advances in Neurotoxicology, Volume 12
ISSN 2468-7480, https://doi.org/10.1016/bs.ant.2024.07.002

1. Introduction

Developmental neurotoxicity (DNT) testing is crucial for protecting public health by identifying chemicals that may adversely affect the developing nervous system. The developing brain is particularly vulnerable to environmental toxicants, and exposure during critical windows of development can contribute to neurodevelopmental disorders such as autism spectrum disorder (ASD), attention deficit hyperactivity disorder (ADHD), and intellectual disabilities (Smirnova et al., 2014). The rising incidence of these disorders, along with the lack of toxicity information for most chemicals, underscores the urgent need for reliable DNT testing strategies (Grandjean and Landrigan, 2006, 2014).

However, current animal-based approaches for DNT testing have several limitations. These include high costs, low throughput, lengthy study durations, and the use of large numbers of animals (Smirnova et al., 2024). Moreover, interspecies differences in brain development and responses to toxicants can limit the extrapolation of animal data to humans (Smirnova et al., 2014). The regulatory guidelines for DNT testing, such as the OECD Test Guideline 426, rely on *in vivo* rodent studies that are not formally validated and may not adequately capture the complexity of human neurodevelopment (Makris et al., 2009; Smirnova and Hartung, 2024).

To address these challenges, there has been a growing interest in developing alternative, human-relevant models for DNT testing. Human brain organoids, derived from pluripotent stem cells, have emerged as a promising tool for this purpose (Pamies et al., 2017; Anderson et al., 2021; Fan et al., 2022; Smirnova and Hartung, 2024). These 3D cell culture systems recapitulate key features of the developing human brain, including diverse cell types, complex architecture, and functional neural networks (Lancaster et al., 2013; Qian et al., 2019; Yoon et al., 2019). By enabling the direct assessment of toxicant effects on human neural tissue, brain organoids offer the potential to overcome the limitations of animal models and provide a more predictive and mechanistically informative approach to DNT testing (Smirnova et al., 2014, 2024).

In this article, we review the current state of and advances in brain organoid technology, discuss their applications in DNT testing, highlight challenges and limitations, and provide recommendations for future research and regulatory acceptance. We also explore the ethical considerations surrounding the use of brain organoids and their potential to revolutionize our understanding of human neurodevelopment and susceptibility to environmental toxicants.

2. The journey to *in vitro* DNT testing

The journey to *in vitro* DNT testing has been driven by the need for more human-relevant, efficient, and predictive approaches to assess the potential risks posed by chemicals to the developing brain. Traditional animal-based methods, such as the Organization for Economic Co-operation and Development (OECD) Test Guideline 426, have several limitations, including high costs, low throughput, and questionable relevance to human health outcomes (Smirnova et al., 2014, 2024). The development of alternative, non-animal methods for DNT testing has been a long-standing goal in the field of toxicology, with the aim of reducing animal use, improving the predictivity of testing strategies, and protecting human health (Bal-Price et al., 2018).

The journey towards *in vitro* DNT testing has been marked by several key milestones and collaborative efforts among researchers, industry partners, and regulatory agencies (Smirnova et al., 2024). In 2005, the first workshop on incorporating *in vitro* alternative methods for DNT testing into international hazard and risk assessment strategies was held in a collaboration between CAAT, CEFIC and ECVAM,[1] setting the stage for future developments (Coecke et al., 2007). Subsequent workshops and conferences, such as the TestSmart DNT series and the OECD/EFSA workshop on DNT, have brought together stakeholders to discuss the state of the science, identify research needs, and develop recommendations for advancing alternative approaches (Lein et al., 2007; Crofton et al., 2011; Bal-Price et al., 2012; Fritsche et al., 2017).

The advent of human induced pluripotent stem cell (hiPSC) technology has been a critical turning point in the journey towards *in vitro* DNT testing. The ability to generate human iPSC-derived neural cells and tissues has provided a physiologically relevant and genetically diverse platform for modeling human brain development and assessing the effects of neurotoxicants (Bal-Price et al., 2010; Bal-Price et al., 2015). The development of 3D brain organoid models, which recapitulate key features of the developing human brain, has further advanced the field by enabling the assessment of DNT in a more complex and organized cellular environment (Lancaster et al., 2013; Pamies et al., 2017; Yoon et al., 2019; Jacob et al., 2021; Adams et al., 2023; Smirnova and Hartung, 2024).

[1] CAAT – Johns Hopkins Center for Alternatives to Animal Testing; CEFIC - the European Chemical Industry Council; ECVAM – European Centre for the Validation of Alternative Methods.

In parallel with the development of human iPSC-based models, efforts have been made to establish *in vitro* assays that cover critical neurodevelopmental processes, such as proliferation, migration, differentiation, synaptogenesis, and network formation (Bal-Price et al., 2015; Sachana et al., 2019). The integration of these assays into testing batteries and the development of adverse outcome pathways (AOPs) (Leist et al., 2021, 2017; Meijer et al., 2017) have provided a mechanistic framework for understanding the sequence of events leading from chemical exposure to adverse neurodevelopmental outcomes (Bal-Price et al., 2015, 2018).

The application of high-throughput screening (HTS) technologies and computational modeling approaches has further enhanced the efficiency and predictive power of *in vitro* DNT testing (Schmidt et al., 2016; Smirnova et al., 2018). The integration of *in vitro* data with *in silico* models, such as quantitative structure-activity relationship (QSAR) and physiologically based pharmacokinetic (PBPK) models, has enabled the extrapolation of *in vitro* results to *in vivo* outcomes and the prioritization of chemicals for further testing (Hartung et al., 2012; Smirnova et al., 2018).

Despite significant progress in the development of *in vitro* DNT testing approaches, several challenges remain to be addressed. These include the need for further standardization and validation of *in vitro* assays, the incorporation of metabolic competence and other supporting cell types into brain organoid models, and the establishment of regulatory acceptance criteria for alternative methods (Smirnova et al., 2024). Ongoing efforts to address these challenges include the development of performance standards (Hartung et al., 2004), the optimization of organoid culture conditions, and the engagement of regulatory agencies in the validation and implementation of *in vitro* DNT testing strategies (Bal-Price et al., 2018; Pamies et al., 2017; Sachana et al., 2019).

In conclusion, the journey towards *in vitro* DNT testing has been a collaborative and iterative process, driven by advances in stem cell biology, high-throughput screening, computational modeling, and regulatory science. It represents a prime example for a strategic development of a non-animal test strategy where initially in 2005 no approach was available to forming a community and resulting in initial OECD *in vitro* test guidance in 2023 (OECD, 2023). While challenges remain, the development of human-relevant, efficient, and predictive *in vitro* approaches holds great promise for transforming the way we assess the potential risks of chemicals to the developing brain and for protecting human health (Fig. 1). Neurospheres, *i.e.*, three-dimensional neuronal cultures are already part of the

	in vivo	*in vitro* battery	brain organoids
Costs	$ 1,4 million	$ X0,000	$ X,000
Animals	1,000+	---	---
Duration	2 years	X months	~3 months
Mechanistic	No	Yes	Yes
Validation	unlikely	in preparation	t.b.d.
Glial cells	+++	---	+++
In preparation		Defined Approaches	Multiplexing Organoid Intelligence

Fig. 1 Cursory comparison of *in vivo* DNT testing, the OECD *in vitro* test battery and the prospect brain organoids.

emerging *in vitro* battery, but more refined brain microphysiological systems (MPS) promise further refinement and multiplexing (Fig. 1). The continued collaboration among researchers, industry partners, and regulatory agencies will be essential to realize the full potential of *in vitro* DNT testing and to ensure its successful implementation in chemical safety assessment.

3. Current state and advances in brain organoid technology

Brain organoids are 3D cell culture systems derived from human pluripotent stem cells (hPSCs), which can be either embryonic stem cells or iPSCs (Lancaster et al., 2013; Paşca et al., 2015; Pamies et al., 2017; Muotri, 2023). The derivation of brain organoids typically involves the differentiation of hPSCs into neuroectoderm, followed by the embedding of the cells in an extracellular matrix scaffold, such as Matrigel (Lancaster and Knoblich, 2014). The cells then self-organize into complex structures that resemble various brain regions, including the cerebral cortex, ventral forebrain, midbrain, and hypothalamus (Qian et al., 2018). Other protocols use bioreactors, which keep organoids in suspension by stirring or shaking (Smirnova and Hartung, 2024).

One of the key advantages of brain organoids is their ability to recapitulate the cellular diversity and organization of the developing human brain (Smirnova and Hartung, 2024). Brain organoids contain a variety of cell types, including neural progenitors, neurons, astrocytes, and oligodendrocytes, which can arrange in a manner that mimics the layered structure of the cerebral cortex (Qian et al., 2019). Moreover, brain organoids exhibit functional properties, such as spontaneous electrical activity and synaptic connectivity, which are essential for modeling neurodevelopmental processes (Trujillo et al., 2019).

The use of human-derived cells is another significant advantage of brain organoids over animal models. By capturing the genetic background of the donor, brain organoids can be used to study individual differences in susceptibility to neurotoxicants and to develop personalized approaches to risk assessment (Modafferi et al., 2021; Smirnova and Hartung, 2024). Additionally, brain organoids derived from patient-specific iPSCs provide a valuable tool for investigating the mechanisms underlying neurodevelopmental disorders and for screening potential therapeutic compounds (Mariani et al., 2015; Urresti et al., 2021).

Recent advances in brain organoid culture methods have further enhanced their utility for DNT testing. The development of spinning bioreactor systems has enabled the generation of larger and more complex organoids with improved oxygen and nutrient diffusion (Qian et al., 2018). The incorporation of vascularization strategies, such as co-culturing with endothelial cells or transplantation into animal hosts, has also been shown to promote organoid maturation and functionality (Mansour et al., 2018). Furthermore, the use of gene-editing technologies, such as CRISPR/Cas9, has allowed for the introduction of disease-associated mutations or reporter genes into brain organoids, expanding their applications in disease modeling and toxicity testing (Birtele et al., 2023; Smirnova and Hartung, 2024).

Advances in characterization methods have also contributed to the growing utility of brain organoids for DNT testing. High-content imaging and automated analysis platforms have enabled the quantitative assessment of morphological and functional endpoints, such as neurite outgrowth, synaptogenesis, and calcium signaling (Pamies et al., 2018; Zhong et al., 2020). The application of omics technologies, including transcriptomics, proteomics, and metabolomics, has provided insights into the molecular mechanisms underlying the effects of neurotoxicants on brain organoids (Smirnova et al., 2015).

In summary, the derivation of brain organoids from human pluripotent stem cells has revolutionized the field of DNT testing by providing a physiologically relevant and genetically diverse model of the developing human brain. Ongoing improvements in culture methods, characterization techniques, and genetic engineering tools are expected to further enhance the predictive power and translational potential of brain organoids for identifying and assessing the risks posed by developmental neurotoxicants.

4. Applications of brain organoids in DNT testing

Brain organoids have emerged as a powerful tool for assessing the effects of toxicants on key neurodevelopmental processes. These 3D models allow for the investigation of critical events such as proliferation, differentiation, migration, and synaptogenesis in a human-relevant context (Pamies et al., 2017; Smirnova and Hartung, 2024). By recapitulating the complex cellular interactions and microenvironment of the developing brain, organoids provide a more comprehensive and physiologically accurate platform for DNT testing compared to traditional 2D cell cultures or animal models (Smirnova et al., 2014, 2024).

Brain organoids offer several advantages over conventional *in vitro* and *in vivo* approaches for DNT assessment. Compared to 2D cell cultures, organoids exhibit higher complexity, with multiple cell types and more authentic cell-cell and cell-matrix interactions (Mansour et al., 2021). This complexity allows for the evaluation of toxicant effects on diverse neural cell populations and their interplay during development. Moreover, brain organoids can be maintained in culture for extended periods, enabling the study of long-term and delayed effects of neurotoxicants (Smirnova and Hartung, 2024).

The human brain is a unique organ and has been shown to exhibit many molecular and cellular differences even compared to non-human primates (Sousa et al., 2017). Compared to animal models, brain organoids provide a more direct representation of human neurodevelopment, avoiding the need for inter-species extrapolation (Smirnova et al., 2024). The use of human cells is particularly relevant given the differences in developmental timelines, metabolism, and susceptibility to toxicants between humans and animals (Smirnova et al., 2014; Masjosthusmann et al., 2018). Additionally, brain organoids can be generated from a large number of individuals, capturing the genetic diversity of human populations and allowing for the identification of vulnerable subgroups (Smirnova and Hartung, 2024).

Several studies have demonstrated the utility of brain organoids for evaluating the effects of known and suspected neurotoxicants. For example, exposure of human brain organoids to the flame retardant PBDE-47 was found to impair neuronal differentiation and neurite outgrowth, providing evidence for its developmental neurotoxicity (Li et al., 2019). Similarly, brain organoids exposed to the pesticide rotenone exhibited oxidative stress, reduced dopaminergic differentiation, and impaired electrophysiological activity, recapitulating the hallmarks of Parkinson's disease (Pamies et al., 2018).

Brain organoids have also been used to investigate the neurodevelopmental effects of drugs and environmental pollutants (Fan et al., 2022). The antidepressant paroxetine was shown to disrupt neurogenesis and synaptic function in human brain organoids at clinically relevant concentrations (Zhong et al., 2020).

The ability to generate brain organoids from patient-derived iPSCs opens up exciting possibilities for personalized toxicology and risk assessment (Smirnova and Hartung, 2024). By capturing the unique genetic background of individuals, patient-derived organoids can be used to identify susceptibility factors and to predict individual responses to neurotoxicants. This approach has the potential to inform personalized risk management strategies and to guide the development of targeted interventions for vulnerable populations.

Moreover, brain organoids derived from patients with neurodevelopmental disorders, such as ASD or schizophrenia, can be used to study the complex interplay between genetic predisposition and environmental exposures in the etiology of these conditions (Mariani et al., 2015). By exposing patient-derived organoids to various neurotoxicants, researchers can investigate how specific genetic variants modulate the response to environmental stressors and contribute to the manifestation of neurodevelopmental phenotypes (Modafferi et al., 2021).

Ellen Fritsche and her colleagues have made significant contributions to the field of developmental neurotoxicity (DNT) testing through the development and application of the neurosphere assay. The neurosphere assay is an *in vitro* method that initially utilized primary fetal rat cortical, and now uses human neural progenitor cells to form three-dimensional neurospheres, which recapitulate key events in neurodevelopment, such as proliferation, differentiation, and migration (Fritsche et al., 2005). Fritsche and her team have demonstrated the utility of the neurosphere assay for assessing the DNT potential of various chemicals, including methylmercury, lead, and polychlorinated biphenyls (PCBs) (Fritsche et al., 2005; Moors et al., 2009; Gassmann et al., 2010). By comparing the effects of these chemicals on neurosphere formation and differentiation with *in vivo* data, they have shown that the assay can predict DNT with high sensitivity and specificity (Moors et al., 2009). Furthermore, Fritsche and colleagues have integrated the neurosphere assay into the OECD DNT testing battery and have contributed to the development of adverse outcome pathways (AOPs) for DNT (Bal-Price et al., 2015). Their work has been instrumental in advancing the use of alternative methods for DNT testing and has paved the way for the development of more comprehensive and predictive *in vitro* approaches to assess the potential risks of chemicals to the developing brain.

In conclusion, brain organoids have emerged as a powerful and versatile tool for DNT testing, offering advantages over traditional *in vitro* and animal models. The ability to assess the effects of toxicants on key neurodevelopmental processes in a human-relevant context, combined with the potential for personalized toxicology using patient-derived organoids, positions brain organoids as a transformative platform for advancing our understanding of DNT and for protecting human health.

5. Challenges and limitations

Despite the remarkable potential of brain organoids for DNT testing, several challenges and limitations need to be addressed to fully realize their utility in toxicological research and regulatory decision-making.

One of the main challenges associated with brain organoids is the variability and reproducibility of the models (Smirnova et al., 2024). The self-organization process that underlies organoid formation can lead to heterogeneity in size, shape, and cellular composition, both within and between batches (Qian et al., 2019). This variability can confound the interpretation of toxicity data and hinder the establishment of robust and reproducible assays. Efforts to standardize organoid generation protocols, implement quality control measures, and develop high-throughput platforms are ongoing to improve the consistency and scalability of brain organoid models (Pamies et al., 2017; Smirnova and Hartung, 2024).

Another limitation of current brain organoid models is the lack of vascularization and other supporting cell types (Smirnova and Hartung, 2024). The absence of blood vessels limits the size and complexity of organoids, as nutrients and oxygen cannot efficiently diffuse into the core of larger structures (Qian et al., 2019). This can lead to necrotic centers and impaired functionality, particularly in long-term cultures. Moreover, the lack of endothelial cells and microglia, which play critical roles in neurodevelopment and neurotoxicity, may limit the physiological relevance of brain organoids (Pamies et al., 2017). Ongoing efforts to incorporate vascularization strategies and to co-culture brain organoids with other relevant cell types aim to address these limitations (Mansour et al., 2018; Smirnova and Hartung, 2024; Pamies et al., 2024).

The immature or fetal-like state of brain organoids presents another challenge for their application in DNT testing (Smirnova et al., 2024). While organoids can recapitulate early stages of neurodevelopment, they often lack the cellular diversity, organization, and functionality of the adult brain (Qian et al., 2019). This limitation may restrict their utility for modeling later stages of development or for predicting long-term consequences of neurotoxicant exposure. Strategies to promote organoid maturation, such as extended culture times, optimized differentiation protocols, and exposure to physiologically relevant stimuli, are being explored to enhance the adult-like features of brain organoids (Smirnova and Hartung, 2024).

While brain organoids can mimic the microenvironment of the brain, they usually lack multiorgan-interactions and hormonal communication. For example, the development of the brain is influenced by a surge of testosterone from the developing testes of a male human fetus (Bakker, 2022) which would be needed to be added to a brain cell culture. This hormonal influence permanently alters the brain and is thought to affect the vulnerability towards neurotoxicants and the incidence of neurological disorders, *e.g.* in ASD, which is about 4 times more common in men compared to women (Schaafsma et al., 2017; Kern et al., 2017). In rodents however, estrogen conveys the masculinization of the male brain, impacting their suitability to study neurotoxicity and neurodevelopment (Arambula and McCarthy, 2020).

Validation of brain organoids against *in vivo* animal and human data is essential to establish their predictive value and regulatory acceptance for DNT testing (Hartung et al., 2024). While initial studies have demonstrated the ability of brain organoids to recapitulate key aspects of neurodevelopment and to respond to known neurotoxicants, more comprehensive validation efforts are needed (Pamies et al., 2018; Zhong et al., 2020). This includes comparing organoid-based assays with existing animal models, evaluating their performance in predicting human DNT outcomes, and establishing a robust set of reference compounds and endpoints for standardization (Smirnova and Hartung, 2024). Collaborative efforts among researchers, industry partners, and regulatory agencies will be crucial to advance the validation and implementation of brain organoid models in DNT testing.

The OECD DNT *In Vitro* Testing Battery, while representing a significant advancement in non-animal approaches for DNT assessment, may benefit from pruning and optimization to enhance its efficiency and practical utility. The current battery consists of multiple assays covering various key neurodevelopmental processes, such as neural progenitor cell proliferation, neuronal and glial differentiation, neurite outgrowth, synaptogenesis, and neuronal network formation (Sachana et al., 2019; Bal-Price et al., 2018). However, running such a comprehensive battery can be time-consuming, resource-intensive, and may not always be necessary for all chemicals or regulatory contexts (Smirnova et al., 2024). Multiplexing these assays within a single brain organoid model could offer a more streamlined and cost-effective approach to DNT testing. Brain organoids have the capacity to recapitulate multiple neurodevelopmental processes simultaneously within a single 3D culture system (Smirnova and Hartung, 2024). By incorporating appropriate readouts and endpoints, brain organoids could potentially capture the key events covered by the individual assays of the OECD DNT *In Vitro* Testing Battery, thereby reducing the need for multiple separate tests (Smirnova et al., 2024). This multiplexed approach could not only improve the efficiency of DNT testing but also provide a more integrated and holistic assessment of neurotoxicant effects on the developing brain. However, further research is needed to validate the performance of multiplexed brain organoid assays against the individual tests of the battery and to establish their predictive value for regulatory decision-making.

In addition to these technical challenges, ethical considerations surrounding the use of human-derived brain organoids must be addressed (Sawai et al., 2022). As organoids become more complex and mature, questions arise regarding their moral status, potential for consciousness, and the implications for informed consent and research oversight (Boers et al., 2019; de Jongh et al., 2022). Engaging in interdisciplinary dialogues and developing appropriate governance frameworks will be essential to ensure the responsible and ethically sound use of brain organoids in toxicological research (Hartung et al., 2024).

In summary, while brain organoids hold immense promise for revolutionizing DNT testing, several challenges and limitations need to be addressed to fully realize their potential. Ongoing efforts to improve the reproducibility, physiological relevance, and validation of brain organoid models, coupled with attention to ethical considerations, will be crucial to advance their application in toxicological research and regulatory decision-making.

6. The importance of quality assurance for DNT testing with brain organoids

The use of brain organoids DNT testing requires adherence to rigorous quality standards and reporting practices to ensure the reliability, reproducibility, and regulatory acceptance of the obtained results. Good Cell Culture Practice (GCCP) is a set of principles that outline the necessary conditions for the proper handling, maintenance, and documentation of cell cultures (Pamies et al., 2022). Adherence to GCCP is essential for minimizing variability and ensuring the quality and consistency of brain organoid cultures used in DNT testing. Similarly, Good In Vitro Reporting Standards (GIVReSt) provide a framework for the accurate and transparent reporting of *in vitro* experimental details, enabling the reproducibility and comparability of DNT studies (Hartung et al., 2019). Test readiness criteria, such as those proposed by Bal-Price et al. (2018), are crucial for evaluating the suitability of brain organoid-based assays for DNT testing. These criteria consider factors such as the biological relevance of the model, the robustness and reproducibility of the assay, and the availability of reference compounds and performance standards. Qualification and validation of brain organoid-based DNT assays are essential steps in establishing their scientific validity and regulatory acceptance. Qualification involves the assessment of the assay's performance characteristics, such as sensitivity, specificity, and reproducibility, while validation entails the demonstration of the assay's relevance and reliability for its intended purpose (Hartung et al., 2013; Pamies et al., 2022). Adherence to GCCP, GIVReSt, test readiness criteria, and rigorous qualification and validation processes will be critical for the successful implementation of brain organoids in DNT testing and their integration into regulatory decision-making frameworks.

7. Future directions and recommendations

To fully harness the potential of brain organoids for DNT testing, several key advancements and recommendations should be considered. These include the incorporation of vasculature, immune cells, and blood-brain barrier (BBB) components, strategies to enhance organoid maturation and aging, the development of high-throughput screening platforms, integration with other emerging technologies, and efforts toward standardization and regulatory acceptance.

The incorporation of vasculature, immune cells, and BBB components into brain organoid models is a critical step toward improving their physiological relevance and predictive value (Smirnova and Hartung, 2024). The development of vascularized organoids, through co-culture with endothelial cells or *in vivo* transplantation, can enhance nutrient and oxygen delivery, allowing for the generation of larger and more complex structures (Mansour et al., 2018). The inclusion of microglia and other immune cells can recapitulate the neuroimmune interactions that play a key role in neurodevelopment and neurotoxicity (Abreu et al., 2018). Moreover, the incorporation of BBB components, such as endothelial cells and pericytes, can enable the study of toxicant transport and distribution in a more physiologically relevant context (Bergmann et al., 2018).

Strategies to enhance the maturation and aging of brain organoids are essential to expand their utility beyond the modeling of early developmental stages (Hartung et al., 2024). Prolonged culture times, optimized differentiation protocols, and exposure to physiologically relevant stimuli, such as neurotransmitters and growth factors, can promote the emergence of more mature neuronal subtypes and complex network dynamics (Qian et al., 2019; Smirnova and Hartung, 2024). The development of "aging" protocols, involving the introduction of age-associated stressors or the use of iPSCs derived from older individuals, can enable the modeling of age-related neurodegenerative processes and their interaction with developmental neurotoxicity (Smirnova and Hartung, 2024).

High-throughput screening (HTS) platforms are critical for the efficient and cost-effective assessment of large chemical libraries for DNT potential (Hartung et al., 2024). The development of miniaturized and automated organoid culture systems, coupled with advanced imaging and analysis tools, can enable the rapid generation and characterization of brain organoids in a format compatible with HTS (Pamies et al., 2017; Smirnova and Hartung, 2024). The incorporation of machine learning algorithms and artificial intelligence can further streamline data analysis and improve the predictive power of organoid-based DNT assays (Smirnova et al., 2015; Hartung et al., 2024).

The integration of brain organoids with other emerging technologies, such as organ-on-a-chip devices, 3D bioprinting, and in silico modeling, can provide a more comprehensive and predictive approach to DNT testing (Smirnova and Hartung, 2024). Organ-on-a-chip platforms can enable the co-culture of brain organoids with other relevant tissues, such as the liver or the placenta, to study the effects of metabolites and the role of the maternal-fetal interface in DNT (Pamies et al., 2017). 3D bioprinting can allow for the precise control over organoid architecture and the incorporation of scaffold materials to enhance structural and functional maturation (Zhuang et al., 2018). *In silico* modeling, including quantitative structure-activity relationship (QSAR) models and physiologically based pharmacokinetic (PBPK) models, can complement organoid-based assays by providing predictions of toxicant fate and transport, and by aiding in the interpretation and extrapolation of *in vitro* data (Smirnova and Hartung, 2024).

Standardization and regulatory acceptance are key priorities for the successful implementation of brain organoids in DNT testing (Hartung et al., 2024). The development of standardized protocols, quality control measures, and performance benchmarks can enhance the reproducibility and comparability of organoid-based assays across different laboratories (Pamies et al., 2017). The establishment of reference compounds, endpoint definitions, and data reporting standards can facilitate the validation and regulatory acceptance of brain organoid models (Aschner et al., 2017; Sachana et al., 2019). Collaborative efforts among researchers, industry partners, and regulatory agencies, such as the OECD and the European Food Safety Authority (EFSA), are essential to harmonize approaches and to develop internationally accepted guidelines for the use of brain organoids in DNT testing (Bal-Price et al., 2018; Fritsche et al., 2017).

Organoid Intelligence (OI) represents an emerging frontier in developmental neurotoxicity (DNT) testing, where brain organoids are integrated with artificial intelligence (AI) to create advanced *in vitro* models of learning and cognition (Smirnova et al., 2023a). The goal of OI is to leverage the self-organizing complexity of brain organoids and the computational power of AI to generate hybrid systems capable of modeling higher-order brain functions (Hartung et al., 2023). This approach could revolutionize DNT testing by enabling the assessment of neurotoxicant effects on cognitive processes that are difficult to study using current methods (Smirnova and Hartung, 2024). OI could also facilitate the development of more predictive and efficient DNT assays by automating data analysis and integrating multi-modal readouts from high-throughput screening platforms (Smirnova et al., 2023b). However, the realization of OI for DNT testing will require significant advancements in organoid culture techniques, AI algorithms, and biohybrid interface technologies, as well as careful consideration of the ethical implications of creating intelligent *in vitro* systems (Hartung et al., 2023, 2024). As the field of OI continues to evolve, it holds immense potential for transforming DNT testing and advancing our understanding of the complex interactions between environmental exposures and neurodevelopment.

In conclusion, the future of brain organoids in DNT testing relies on the continued advancement of organoid culture methods, the integration with emerging technologies, and the establishment of standardized and validated approaches. By incorporating vasculature, immune cells, and BBB components, enhancing organoid maturation and aging, developing high-throughput screening platforms, and integrating with other cutting-edge tools, brain organoids can provide an increasingly comprehensive and physiologically relevant model for assessing the DNT potential of chemicals. Collaborative efforts toward standardization and regulatory acceptance will be crucial to realize the full potential of brain organoids in protecting human health from the adverse effects of developmental neurotoxicants.

8. Ethical considerations

The use of human brain organoids in DNT testing raises important ethical questions that must be carefully considered and addressed. These include the moral status of brain organoids, the potential for consciousness, selection of donors, informed consent from cell donors, and the need for appropriate oversight frameworks.

The moral status of brain organoids is a complex issue that depends on various factors, such as their cellular composition, level of organization, and potential for consciousness (Lavazza and Massimini, 2018; Sawai et al., 2022). As brain organoids become more sophisticated and capable of recapitulating higher-order neural functions, concerns arise about their moral considerability and the ethical obligations of researchers (Farahany et al., 2018). While current brain organoid models are unlikely to possess consciousness or sentience, the possibility of these capacities emerging in future models cannot be ruled out (Lavazza, 2021a,b). Engaging in interdisciplinary dialogues among scientists, ethicists, and policymakers is essential to proactively address these issues and develop appropriate guidelines for the ethical use of brain organoids (Sawai et al., 2022; Hartung et al., 2024).

Donors should be selected with care, as sex and gender as well the genetic background of an individual can influence their vulnerability to neurotoxicants, while increasing diversity will improve the progression towards precision medicine (Ghosh et al., 2022). Informed consent from cell donors is a critical ethical requirement for the generation and use of brain organoids in research (Boers et al., 2019). Donors should be provided with clear and comprehensive information about the intended use of their cells, including the potential for organoid development, long-term storage, and sharing with other researchers (Bollinger et al., 2021). Special considerations may apply when using cells from vulnerable populations, such as children or individuals with cognitive impairments (Boers and Bredenoord, 2018). The development of dynamic consent models, which allow donors to update their preferences over time, can help ensure that their wishes are respected as research progresses (Boers et al., 2019).

The use of patient-derived iPSCs for generating brain organoids raises additional ethical considerations. Donors should be informed about the potential for incidental findings, such as the discovery of disease-related genetic variants, and the implications for their health and that of their family members (de Jongh et al., 2022). Researchers should have clear policies in place for managing and communicating such findings, and for protecting the privacy and confidentiality of donor information (Hartung et al., 2024).

As brain organoid research becomes more complex and sophisticated, there is a need for appropriate oversight frameworks to ensure responsible and ethically sound practices (Hyun et al., 2020; Sawai et al., 2022). Existing oversight mechanisms, such as institutional review boards (IRBs) and stem cell research oversight (SCRO) committees, may require adaptations to address the unique challenges posed by brain organoids (Farahany et al., 2018). The development of specialized guidelines and review processes for brain organoid research can help ensure that ethical considerations are systematically addressed and that research is conducted in accordance with societal values and expectations (Hyun et al., 2020; Hartung et al., 2024).

Particular attention should be given to the ethical implications of using brain organoids for modeling neurological and psychiatric disorders, as well as for developing personalized therapies (Smirnova and Hartung, 2024). The use of patient-derived organoids in these contexts raises questions about the ownership and control of biological materials, the return of research results to participants, and the equitable access to potential benefits (Boers et al., 2019). Engaging patient communities and other stakeholders in the design and conduct of brain organoid research can help ensure that these issues are addressed in a transparent and inclusive manner (Bollinger et al., 2021).

In conclusion, the ethical considerations surrounding the use of human brain organoids in DNT testing are complex and multifaceted. Addressing these issues requires ongoing interdisciplinary collaboration among scientists, ethicists, policymakers, and the public. The development of appropriate guidelines, oversight mechanisms, and public engagement strategies will be essential to ensure that brain organoid research advances in a responsible and ethically sound manner, while realizing its potential to protect human health and well-being.

9. Conclusions

Brain organoids hold immense potential to revolutionize DNT testing by providing a human-relevant, efficient, and predictive platform for assessing the effects of chemicals on the developing brain. By recapitulating key features of human neurodevelopment, including cellular diversity, complex organization, and functional activity, brain organoids offer a powerful tool to overcome the limitations of traditional animal-based testing approaches (Pamies et al., 2017; Smirnova et al., 2014, 2024).

The ability to generate brain organoids from human pluripotent stem cells, including patient-derived iPSCs, enables the modeling of genetically diverse populations and the investigation of gene-environment interactions in DNT (Modafferi et al., 2021; Smirnova and Hartung, 2024). This approach opens up new avenues for personalized toxicology and risk assessment, allowing for the identification of susceptible individuals and the development of targeted prevention and intervention strategies (Smirnova et al., 2024; Suciu et al., 2023).

The integration of brain organoids with high-throughput screening technologies, advanced imaging and analytical methods, and computational modeling tools can further enhance the efficiency and predictive power of DNT testing (Pamies et al., 2017; Zhong et al., 2020). The development of standardized protocols, performance benchmarks, and validation criteria will be essential to establish the reliability and regulatory acceptance of brain organoid-based DNT assays (Bal-Price et al., 2018; Fritsche et al., 2017).

However, realizing the full potential of brain organoids in DNT testing will require continued research efforts to address current limitations and challenges. These include improving the reproducibility and scalability of organoid generation, incorporating vascularization and other supporting cell types, enhancing organoid maturation and aging, and establishing robust validation against *in vivo* animal and human data (Qian et al., 2019; Smirnova and Hartung, 2024).

Moreover, the ethical considerations surrounding the use of human-derived brain organoids, such as their moral status, potential for consciousness, and informed consent requirements, must be proactively addressed through interdisciplinary dialogue and the development of appropriate governance frameworks (Farahany et al., 2018; Hartung et al., 2024).

Despite these challenges, the promise of brain organoids in transforming DNT testing and advancing human health protection is immense. By providing a more human-relevant and mechanistically informative approach, brain organoids can help fill critical knowledge gaps in our understanding of DNT and support the development of safer and more sustainable chemicals and products (Smirnova et al., 2014, 2024).

Ultimately, the vision for the future of DNT testing is one that leverages the power of brain organoids, along with other cutting-edge technologies and innovative approaches, to deliver a more efficient, predictive, and human-relevant assessment of the risks posed by developmental neurotoxicants (Smirnova and Hartung, 2024). This vision calls for a collaborative and interdisciplinary effort among researchers, industry partners, regulators, and other stakeholders to advance the science and ethics of brain organoid-based DNT testing and to translate this knowledge into effective strategies for protecting the health and well-being of current and future generations.

In conclusion, brain organoids represent a groundbreaking tool for advancing DNT testing and improving our understanding of the complex interactions between environmental exposures and neurodevelopment. By harnessing the potential of this innovative technology, while addressing its limitations and ethical implications, we can move towards a more human-relevant, efficient, and predictive approach to safeguarding brain health and promoting sustainable chemical innovation.

Acknowledgments

The authors acknowledge funding from a Johns Hopkins Discovery award, the Wendy Klag Center at the Johns Hopkins University and a Johns Hopkins SURPASS award. Funding by the US Environmental Protection Agency (EPA) EPA-STAR grant (R83950501), the FDA-JHU Center of Excellence in Risk Sciences (CERSI) BrainMixTox Project, the Maryland Stem Cell Research Fund, and the Combining advances in Genomics and Environmental science to accelerate Actionable Research in ASD (GEARS) Network (R01ES034554); Ladd-Acosta, Volk) is gratefully appreciated. M.S. has received funding from the Deutsche Forschungsgemeinschaft (DFG, 507269789) and the Alternatives Research & Development Foundation (ARDF).

Author contributions

TH: Conceptualization, Writing – Original Draft Preparation, All: Writing – Review & Editing.

Conflict of interest

T.H. is named inventor on a patent by Johns Hopkins University on the production of mini-brains (also called BrainSpheres), which is licensed to Axo-Sim, New Orleans, LA, USA. He is a shareholder of and he and L.S. are consultants for AxoSim, New Orleans; T.H. is also a consultant American Type Culture Collection (ATCC), InSphero, Zurich, Switzerland, Crown Biosciences, San Diego, CA, and was until recently consultant for AstraZeneca on advanced cell culture methods.

References

Abreu, C.M., Gama, L., Krasemann, S., Chesnut, M., Odwin-Dacosta, S., Hogberg, H.T., et al., 2018. Microglia increase inflammatory responses in iPSC-derived human BrainSpheres. Front. Microbiol. 9, 2766. https://doi.org/10.3389/fmicb.2018.02766.

Adams, J.W., Negraes, P.D., Truong, J., Tran, T., Szeto, R.A., Guerra, B.S., et al., 2023. Impact of alcohol exposure on neural development and network formation in human cortical organoids. Mol. Psych. 28 (4), 1571–1584. https://doi.org/10.1038/s41380-022-01862-7.

Anderson, W.A., Bosak, A., Hogberg, H.T., Hartung, T., Moore, M.J., 2021. Advances in 3D neuronal microphysiological systems: towards a functional nervous system on a chip. Vitro Cell. Develop. Biol. Anim. 57, 191–206. https://doi.org/10.1007/s11626-020-00532-8.

Arambula, S.E., McCarthy, M.M., 2020. Neuroendocrine-immune crosstalk shapes sex-specific brain development. Endocrinology 161, bqaa055. https://doi.org/10.1210/endocr/bqaa055.

Aschner, M., Ceccatelli, S., Daneshian, M., Fritsche, E., Hasiwa, N., Hartung, T., et al., 2017. Reference compounds for alternative test methods to indicate developmental neurotoxicity (DNT) potential of chemicals: example lists and criteria for their selection and use. ALTEX 34 (1), 49–74. https://doi.org/10.14573/altex.1604201.

Bakker, J., 2022. The role of steroid hormones in the sexual differentiation of the human brain. J. Neuroendocrinol. 34, 1–11. https://doi.org/10.1111/jne.13050.

Bal-Price, A., Hogberg, H.T., Crofton, K.M., Daneshian, M., FitzGerald, R.E., Fritsche, E., et al., 2018. Recommendation on test readiness criteria for new approach methods in toxicology: exemplified for developmental neurotoxicity. ALTEX 35 (3), 306–352. https://doi.org/10.14573/altex.1712081.

Bal-Price, A., Crofton, K.M., Leist, M., Allen, S., Arand, M., Buetler, T., et al., 2015. International STakeholder NETwork (ISTNET): creating a developmental neurotoxicity (DNT) testing road map for regulatory purposes. Arch. Toxicol. 89 (2), 269–287. https://doi.org/10.1007/s00204-015-1464-2.

Bal-Price, A.K., Coecke, S., Costa, L., Crofton, K.M., Fritsche, E., Goldberg, A., et al., 2012. Advancing the science of developmental neurotoxicity (DNT): testing for better safety evaluation. ALTEX 29 (2), 202–215. https://doi.org/10.14573/altex.2012.2.202.

Bal-Price, A., Hogberg, H.T., Buzanska, L., Lenas, P., van Vliet, E., Hartung, T., 2010. In vitro developmental neurotoxicity (DNT) testing: relevant models and endpoints. Neurotoxicology 31 (5), 545–554. https://doi.org/10.1016/j.neuro.2009.11.006.

Bergmann, S., Lawler, S.E., Qu, Y., Fadzen, C.M., Wolfe, J.M., Regan, M.S., et al., 2018. Blood–brain-barrier organoids for investigating the permeability of CNS therapeutics. Nat. Protoc. 13 (12), 2827–2843. https://doi.org/10.1038/s41596-018-0066-x.

Birtele, M., Del Dosso, A., Xu, T., et al., 2023. Non-synaptic function of the autism spectrum disorder-associated gene SYNGAP1 in cortical neurogenesis. Nat. Neurosci. 26, 2090–2103. https://doi.org/10.1038/s41593-023-01477-3.

Boers, S.N., van Delden, J.J.M., Bredenoord, A.L., 2019. Organoids as hybrids: ethical implications for the exchange of human tissues. J. Med. Ethics 45 (2), 131–139. https://doi.org/10.1136/medethics-2018-104846.

Boers, S.N., Bredenoord, A.L., 2018. Consent for governance in the ethical use of organoids. Nat. Cell Biol. 20 (6), 642–645. https://doi.org/10.1038/s41556-018-0110-y.

Bollinger, J., May, E., Mathews, D., Donowitz, M., Sugarman, J., 2021. Patients' perspectives on the derivation and use of organoids. Stem. Cell Rep. 16 (8), 1874–1883. https://doi.org/10.1016/j.stemcr.2021.07.004.

Coecke, S., Goldberg, A.M., Allen, S., Buzanska, L., Calamandrei, G., Crofton, K., et al., 2007. Workgroup report: incorporating in vitro alternative methods for developmental neurotoxicity into international hazard and risk assessment strategies. Environ. Health Perspect. 115 (6), 924–931. https://doi.org/10.1289/ehp.9427.

Crofton, K.M., Mundy, W.R., Lein, P.J., Bal-Price, A., Coecke, S., Seiler, A.E.M., et al., 2011. Developmental neurotoxicity testing: recommendations for developing alternative methods for the screening and prioritization of chemicals. ALTEX 28 (1), 9–15. https://doi.org/10.14573/altex.2011.1.009.

de Jongh, D., Massey, E.K., the VANGUARD consortium, Bunnik, E.M., 2022. Organoids: a systematic review of ethical issues. Stem Cell Res. Ther. 13, 337. https://doi.org/10.1186/s13287-022-02950-9.

Fan, P., Wang, Y., Xu, M., Han, X., Liu, Y., 2022 Feb 9. The application of brain organoids in assessing neural toxicity. Front. Mol. Neurosci. 15, 799397. https://doi.org/10.3389/fnmol.2022.799397.

Farahany, N.A., Greely, H.T., Hyman, S., Koch, C., Grady, C., Paşca, S.P., et al., 2018. The ethics of experimenting with human brain tissue. Nature 556 (7702), 429–432. https://doi.org/10.1038/d41586-018-04813-x.

Fritsche, E., Crofton, K.M., Hernandez, A.F., Hougaard Bennekou, S., Leist, M., Bal-Price, A., et al., 2017. OECD/EFSA workshop on developmental neurotoxicity (DNT): the use of non-animal test methods for regulatory purposes. ALTEX 34 (2), 311–315. https://doi.org/10.14573/altex.1701171.

Fritsche, E., Cline, J.E., Nguyen, N.-H., Scanlan, T.S., Abel, J., 2005. Polychlorinated biphenyls disturb differentiation of normal human neural progenitor cells: clue for involvement of thyroid hormone receptors. Environ. Health Perspect. 113 (7), 871–876. https://doi.org/10.1289/ehp.7793.

Ghosh, S., Nehme, R., Barrett, L.E., 2022 Nov 26. Greater genetic diversity is needed in human pluripotent stem cell models. Nat. Commun. 13 (1), 7301. https://doi.org/10.1038/s41467-022-34940-z.

Grandjean, P., Landrigan, P.J., 2006. Developmental neurotoxicity of industrial chemicals. Lancet 368 (9553), 2167–2178. https://doi.org/10.1016/S0140-6736(06)69665-7.

Grandjean, P., Landrigan, P.J., 2014. Neurobehavioural effects of developmental toxicity. Lancet Neurol. 13 (3), 330–338. https://doi.org/10.1016/S1474-4422(13)70278-3.

Gassmann, K., Abel, J., Bothe, H., Haarmann-Stemmann, T., Merk, H.F., Quasthoff, K.N., et al., 2010. Species-specific differential AhR expression protects human neural progenitor cells against developmental neurotoxicity of PAHs. Environ. Health Perspect. 118 (11), 1571–1577. https://doi.org/10.1289/ehp.0901545.

Hartung, T., Morales Pantoja, I.E., Smirnova, L., 2023. Brain organoids and organoid intelligence (OI) from ethical, legal, and social points of view (accepted). Front. Artif. Intell. https://doi.org/10.3389/frai.2023.1307613.

Hartung, T., Morales Pantoja, I.E., Smirnova, L., 2024. Brain organoids and Organoid Intelligence (OI) from ethical, legal, and social points of view. Front. Artif. Intell. 6, 1307613. https://doi.org/10.3389/frai.2023.1307613.

Hartung, T., de Vries, R., Hoffmann, S., Hogberg, H.T., Smirnova, L., Tsaioun, K., et al., 2019. Toward good in vitro reporting standards. ALTEX 36 (1), 3–17. https://doi.org/10.14573/altex.1812191.

Hartung, T., Hoffmann, S., Stephens, M., 2013. Mechanistic validation. ALTEX 30 (2), 119–130. https://doi.org/10.14573/altex.2013.2.119.

Hartung, T., van Vliet, E., Jaworska, J., Bonilla, L., Skinner, N., Thomas, R., 2012. Systems toxicology. ALTEX 29 (2), 119–128. https://doi.org/10.14573/altex.2012.2.119.

Hartung, T., Bremer, S., Casati, S., Coecke, S., Corvi, R., Fortaner, S., et al., 2004. A modular approach to the ECVAM principles on test validity. ATLA: Alter. Lab. Anim. 32 (5), 467–472. https://doi.org/10.1177/026119290403200503.

Hyun, I., Scharf-Deering, J.C., Lunshof, J.E., 2020. Ethical issues related to brain organoid research. Brain Res. 1732, 146653. https://doi.org/10.1016/j.brainres.2020.146653.

Jacob, F., Schnoll, J.G., Song, H., Ming, G.L., 2021. Building the brain from scratch: engineering region-specific brain organoids from human stem cells to study neural development and disease. Curr. Top. Dev. Biol. 142, 477–530. https://doi.org/10.1016/bs.ctdb.2020.12.011.

Kern, J.K., Geier, D.A., Homme, K.G., King, P.G., Bjørklund, G., Chirumbolo, S., et al., 2017. Developmental neurotoxicants and the vulnerable male brain: a systematic review of suspected neurotoxicants that disproportionally affect males. Acta Neurobiol. Exp. (Wars). 77, 269–296. https://doi.org/10.21307/ane-2017-061.

Lancaster, M.A., Renner, M., Martin, C.-A., Wenzel, D., Bicknell, L.S., Hurles, M.E., et al., 2013. Cerebral organoids model human brain development and microcephaly. Nature 501 (7467), 373–379. https://doi.org/10.1038/nature12517.

Lancaster, M.A., Knoblich, J.A., 2014. Generation of cerebral organoids from human pluripotent stem cells. Nat. Protoc. 9 (10), 2329–2340. https://doi.org/10.1038/nprot.2014.158.

Lavazza, A., 2021a. Potential ethical problems with human cerebral organoids: consciousness and moral status of future brains in a dish. Brain Res. 1750, 147146. https://doi.org/10.1016/j.brainres.2020.147146.

Lavazza, A., 2021b. Consciousnessoids: clues and insights from human cerebral organoids for the study of consciousness. Neurosci. Conscious. 7 (2), niab029. https://doi.org/10.1093/nc/niab029.

Lavazza, A., Massimini, M., 2018. Cerebral organoids: ethical issues and consciousness assessment. J. Med. Ethics 44 (9), 606–610. https://doi.org/10.1136/medethics-2017-104555.

Lein, P., Locke, P., Goldberg, A., 2007. Meeting report: alternatives for developmental neurotoxicity testing. Environ. Health Perspect. 115 (5), 764–768. https://doi.org/10.1289/ehp.9841.

Leist, M., Ghallab, A., Graepel, R., Marchan, R., Hassan, R., Bennekou Hougaard, S., et al., 2021. Cellular complexity in brain organoids: current progress and unsolved issues. Semin. Cell Dev. Biol. 111, 32–39. https://doi.org/10.1016/j.semcdb.2020.05.013.

Leist M., Ghallab A., Graepel R., Marchan R., Hassan R., Hougaard, Bennekou S., Limonciel A., Vinken M., Schildknecht S., Waldmann T., Danen E., van Ravenzwaay B., Kamp H., Gardner I., Godoy P., Bois F.Y., Braeuning A., Reif R., Oesch F., Drasdo D., Höhme S., Schwarz M., Hartung T., Braunbeck T., Beltman J., Vrieling H., Sanz F., Forsby A., Gadaleta D., Fisher C., Kelm J., Fluri D., Ecker G., Zdrazil B., Terron A., Jennings P., van der Burg B., Dooley S., Meijer A.H., Willighagen E., Martens M., Evelo C., Mombelli E., Taboureau O., Mantovani A., Hardy B., Koch B., Escher S., van Thriel C., Cadenas C., Kroese D., van de Water B. and Hengstler J.G. Adverse outcome pathways: opportunities, limitations and open questions. *Arch. Toxicol.* 2017, 31:221–229. https://doi.org/10.1007/s00204-017-2045-3.

Li, P., Ma, R., Dong, L., Liu, L., Zhou, G., Tian, Z., Zhao, Q., Xia, T., Zhang, S., Wang, A., 2019. Autophagy impairment contributes to PBDE-47-induced developmental neurotoxicity and its relationship with apoptosis. Theranostics 9 (15), 4375–4390. https://doi.org/10.7150/thno.33688. Jun 9.

Masjosthusmann, S., Becker, D., Petzuch, B., Klose, J., Siebert, C., Deenen, R., et al., 2018. A transcriptome comparison of time-matched developing human, mouse and rat neural progenitor cells reveals human uniqueness. Toxicol. Appl. Pharmacol. 354, 40–55. https://doi.org/10.1016/j.taap.2018.05.009.

Makris, S.L., Raffaele, K., Allen, S., Bowers, W.J., Hass, U., Alleva, E., et al., 2009. A retrospective performance assessment of the developmental neurotoxicity study in support of OECD test guideline 426. Environ. Health Perspect. 117 (1), 17–25. https://doi.org/10.1289/ehp.11447.

Mansour, A.A., Gonçalves, J.T., Bloyd, C.W., Li, H., Fernandes, S., Quang, D., et al., 2018. An in vivo model of functional and vascularized human brain organoids. Nat. Biotechnol. 36 (5), 432–441. https://doi.org/10.1038/nbt.4127.

Mansour, A.A., Schafer, S.T., Gage, F.H., 2021. Cellular complexity in brain organoids: Current progress and unsolved issues. Semin. Cell Dev. Biol. 111, 32–39. https://doi.org/10.1016/j.semcdb.2020.05.013. Mar.

Mariani, J., Coppola, G., Zhang, P., Abyzov, A., Provini, L., Tomasini, L., et al., 2015. FOXG1-dependent dysregulation of GABA/glutamate neuron differentiation in autism spectrum disorders. Cell 162 (2), 375–390. https://doi.org/10.1016/j.cell.2015.06.034.

Meijer, A.H., Willighagen, E., Martens, M., Evelo, C., Mombelli, E., Taboureau, O., et al., 2017. Adverse outcome pathways: opportunities, limitations and open questions. Arch. Toxicol. 31, 221–229. https://doi.org/10.1007/s00204-017-2045-3.

Modafferi, S., Zhong, X., Kleensang, A., Murata, Y., Fagiani, F., Pamies, D., et al., 2021. Gene-environment interactions in developmental neurotoxicity: a case study of synergy between chlorpyrifos and CHD8 knockout in human BrainSpheres. Environ. Health Perspect. 129 (7), 77001. https://doi.org/10.1289/EHP8580.

Moors, M., Rockel, T.D., Abel, J., Cline, J.E., Gassmann, K., Schreiber, T., et al., 2009. Human neurospheres as three-dimensional cellular systems for developmental neurotoxicity testing. Environ. Health Perspect. 117 (7), 1131–1138. https://doi.org/10.1289/ehp.0800207.

Muotri, A.R., 2023. Brain model technology and its implications. Camb. Q. Healthc. Ethics 1–5. https://doi.org/10.1017/S096318012300018X.

OECD, 2023. Initial recommendations on evaluation of data from the developmental neurotoxicity (DNT) in-vitro testing battery. Series on Testing and Assessment No. 377. https://one.oecd.org/document/ENV/CBC/MONO(2023)13/en/pdf.

Pamies, D., Leist, M., Coecke, S., Bowe, G., Allen, D., Gstraunthaler, G., et al., 2022. Guidance document on good cell and tissue culture practice 2.0 (GCCP 2.0). ALTEX 39 (1), 30–70. https://doi.org/10.14573/altex.2111011.

Pamies, D., Block, K., Lau, P., Gribaldo, L., Pardo, C.A., Barreras, P., et al., 2018. Rotenone exerts developmental neurotoxicity in a human brain spheroid model. Toxicol. Appl. Pharmacol. 354, 101–114. https://doi.org/10.1016/j.taap.2018.02.003.

Pamies, D., Barreras, P., Block, K., Makri, G., Kumar, A., Wiersma, D., et al., 2017. A human brain microphysiological system derived from induced pluripotent stem cells to study neurological diseases and toxicity. ALTEX 34 (3), 362–376. https://doi.org/10.14573/altex.1609122.

Pamies, D., Ekert J., Zurich, M.-G., Frey, O., Werner, S., Pergiovanni, M., Freedman, B., Teo A.K.K., Erfurth, H., Reyes, D.R., Loskill, P., Candarlioglu, P., Suter-Dick, L., Wang, S., Hartung, T., Coecke, S., Stacey, G., Wagegg, B.A., Dehne, E.-M., Pistollato, F. and Leist, M. Recommendations on fit-for-purpose criteria to establish quality management for Microphysiological Systems (MPS) and for monitoring of their reproducibility. Stem Cell

Reports, Recommendations on fit-for-purpose criteria to establish quality management for microphysiological systems and for monitoring their reproducibility, *Stem Cell Rep.* 2024, 19:604-617. https://doi.org/10.1016/j.stemcr.2024.03.009.

Paşca, A.M., Sloan, S.A., Clarke, L.E., Tian, Y., Makinson, C.D., Huber, N., et al., 2015. Functional cortical neurons and astrocytes from human pluripotent stem cells in 3D culture. Nat. Methods 12 (7), 671–678. https://doi.org/10.1038/nmeth.3415.

Qian, X., Song, H., Ming, G.-L., 2019. Brain organoids: advances, applications and challenges. Development 146 (8), dev166074. https://doi.org/10.1242/dev.166074.

Qian, X., Jacob, F., Song, M.M., Nguyen, H.N., Song, H., Ming, G.-L., 2018. Generation of human brain region–specific organoids using a miniaturized spinning bioreactor. Nat. Protoc. 13 (3), 565–580. https://doi.org/10.1038/nprot.2017.152.

Sachana, M., Bal-Price, A., Crofton, K.M., Bennekou, S.H., Shafer, T.J., Behl, M., et al., 2019. International regulatory and scientific effort for improved developmental neurotoxicity testing. Toxicol. Sci. 167 (1), 45–57. https://doi.org/10.1093/toxsci/kfy211.

Sawai, T., Hayashi, Y., Niikawa, T., Shepherd, J., Thomas, E., Lee, T., et al., 2022. Mapping the ethical issues of brain organoid research and application. AJOB Neurosci. 13 (2-3), 81–94. https://doi.org/10.1080/21507740.2021.2018345.

Schaafsma, S.M., Gagnidze, K., Reyes, A., Norstedt, N., Månsson, K., Francis, K., et al., 2017. Sex-specific gene–environment interactions underlying ASD-like behaviors. Proc. Natl Acad. Sci. 114, 1383–1388.

Schmidt, B.Z., Lehmann, M., Gutbier, S., Nembo, E., Noel, S., Smirnova, L., et al., 2016. In vitro acute and developmental neurotoxicity screening: an overview of cellular platforms and high-throughput technical possibilities. Arch. Toxicol. 91 (1), 1–33. https://doi.org/10.1007/s00204-016-1805-9.

Smirnova, L., Hogberg, H.T., Leist, M., Hartung, T., 2024. Revolutionizing developmental neurotoxicity testing - a journey from animal models to advanced in vitro systems. ALTEX 2024 41, 152–178. https://doi.org/10.14573/altex.2403281.

Smirnova, L., Hartung, T., 2024. The promise and potential of brain organoids. Adv. Healthc. Mater., 2302745. https://doi.org/10.1002/adhm.202302745.

Smirnova, L., Morales Pantoja, I.E., Hartung, T., 2023a. Organoid intelligence (OI): the new frontier in biocomputing and intelligence-in-a-dish. Front. Sci. 1, 1017235. https://doi.org/10.3389/fsci.2023.1017235.

Smirnova, L., Morales Pantoja, I.E., Hartung, T., 2023b. Organoid Intelligence (OI) – the ultimate functionality of a brain microphysiological system. ALTEX 40, 191–203. https://doi.org/10.14573/altex.2303261.

Smirnova, L., Kleinstreuer, N., Corvi, R., Levchenko, A., Fitzpatrick, S.C., Hartung, T., 2018. 3S - Systematic, systemic, and systems biology and toxicology. ALTEX 35 (2), 139–162. https://doi.org/10.14573/altex.1804051.

Smirnova, L., Seiler, A.E.M., Luch, A., 2015. microRNA profiling as tool for developmental neurotoxicity testing (DNT). 20.9.1-22. Curr. Protoc. Toxicol. 64. https://doi.org/10.1002/0471140856.tx2009s64.

Smirnova, L., Hogberg, H.T., Leist, M., Hartung, T., 2014. Developmental neurotoxicity - challenges in the 21st century and in vitro opportunities. ALTEX 31 (2), 129–156. https://doi.org/10.14573/altex.1403271.

Sousa, A.M.M., Zhu, Y., Raghanti, M.A., Kitchen, R.R., Onorati, M., Tebbenkamp, A.T.N., et al., 2017. Molecular and cellular reorganization of neural circuits in the human lineage. Science 1032, 1027–1032. https://doi.org/10.1126/science.aan3456.

Suciu I., Pamies D., Peruzzo R., Wirtz P.H., Smirnova L., Pallocca G., Hauck C., Cronin M.T.D., Hengstler J.G., Brunner T., Hartung T., Amelio I. and Leist M. GxE interactions as a basis for toxicological uncertainty. Archives Toxicology 2023, 97:2035-2049. https://doi.org/10.1007/s00204-023-03500-9.

Trujillo, C.A., Gao, R., Negraes, P.D., Gu, J., Buchanan, J., Preissl, S., et al., 2019. Complex oscillatory waves emerging from cortical organoids model early human brain network development. Cell Stem. Cell 25 (4), 558–569.e7. https://doi.org/10.1016/j.stem.2019.08.002.

Urresti, J., Zhang, P., Moran-Losada, P., et al., 2021. Cortical organoids model early brain development disrupted by 16p11.2 copy number variants in autism. Mol. Psych. 26, 7560–7580. https://doi.org/10.1038/s41380-021-01243-6.

Yoon, S.J., Elahi, L.S., Paşca, A.M., Marton, R.M., Gordon, A., Revah, O., et al., 2019. Reliability of human cortical organoid generation. Nat. Methods 16 (1), 75–78. https://doi.org/10.1038/s41592-018-0255-0.

Zhong, X., Harris, G., Smirnova, L., Zufferey, V., de Cássia da Silveira, E., Sá, R., et al., 2020. Antidepressant paroxetine exerts developmental neurotoxicity in an iPSC-derived 3D human brain model. Front. Cell. Neurosci. 14, 25. https://doi.org/10.3389/fncel.2020.00025.

Zhuang, P., Sun, A.X., An, J., Chua, C.K., Chew, S.Y., 2018. 3D neural tissue models: from spheroids to bioprinting. Biomaterials 154, 113–133. https://doi.org/10.1016/j.biomaterials.2017.10.

CHAPTER FOUR

Self-organizing human neuronal cultures in the modeling of environmental impacts on learning and intelligence

Thomas Hartung[a,b,c,*], **Jack R. Thornton**[a], **and Lena Smirnova**[a]
[a]Center for Alternatives to Animal Testing (CAAT), Bloomberg School of Public Health and Whiting School of Engineering, Johns Hopkins University, Baltimore, MD, United States
[b]Doerenkamp-Zbinden Chair for Evidence-based Toxicology, Baltimore, MD, United States
[c]CAAT Europe, University of Konstanz, Konstanz, Germany
*Corresponding author. e-mail address: thartun1@jhu.edu

Contents

Advances in Neurotoxicology, Volume 12
ISSN 2468-7480, https://doi.org/10.1016/bs.ant.2024.09.001

Abstract

The increase in neurodevelopmental disorders such as autism and ADHD calls for better models to study the environmental influences on brain development. Traditional animal models and simple cell cultures have limitations. Our study focuses on advanced three-dimensional brain organoids derived from human induced pluripotent stem cells, which more accurately replicate human brain development. These models enable us to investigate the effects of environmental factors, including toxicants and nutrients, on learning and intelligence.

We highlight methodologies for simulating real-life environmental exposures and integrating omics technologies to elucidate the molecular impacts. Examples of neurodevelopmental disorders provide insights into these processes. The emerging concept of Organoid Intelligence, demonstrated by brain organoids exhibiting learning and memory capabilities, underscores the potential of these models to advance public health strategies and interventions.

1. Introduction

Studying environmental impacts on neurodevelopment is critical given the continuing increase in neurodevelopmental disorders, which can be largely explained by changes in lifestyle factors and exposures. According to the latest data from 2023, 1 in 36 children in the US aged 8 years is diagnosed with ASD (Maenner et al., 2023). Genetic factors alone cannot account for this rapid increase, and there is no evidence suggesting an infectious component. The development of the human brain is a complex, finely-tuned process that begins before birth and continues into early adulthood. This developmental trajectory makes the brain uniquely susceptible to environmental influences, which can range from chemical exposures to nutritional deficits. Understanding how environmental factors affect neurodevelopment is crucial for public health, as early-life exposures can have lasting effects, potentially leading to neurodevelopmental disorders such as autism spectrum disorder, attention-deficit hyperactivity disorder, and reduced cognitive abilities.

We will start with an overview of current methods in neuronal culture models, focusing on the potential of stem-cell-derived brain organoids (Smirnova and Hartung, 2024). Traditionally, the study of neurodevelopmental impacts has relied on animal models and simple two-dimensional cell cultures. While these methods have provided valuable insights, they come with significant limitations (Hartung et al., 2024). Animal models often fail to fully replicate human biological processes due to interspecies differences, and 2D cultures lack the complex cell interactions found in living organisms (Alépée et al., 2014). These limitations necessitate the development of more accurate and human-relevant models to study the intricate processes of human neurodevelopment and the influences that can alter its course.

Self-organizing human neuronal cultures, particularly brain organoids, represent a significant advancement in neuroscience research (Adlakha, 2023; Smirnova and Hartung, 2024). These three-dimensional cultures are derived from human pluripotent stem cells and mimic the development and organization of the human brain more closely than previous models. They provide a dynamic environment where neurons can grow, differentiate, and interact in ways that closely resemble *in vivo* neurodevelopment.

The advantages of using self-organizing cultures include the ability to observe the development of brain structures and complex neural networks over time, offering a unique window into the cellular and molecular mechanisms that underlie human brain development. Furthermore, these models allow for the direct study of the effects of environmental toxins, nutritional factors, and genetic variations in a controlled setting, providing insights that are both profound and directly relevant to human health.

As we advance into an era where the impact of environmental factors on health is increasingly recognized, calling now for a Human Exposome Project (Hartung, 2023). The development and refinement of self-organizing human neuronal cultures stand out as pivotal innovations. These models not only enhance our understanding of neurodevelopmental processes but also hold the potential to transform approaches to disease prevention and intervention, ultimately contributing to the safeguarding of future generations.

2. Self-organizing neural cultures

2.1 Definition and characteristics of self-organizing human neural cultures

Self-organizing human neuronal cultures represent a class of advanced *in vitro* systems derived from human pluripotent stem cells (hPSCs). These cultures autonomously form complex, three-dimensional structures that mimic several aspects of human brain development, such as the formation of distinct neuronal layers and regions with functional neuronal networks (Lancaster and Knoblich, 2014; Pasca et al., 2015). The key characteristic of these cultures is their ability to self-assemble into organoid structures that exhibit cellular diversity and architecture similar to *in vivo* conditions. This dynamic environment allows neurons and glial cells to interact and mature over time, closely resembling the developmental processes of the human brain (Di Lullo and Kriegstein, 2017).

2.2 Advantages over traditional neural culture methods

Self-organizing human neural cultures represent a significant advancement over traditional two-dimensional cell culture methods, offering multiple advantages that enhance their utility in neuroscience research (Fig. 1). Primarily, these three-dimensional cultures provide a physiologically relevant model that closely mimics the *in vivo* environment of the human brain due to their complex three-dimensional structure and the integration of various cell types, which more accurately replicate the cellular interactions and microenvironments found within the brain (Quadrato et al., 2017). This complexity facilitates the study of fundamental neurobiological

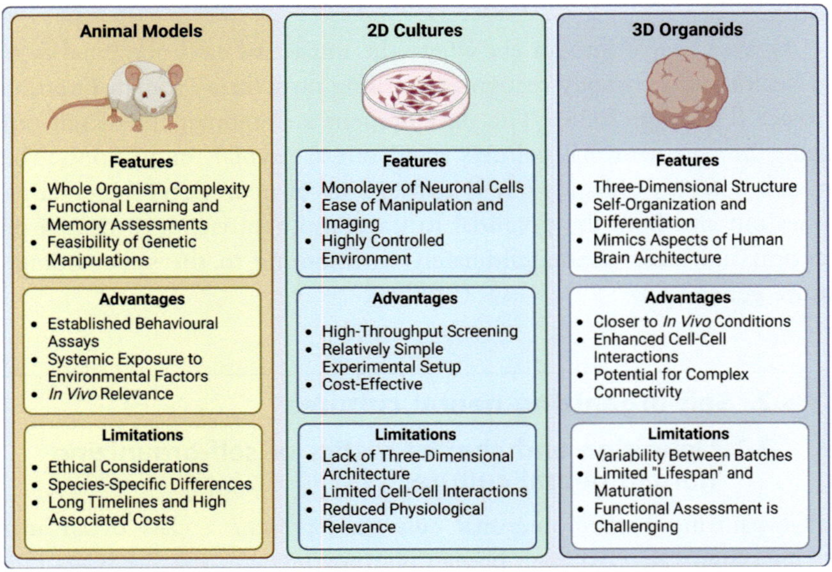

Fig. 1 The key differences between three model systems: *in vivo* animal models, simple *in vitro* 2D cell cultures, and complex *in vitro* 3D organoids. *Created in BioRender.*

processes, including neuronal development, synaptogenesis, and network formation, thereby increasing the biological relevance and translational potential of research findings (Trujillo et al., 2019).

Furthermore, self-organizing cultures have the advantage of longevity; they can be maintained for extended periods, allowing for the investigation of chronic processes such as aging, neurodegeneration, and long-term synaptic stability. This capability enables researchers to observe the progression of neurodegenerative diseases or the long-term effects of chronic exposure to environmental toxicants, providing insights not feasible with traditional short-lived two-dimensional cultures (Kelava and Lancaster, 2016).

Additionally, self-organizing human neuronal cultures are highly amenable to various experimental manipulations. They can be genetically engineered to express reporter genes or to knock down specific genes of interest, facilitating the study of gene function in neurodevelopment or pathology. Risk genes for certain diseases can be introduced into healthy donor genomes or repaired in those of patient donors. These cultures are also compatible with high-resolution imaging techniques, allowing for detailed structural and functional analyses at the cellular and network levels. This compatibility is particularly beneficial for high-throughput screening applications and detailed morphological assessments, which are crucial for drug development and disease modeling (Mansour et al., 2018).

These cultures offer an unprecedented opportunity to study human brain development and diseases under controlled laboratory conditions. By providing a model that better represents human physiology, self-organizing neuronal cultures enable a more accurate simulation of human neuropathological conditions, potentially reducing the translational gap between preclinical studies and clinical applications. This advancement is crucial for developing more effective therapies and gaining deeper insights into the complex mechanisms underlying human brain function and its disruption in various neurological disorders.

2.3 Current methodologies for developing self-organizing cultures

The development of self-organizing human neuronal cultures is a sophisticated process that begins with the differentiation of PSCs into neural liniages. This differentiation is typically induced using a cocktail of growth factors and small molecules that simulate the developmental cues present during early brain development. These factors influence the stem cells to forego their pluripotent state and begin transitioning into cells that possess the characteristics of neural tissue (Sloan et al., 2018).

Differentiation is supported by three-dimensional matrices or scaffolds designed to support the growth and organization of neural tissues by mimicking the extracellular matrix of the brain. Alternatively, the organoids are kept in suspension by shaking (Pamies et al., 2017; Fig. 2) or stirring (Qian et al., 2018) to continue their development, differentiating into a variety of neuronal and glial subtypes. Over time, these cells self-organize to form complex, structured neural networks that exhibit properties similar to those observed *in vivo*. This spontaneous organization is crucial for studying various aspects of brain development, neural connectivity, and pathology under controlled conditions.

Recent advancements in cell culture technologies have focused on enhancing the reproducibility and scalability of cell cultures, particularly through the integration of microfluidic devices and bioreactors. These technologies aim to create more physiologically relevant models that can be used for various applications, including drug discovery, tissue engineering, and disease modeling (Roth et al., 2021).

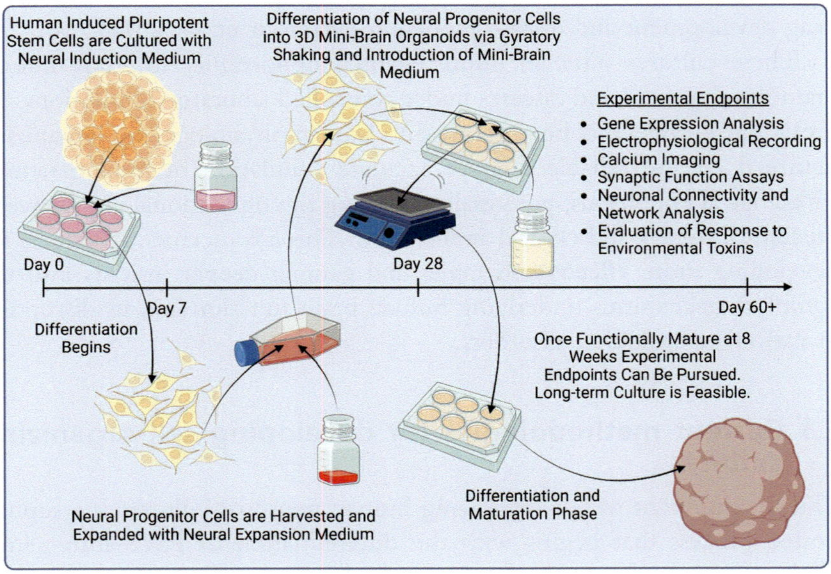

Fig. 2 Timeline and methodology for differentiating human induced pluripotent stem cells (iPSCs) into 3D organoids. The figure depicts the process of generating neural organoids in our laboratory (Pamies et al., 2017). *Created in BioRender.*

Microfluidic devices control the cellular microenvironment with high precision, allowing manipulation of small fluid volumes and continuous perfusion of culture media, beneficial for long-term cell cultivation. Microfluidic systems can also protect cells from shear stress by employing low flow rates or specific channel designs (Tehranirokh et al., 2013). Additionally, microfluidic cell culture can be used for co-culture of different cell types, offering a common medium that promotes the growth of various cells. The use of Polydimethylsiloxane (PDMS) in microfluidic chips is common due to its biocompatibility and transparency, although there are efforts to explore other materials like polystyrene or polylactic acid. Microfluidic systems in biomanufacturing show promise in miniaturizing and improving existing processes by precisely controlling microenvironmental conditions, feeding, changing culture media, and characterizing end products. The use of microfluidic devices is expected to continue growing, with the global bio-based materials market projected to reach about USD 87 Billion by 2026.[1]

Bioreactors provide a three-dimensional environment representative of *in vivo* conditions, managing mechanical forces and fluid shear stress to mimic the physical conditions of the developing human brain. Microfluidic systems allow for the precise control of the culture environment by regulating factors like fluid flow, which can influence nutrient delivery and waste removal. These systems are essential for forming larger, more complex organoids that can provide more physiologically relevant data (Qian et al., 2018). Bioreactors provide a high surface-to-volume ratio suitable for large-scale high-throughput production and can be integrated with microfluidic devices to create dynamic cell culture systems (Huang et al., 2022). They can control environmental conditions such as gas composition and temperature, which are key parameters for cell production and end-product quality. Microfluidic bioreactors can also generate different sizes of droplets for encapsulation and particle mixing, useful for bioprinting and biomaterial production.

Furthermore, co-culture systems represent a significant advancement in modeling brain environments. Incorporating various cell types such as microglia and endothelial cells into neuronal cultures creates a more comprehensive model of the brain's cellular ecosystem. Microglia participate in immune responses and phagocytosis, while endothelial cells

[1] https://starfishmedical.com/blog/integrating-microfluidics-into-biomanufacturing/

contribute to the formation of the blood-brain barrier, a critical component of brain function and integrity. Co-culturing these cells with neurons and glia mimics their interactions and functionalities, providing deeper insights into cellular responses to inflammation, injury, and disease within a brain-like context (Pham et al., 2018; Xu et al., 2021).

These methodologies collectively enhance our ability to generate robust, reproducible, and physiologically relevant human neuronal cultures, instrumental for advancing our understanding of neurobiology, investigating disease mechanisms, and developing therapeutic strategies. The incorporation of microfluidic devices and bioreactors into cell culture methodologies has significantly contributed to the reproducibility and scalability of these cultures. These advanced technologies enable precise control over the culture conditions, facilitating the creation of more physiologically relevant cell culture models for a wide range of applications (Stacey, 2012).

3. Environmental impacts on neurodevelopment

3.1 Overview of environmental factors known to influence neurodevelopment

The neurodevelopmental process is highly susceptible to various environmental factors that can significantly alter its course. These factors include a broad range of chemical and non-chemical elements, from industrial pollutants and household chemicals to dietary deficiencies and enriched environments. Among chemical exposures, heavy metals such as lead and mercury are notoriously detrimental to neurodevelopment, disrupting neuronal and cognitive functions even at low levels of exposure (Grandjean and Landrigan, 2014). Similarly, exposure to pesticides and polychlorinated biphenyls (PCBs) has been linked to various developmental deficits and disorders, reflecting their pervasive impact on pediatric health (Eskenazi et al., 2007).

Dietary factors also play a critical role in shaping neurodevelopment. Nutrients such as essential fatty acids are vital for brain development, influencing synaptic function and plasticity (Innis, 2007; Flanagan et al., 2020). Likewise, iodine deficiency is the leading preventable cause of intellectual disabilities in children worldwide, underscoring the importance of adequate nutrition for cognitive development (Zimmermann, 2009). These factors, combined with vitamins such as B_{12} and folate, are fundamental for proper brain development and function, and deficiencies can have long-lasting effects on cognitive abilities and neurological health (Bhatnagar and Taneja, 2001).

3.2 Mechanisms of environmental impact at the cellular and molecular levels

The impact of environmental factors on neurodevelopment is mediated through various cellular and molecular mechanisms that alter the fundamental processes of brain development. Lead, for example, is known to interfere with the homeostasis of calcium, and antogonise NMDA receptor, leading to impaired neurotransmitter release and synaptic plasticity, essential for learning and memory (Bellinger, 2004). This metal also induces oxidative stress and apoptosis, contributing to neurodegenerative processes (Lidsky and Schneider, 2003).

Organic pollutants such as PCBs exert their effects through the disruption of thyroid hormone signaling, which is crucial for brain development. PCBs mimic or inhibit thyroid hormone actions, leading to alterations in gene expression patterns that are critical for neuronal differentiation and maturation (Zoeller, 2007). These disruptions can lead to structural abnormalities in the brain and long-term functional deficits, impacting cognitive and motor abilities (Schantz et al., 2003).

Additionally, air pollutants including particulate matter and noxious gases can induce neuroinflammation and oxidative stress, which are implicated in various neurodegenerative diseases and developmental disorders. These pollutants have been shown to alter brain development through the induction of inflammatory cytokines, disruption of blood-brain barrier integrity, and direct toxicity to neural cells, thereby affecting cognitive functions (Block and Calderón-Garcidueñas, 2009).

3.3 Case studies demonstrating significant environmental effects on neural cultures linked to public health concerns

Case study 1: Lead exposure and cognitive decline

Research has shown that even low levels of lead exposure during early development can lead to significant cognitive deficits. A study utilizing neuronal cultures demonstrated that lead alters calcium signaling and impairs neuronal connectivity, providing a cellular basis for the observed declines in IQ and executive function in exposed children (Lidsky and Schneider, 2003). Boyle et al. (2021) estimated that early childhood lead exposure in the US from 1999 to 2010 resulted in a loss of about $46.2 billion annually.

Case study 2: Pesticides and neurodevelopmental disorders

Exposure to common pesticides like organophosphates has been linked to neurodevelopmental disorders such as ADHD and autism. *In vitro* studies using human neuronal cultures have shown that these chemicals disrupt synaptic development and neural network formation by inhibiting acetylcholinesterase, leading to excessive accumulation of acetylcholine in synaptic junctions (Eskenazi et al., 2007). Bellanger et al. (2015) estimated that organophosphate exposure had a 70–100% probability of 13 million lost IQ points and 59,300 intellectual disability cases, costing €146 billion annually.

Case study 3: Air pollution and neuroinflammation

Studies have highlighted the role of ultrafine particles from air pollution in promoting neuroinflammation and oxidative stress, critical in the pathogenesis of neurodegenerative diseases. Using microglial and neuronal co-cultures, researchers found that exposure to diesel exhaust particles promotes microglial activation and the release of pro-inflammatory cytokines, which in turn affects neuronal integrity (Block and Calderón-Garcidueñas, 2009). Gawryluk et al. (2023) showed that brief diesel exhaust exposure acutely impairs functional brain connectivity in humans. Webb et al. (2018) suggest significant economic implications for IQ loss due to environmental pollutants like diesel exhaust.

3.4 Historical development of the DNT strategy and community by CAAT and ECVAM

Over the past two decades, significant advancements in developmental neurotoxicity (DNT) testing have been achieved through the collaborative efforts initiated by the Center for Alternatives to Animal Testing (CAAT) and the European Centre for the Validation of Alternative Methods (ECVAM) (Smirnova et al., 2024). These efforts, supported by numerous institutions, have driven the development and implementation of innovative, animal-free methods that enhance the predictability and ethical standards of neurotoxicity assessments. Technological advancements in cell culture technologies and computational biology have further contributed to this groundbreaking movement.

Notably, the use of human-derived cells in DNT testing became more prevalent, offering more relevant physiological insights compared to traditional animal models. This change has been largely motivated by the need for greater accuracy and more ethically appropriate experimental methods. One of the most significant impacts has been the integration of these new DNT methods into regulatory frameworks, with the OECD (2023) providing initial guidance in 2023. The establishment of a global DNT community has been a pivotal outcome, fostering an environment of knowledge exchange and collaboration through workshops, conferences, and training sessions. These platforms have not only spread awareness of new methodologies but also sparked innovation through cross-disciplinary partnerships. The DNT community continues to expand, incorporating more diverse scientific perspectives and technological approaches.

The ongoing development of more sophisticated models, such as organ-on-a-chip and multi-omics approaches, promises to further refine our understanding of neurotoxic effects. The push for global regulatory harmonization remains a key agenda, crucial for the universal acceptance of alternative DNT methods. The historical development of the DNT strategy steered by CAAT and ECVAM represents a fundamental shift in toxicological testing, maturing from a nascent endeavor to reduce animal testing into a robust, innovative field that not only enhances scientific accuracy but also upholds ethical standards (Smirnova et al., 2024).

4. Modeling learning and intelligence

Microphysiological Systems (MPS) is an umbrella term for organoids, organ-on-chip systems, and similar constructs (Marx et al., 2016, 2020). These systems employ bioengineering to emulate aspects of organ architecture and function. The ultimate functionality of the brain is cognition and intelligence. Under the term Organoid Intelligence (OI), efforts to implement basic forms of learning and memory *in vitro* are summarized (Fig. 3) (Smirnova et al., 2023).

4.1 Theoretical frameworks for assessing learning and intelligence *in vitro*

The assessment of learning and intelligence *in vitro* requires the adaptation of theoretical frameworks traditionally applied *in vivo*. These frameworks must account for the cellular and molecular foundations of cognitive functions, such as synaptic plasticity, neural connectivity, and network dynamics, which are indicative of learning and memory processes (Fig. 3).

1. *Synaptic Plasticity Models*: Synaptic plasticity is fundamental to learning and memory. *In vitro* systems utilize models of synaptic strength modulation, such as long-term potentiation (LTP) and long-term depression (LTD), to study these processes at the synaptic level (Bliss and Collingridge, 1993; Malenka and Bear, 2004).

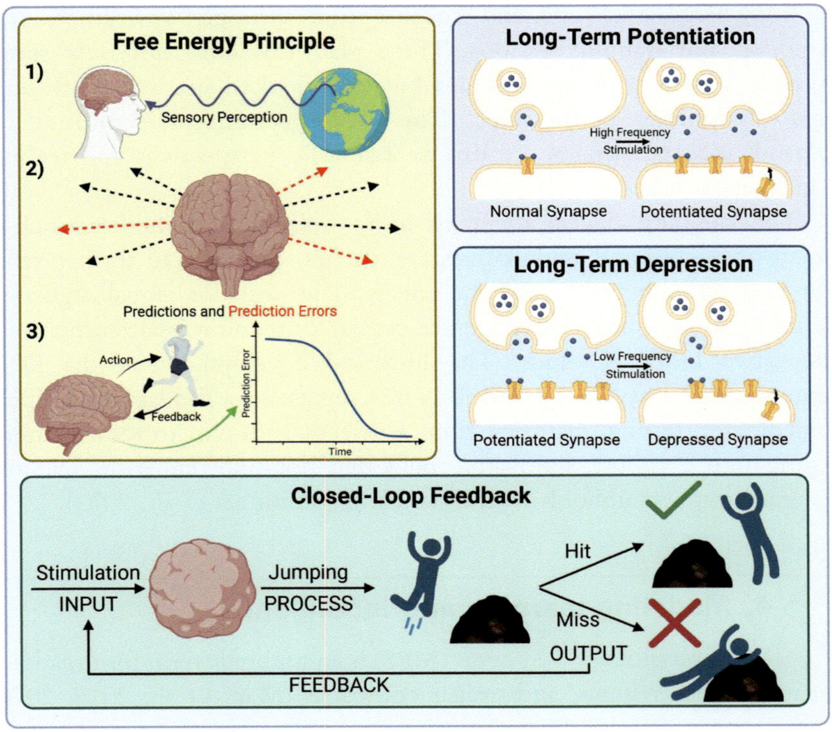

Fig. 3 Depicting the interrelated concepts of the free energy principle, closed-loop feedback learning models, long-term potentiation (LTP), and long-term depression (LTD) in neural processes. *Created in BioRender.*

2. *Neural Network Oscillations*: Cognitive functions are also associated with specific patterns of neural oscillations, such as theta and gamma waves, which can be measured in neuronal cultures. These oscillations reflect the coordinated activity that underpins complex cognitive tasks (Buzsáki and Draguhn, 2004; Colgin, 2016).
3. *Information Processing Models*: Theoretical models that simulate how information is processed within neural networks are used to extrapolate how organoids and cultured neurons might handle learning and intelligence-related tasks (Kriegeskorte, 2015; Yamins and DiCarlo, 2016).
4. *Computational Neuroscience*: Integrating computational models that simulate neural development and functionality may assist in predicting how neural networks might develop *in vitro,* enabling a deeper understanding of both the theoretical and the experimental elements of learning and intelligence (Deco et al., 2008; Eliasmith et al., 2012).

4.2 Techniques for measuring cognitive functions in organoid and neural culture models

Advancements in bioengineering and neurotechnology have enabled the development of several techniques to measure cognitive functions *in vitro*, each offering unique insights into the neural underpinnings of learning and intelligence (Fig. 4).

1. Electrophysiology: Techniques such as multi-electrode arrays (MEAs) allow for the recording of electrical activity from neurons and neural networks in real time, providing data on neural connectivity and activity patterns (Obien et al., 2015; Spira and Hai, 2013).
2. Calcium Imaging: Used to monitor the intracellular calcium levels that fluctuate with neuron firing, thus allowing researchers to visualize and quantify neural activity across different regions of brain organoids (Ahrens et al., 2013; Yang and Yuste, 2017).
3. Optogenetics: This allows controlling neuron activity with light and can be used to study the causal relationships between neural circuit activity and behavior outcomes, albeit in a simplified *in vitro* environment (Deisseroth, 2011; Boyden et al., 2005).
4. Virtual Gaming Environments: Kagan et al. (2022) first used the computer game Pong to demonstrate that neuronal cultures can be

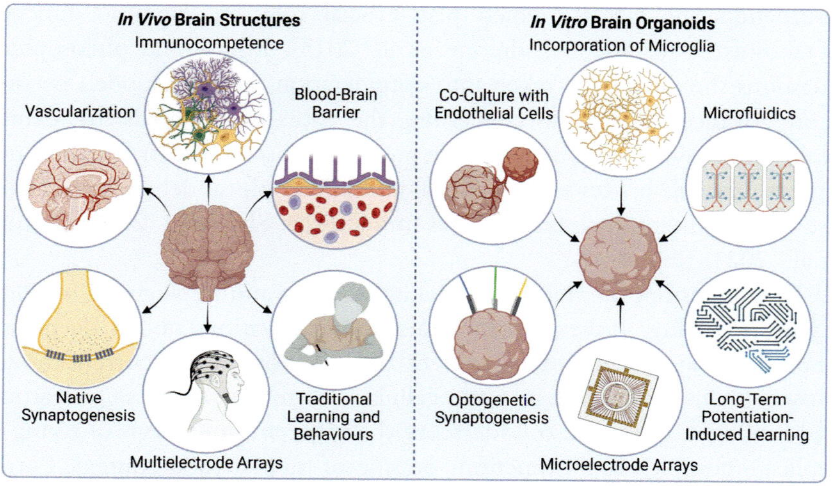

Fig. 4 Comparison of *in vivo* brain structures and *in vitro* brain organoids. Emphasizing technological advancements and ongoing efforts to mimic complex brain functions and structures *in vitro. Created in BioRender.*

trained (Kagan et al., 2022, Smirnova and Hartung, 2022). While these pure neuronal cultures were only able to improve during a given training session, the concept is now pursued within the Organoid Intelligence (OI) community aiming to demonstrate long-term memory (Fig. 3).

4.3 Challenges and limitations in modeling complex brain functions

Despite significant progress, modeling learning and intelligence *in vitro* faces several challenges and limitations. The human brain's complexity, with its billions of interconnected neurons and synapses, is a challenge to replicate *in vitro*. Organoids and neuronal cultures can only mimic a fraction of this complexity, leading to oversimplified models of cognitive processes that do not yet fully encapsulate the dynamic nature of neural networks (Südhof, 2017; Bassett and Gazzaniga, 2011). Furthermore, the stability, longevity, and reproducibility of human brain organoids remain a challenging obstacle to overcome (Paşca, 2018; Qian et al., 2019; Quadrato et al., 2017).

An additional concern is the absence of a systemic context. *In vitro* models lack the systemic physiological interactions present in an intact organism, such as those provided by the vascular system, immune system, and various neuroendocrine factors, which play crucial roles in cognitive function (Trakoshis et al., 2020; Andreone et al., 2015). Ethical and philosophical questions should also be taken into consideration, as these models become more advanced, the ethics concerning the status of organoids that exhibit brain-like activity begin to arise. The potential for sentience or consciousness in these models invites serious ethical and philosophical debates about their use and treatment (Lavazza and Massimini, 2018; Shepherd, 2018; Hartung et al., 2024) (Fig. 4).

In conclusion, the field of modeling learning and intelligence *in vitro* is rapidly evolving, driven by technological innovations and a deepening understanding of neural development and function. While current models provide valuable insights into the cellular and molecular bases of learning and intelligence, they also underscore the inherent limitations of trying to replicate complex brain functions outside of their natural context. Future advancements will likely focus on increasing the complexity and physiological relevance of these models, alongside ongoing discussions about their ethical implications.

5. Organoid intelligence: a new frontier in biocomputing and cognition research

The concept of Organoid Intelligence (OI) (Smirnova et al., 2023) represents a groundbreaking approach to understanding and emulating human cognition by harnessing the computational capabilities of brain organoids (Magliaro and Ahluwalia, 2023; Miller, 2023). A community-forming workshop (Morales Pantoja et al., 2023) led to the Baltimore Declaration toward the Exploration of OI (Hartung et al., 2023). By integrating advancements in stem cell biology, bioengineering, and artificial intelligence, OI aims to create hybrid biocomputing systems that exhibit learning, memory, and problem-solving abilities. Early proof-of-concept studies have demonstrated the potential of brain organoids to function as living neural networks capable of processing information and responding to stimuli. For example, researchers have used brain organoids to control robotic devices and perform simple computational tasks, showcasing their capacity for learning and adaptation (Kagan et al., 2022; Shen, 2018).

Emerging approaches in OI focus on developing brain-inspired computing architectures, such as neuromorphic engineering and spiking neural networks, which aim to mimic the efficiency and flexibility of biological neural systems. Hybrid biocomputing systems that integrate brain organoids with electronic components are also being explored as a way to leverage the strengths of both biological and artificial intelligence (Smirnova et al., 2023).

OI technologies hold immense potential for advancing our understanding of complex cognitive processes, such as perception, decision-making, memory formation, and emotion. By modeling these processes in brain organoids, researchers can gain insights into the neural basis of human cognition and develop novel approaches for studying and treating neurological and psychiatric disorders (Quirion, 2023; Smirnova et al., 2023).

However, the development of OI also raises important ethical and societal questions, particularly regarding the moral status and rights of brain organoids. As these systems become more complex and potentially capable of exhibiting consciousness and sentience, there is a need for robust ethical guidelines and oversight to ensure their responsible development and use (Kinderlerer, 2023; Boyd and Lipshitz, 2024; Hartung et al., 2024). Public engagement and dialogue are also critical for shaping the future direction of OI research and addressing concerns about the risks and benefits of these technologies. Efforts to promote transparency, incorporate diverse perspectives, and foster interdisciplinary collaboration will be essential for realizing the full potential of OI while navigating its ethical and societal implications (Smirnova et al., 2023).

Looking ahead, key challenges for advancing OI include scaling up these systems to increase their complexity and functionality, enabling high-throughput experimentation, and developing real-world applications in areas such as personalized medicine and drug discovery (Smirnova et al., 2023). Especially, the study of neurodevelopmental disorders including developmental neurotoxicity lends itself to deploying a cognitive functional endpoint. Addressing technical hurdles related to reproducibility and standardization, as well as ensuring the responsible development and use of OI technologies, will be critical for their successful translation into practice (Smirnova et al., 2023).

In conclusion, Organoid Intelligence represents a new frontier in biocomputing and cognition research, offering unprecedented opportunities for understanding and emulating human brain function. While still in its early stages, OI has already demonstrated significant progress and holds immense potential for transforming multiple fields, from neuroscience and artificial intelligence to medicine and robotics. As this exciting area continues to evolve, it will be crucial to engage in proactive and inclusive dialogue to ensure that OI technologies are developed and applied responsibly and beneficially for science, technology, and society as a whole.

6. Use of human neuronal cultures in environmental research

Human neuronal cultures, particularly self-organizing brain organoids, provide innovative platforms for simulating environmental exposures and investigating their effects on neurodevelopment (Lee and Sun, 2022; Smirnova and Hartung, 2024). These cultures replicate key aspects of human brain development, offering a relevant and dynamic model to study the impact of environmental toxicants, pollutants, and other factors on neural cells, with first examples for pesticides (Pamies et al., 2018; Modafferi et al., 2021).

Detailed exposure protocols involve dosing regimens that mimic real-life exposure scenarios to toxicants such as heavy metals, pesticides, or endocrine disruptors. These protocols consider critical developmental windows and dose-response relationships to ensure biologically relevant exposures. In addition to chemical toxicants, these cultures can be used to study the effects of non-chemical stressors, including hypoxia, nutritional deficiencies, and physical disturbances, providing a comprehensive view of environmental impacts on neurodevelopment (Kim and Chang, 2023). Furthermore, the incorporation of real-world complexities, such as mixtures of multiple pollutants and their cumulative effects, is crucial for enhancing the ecological validity of these studies.

Following exposure, a variety of techniques, such as live-cell imaging, electrophysiological recording, and molecular assays, are employed to assess changes in neuronal viability, functionality, and morphology. These assessments provide insights into the mechanisms underlying environmental toxicity and its consequences for neural development and function (Zheng et al., 2021; Mulders et al., 2023). Advanced imaging techniques, including single-cell RNA sequencing and spatial transcriptomics, offer detailed cellular and subcellular analyses, enabling a greater understanding of the heterogeneous responses within neuronal cultures.

The integration of omics technologies, such as genomics, proteomics, and metabolomics, plays a crucial role in elucidating the biochemical and genetic impacts of environmental factors on neuronal cultures. Sequencing technologies help identify genetic and transcriptional changes in neurons exposed to environmental toxins, revealing genes that may confer susceptibility or resistance to toxic effects (Murtaza et al., 2020; Cappuccio et al., 2023). Protein expression profiling can uncover signaling pathways disrupted by environmental factors, aiding in the identification of potential biomarkers for neurotoxicity (Lee and Sun, 2022). Furthermore, analysis of metabolic changes in response to environmental exposures can provide clues about the affected biochemical pathways (Van Vliet et al., 2008), offering insights into the mechanisms of toxicity and potential metabolic interventions (Sillé and Hartung, 2014). Moreover, integrating multiomics data with computational models can further elucidate complex interactions and allow the prediction of long-term outcomes as pursued for example in VHP4Safety.[2]

Human neural cultures promise to advance our understanding of the environmental causes of neurodevelopmental disorders. Studies using 3D neural cultures (Pamies et al., 2018; Modafferi et al., 2021) have linked

[2] https://www.sciencrew.com/c/6586?title=VHP4Safety

pesticide exposure during early development to disruptions in neural connectivity and synaptic function, which are characteristic of autism spectrum disorders (ASD). Research involving organoids has shown that exposure to toxicants during critical developmental periods can result in the abnormal formation of neural circuits commonly associated with psychiatric disorders (Dixon and Muotri, 2023). Brain organoids from patients with attention deficit hyperactivity disorder (ADHD) show disease-specific changes (Zhang et al., 2023), but susceptibility to environmental chemicals has not been reported. Investigating epigenetic modifications and their heritability across generations may provide further insight into the long-term neuroactive effects of environmental exposures.

In conclusion, the use of human neural cultures in environmental research represents a significant advancement in our ability to understand and mitigate the impacts of environmental factors on neurodevelopment. By incorporating sophisticated methodologies and omics technologies, researchers can simulate exposures and study their effects in a controlled, yet biologically relevant, setting. These efforts not only enhance our understanding of the etiology of neurodevelopmental disorders but also support the development of strategies to prevent or counteract these adverse effects. This underscores the importance of using advanced *in vitro* human models to deepen our understanding of environmental influences on brain development, providing a foundation for future research and policymaking to safeguard public health. In addition, collaborative efforts between academia, industry, and regulatory agencies are essential to translating these findings into effective public health interventions and policies.

7. Ethical, legal, and social implications

The use of human-derived neuronal cultures, especially organoids that may mimic aspects of neural activity found in sentient beings, raises profound ethical concerns (Hartung et al., 2024). These concerns revolve around issues of consent, the moral status of cerebral organoids, and the potential for consciousness. Obtaining clear consent from donors is crucial, particularly given the sensitive nature of brain-like tissue research. As organoids become increasingly complex, questions arise about their status as potential sentient entities and the ethical implications of such possibilities, including whether they can experience pain or distress. Guidelines for handling and disposing of human-derived neuronal cultures must balance scientific utility with respect for the human origin of these cells (Hyun et al., 2020). Furthermore, transparency when gaining informed consent about the potential uses of donated tissues in complex brain organoid modeling is necessary to uphold ethical standards.

The development and use of human brain models are subject to an evolving regulatory landscape that seeks to ensure research progresses within ethical and legal bounds. International organizations, such as the International Society for Stem Cell Research (ISSCR), have developed guidelines addressing ethical concerns associated with human stem cell-derived models and recommending oversight mechanisms to evaluate the ethical implications of new technologies (ISSCR, 2021). Different countries have varying regulations governing stem cell research, which can impact the speed and direction of scientific advances. As scientific understanding and societal attitudes evolve, regulatory frameworks are periodically updated to reflect new ethical considerations and scientific capabilities, ensuring that research remains responsible and socially acceptable (Hyun et al., 2020).

Public perception of human-derived neural cultures influences societal acceptance and the funding landscape of research. Misunderstandings about the capabilities and purposes of these models can lead to either unwarranted fears or exaggerated expectations. Effective communication strategies are essential to educating the public about the realities and benefits of neuronal culture research, with transparent sharing of research purposes, potential, and limitations helping to alleviate concerns and foster support. Highlighting the potential societal benefits, such as understanding neurodegenerative diseases or developmental disorders, can help align public perceptions with the goals of scientific research. Engaging the public in ethical discussions about brain model research can ensure broader societal values are reflected in how science progresses in this field. Including diverse public voices and stakeholders in these discussions can enhance the legitimacy and acceptance of research practices and outcomes.

In conclusion, the ethical, legal, and social implications of using human-derived neural cultures are as complex as the scientific questions such research aims to answer. Addressing these implications requires a multifaceted approach involving ethical scrutiny, regulatory vigilance, and ongoing dialogue with the public. As this field evolves, so too must our approaches to these challenges, ensuring that research benefits society while respecting fundamental ethical principles. Continuous ethical education for researchers and fostering interdisciplinary collaborations between ethicists, scientists, and policymakers would further support responsible innovation and application in this rapidly advancing field.

8. Future directions and innovations

Advances in biotechnology and neuroscience are set to revolutionize the capabilities of self-organizing neural cultures, promising deeper insights into brain development, disease mechanisms, and the effects of environmental factors on neurodevelopment. Current research projects are pushing the boundaries of what these models can achieve. One major area of innovation is the integration of brain organoids with organ-on-a-chip technology, allowing for the simulation of blood flow and the blood-brain barrier. This integration enhances the physiological relevance of *in vitro* models by closely mimicking *in vivo* conditions. Microfluidics can precisely control the local environment, facilitating the study of neuronal responses to gradients of chemicals or shear stress (Li et al., 2024). This technology is essential for high-throughput screening applications, accelerating the testing of numerous environmental conditions or therapeutic agents, and making it feasible to explore vast combinatorial spaces that were previously impractical (Fig. 4) (Roth et al., 2021).

New imaging technologies, such as light-sheet fluorescence microscopy combined with sophisticated computational models and machine learning, are expanding our ability to visualize and analyze complex neural networks in real-time and three dimensions (Stelzer et al., 2021; Daetwyler and Fiolka, 2023). These advancements provide deeper insights into brain structure and function, enabling more detailed and accurate studies. Furthermore, the concept of Organoid Intelligence (OI) is at the forefront of biocomputing and cognition research. By harnessing the computational capabilities of brain organoids, we are developing hybrid systems that exhibit learning, memory, and problem-solving abilities. These systems hold potential for applications in personalized medicine, drug discovery, and understanding complex cognitive processes. The integration of modeling approaches into microphysiological systems (Smirnova et al., 2018) has tremendous potential to interpret, optimize, and ultimately model their behavior as virtual experiments.

The refinement of self-organizing cultures has significant potential to influence public health policies and environmental regulations. Enhanced *in vitro* models promise more accurate, human-relevant data regarding the neurotoxicity of new chemicals and pollutants before they are widely used, supporting preventive strategies in public health. Organoids derived from individual patients can be used to tailor medical treatments, particularly for neurodegenerative diseases, aligning with the growing trend toward personalized medicine in policymaking (Koo et al., 2019). As these models become more validated and accepted, they can serve as new standards in regulatory science, potentially reducing the reliance on animal testing and speeding up the adoption of safer environmental and public health practices (Sun et al., 2021).

The future of neuro-environmental research using advanced *in vitro* models is envisioned to be transformative, with several key areas of potential impact. Advanced models could simulate not only genetic factors but also environmental influences, providing a holistic view of complex diseases like autism, schizophrenia, and Alzheimer's disease (Smirnova et al., 2014, 2024; Xu and Wen, 2021). Expanding the application of neuronal cultures to ecotoxicology could help assess the impact of environmental stressors on biodiversity, particularly the neurodevelopmental aspects of wildlife exposed to pollutants. Continued development of *in vitro* models aligns with ethical research practices by reducing animal use, driving the scientific community toward more humane and sustainable research methodologies (Smirnova and Hartung, 2024).

The need for validation of self-organizing human neuronal cultures is paramount to ensuring their reliability and reproducibility in scientific research and practical applications. This is critical for their possible regulatory use (Leist et al., 2012). Validated models are essential for regulatory acceptance, as they must demonstrate consistent and predictive results across different laboratories and studies. This consistency is crucial for these models to be used in public health policymaking and the development of new therapeutic interventions. In the absence of gold-standard animal models, this must be mainly based on mechanistic validation (Hartung et al., 2013). Such mechanistic validation involves rigorous testing to confirm that these models accurately replicate human brain development and respond to environmental factors in ways that are consistent with *in vivo* conditions. This process includes comparing the responses of these cultures to known neurodevelopmental pathways and outcomes, ensuring that observed effects are not artifacts of the *in vitro* environment but truly reflect biological processes. As emerging technologies like organ-on-a-chip and advanced imaging techniques are integrated into these models, continuous validation ensures that these innovations enhance rather than compromise the model's accuracy and utility. Robust validation not only boosts confidence in these models within the scientific community but also facilitates their adoption in regulatory frameworks, ultimately contributing to safer and more effective health interventions and environmental regulations.

In conclusion, the future directions and innovations in the use of self-organizing human neural cultures are poised to significantly advance our understanding of the human brain, improve our ability to prevent and treat neurological disease, and enhance the sustainability and ethical standards of neurological research. The continuous validation of these developing approaches is critical for their regulatory use as well as business decisions in drug development. As these technologies mature, their integration into research and policy frameworks will likely reshape our approach to neuroenvironmental and public health challenges.

9. Conclusions

This chapter has explored the profound capabilities and potential applications of self-organizing human neural cultures in environmental research, highlighting their ability to assess and understand the effects of environmental factors on neurodevelopment. Neural organoids offer a dynamic, three-dimensional matrix where neurons and glia cells can interact and organize in ways that closely mimic *in vivo* conditions, providing a more accurate representation of brain development compared to traditional 2D cultures (Smirnova and Hartung, 2024). Environmental factors such as toxicants, pollutants, drugs, and dietary elements significantly influence neurodevelopment by disrupting cellular and molecular pathways, leading to lasting effects on learning and intelligence. The use of organoid and other 3D culture models in studying cognitive functions illustrates the complexity of brain functions and highlights both the capabilities and limitations of current methodologies.

The deployment of human-derived neural models raises important ethical and legal considerations, including the possible regulation of such research and public perception of these technologies. Emerging technologies such as single-cell RNA sequencing, CRISPR-based gene editing, and microfluidic techniques promise to further enhance the fidelity and utility of these models, potentially influencing public health policies and environmental regulations significantly. The integration of self-organizing human neuronal cultures into environmental health research offers transformative potential to both understand and mitigate the impacts of environmental factors on human health. These models serve as a bridge between molecular mechanisms and real-world exposures, providing insights into how specific environmental agents can lead to neurodevelopmental disorders. By offering a platform to study the onset and progression of neurological conditions at an individual level—almost as a "living biopsy" of the donor—these models aid in devising strategies that might prevent or mitigate these conditions in vulnerable populations understanding inter-individual differences.

The detailed insights gained from these studies can inform policymakers, helping craft more effective environmental regulations that protect public health without stifling technological and industrial innovation. Ultimately, the use of advanced human neuronal models in research aligns with broader public health goals, such as reducing the prevalence of neurodevelopmental disorders and improving the quality of life. The work being done with self-organizing human neuronal cultures stands at the frontier of environmental health research, embodying the synthesis of innovative scientific methodologies and real-world applications. As this field continues to evolve, it holds the promise of significant advancements in our understanding of the intricate interplay between environmental factors and human health, potentially leading to more resilient public health frameworks and healthier future generations.

Acknowledgments

The authors acknowledge funding from the US Environmental Protection Agency (EPA) EPA-STAR grant (RD83950501) awarded by the U.S EPA to LS. It had not been formally reviewed by EPA. The views epxressed in the publication are solely those of LS and co-authors and do not necessarily refect theose of the Agency. EPA does not endorse any products or commercial services mentioned in this publication. This work was funded by the Combining advances in Genomics and Environmental science to accelerate Actionable Research in ASD (GEARS) Network (R01ES034554); Ladd-Acosta, Volk. The work was also supported by a Johns Hopkins Discovery award and two Johns Hopkins SURPASS awards.

Author Contributions

TH: Conceptualization, Writing—Original Draft Preparation, All: Writing—Review & Editing.

Conflict of Interest

T.H. is named inventor on a patent by Johns Hopkins University on the production of mini-brains (also called BrainSpheres), which is licensed to Axo-Sim, New Orleans, LA, USA. He is a shareholder of and he and L.S. are consultants for AxoSim, New Orleans; T.H. is also a consultant American Type Culture Collection (ATCC), InSphero, Zurich, Switzerland, Crown Biosciences, SanDiego, CA, and was until recently consultant for AstraZeneca on advanced cell culture methods.

References

Adlakha, Y.K., 2023. Human 3D brain organoids: steering the demolecularization of brain and neurological diseases. Cell Death Discov. 9, 221. https://doi.org/10.1038/s41420-023-01523-w.

Ahrens, M.B., Orger, M.B., Robson, D.N., Li, J.M., Keller, P.J., 2013. Whole-brain functional imaging at cellular resolution using light-sheet microscopy. Nat. Methods 10 (5), 413–420.

Alépée, N., Bahinski, T., Daneshian, M., De Wever, B., Fritsche, E., Goldberg, A., et al., 2014. State-of-the-art of 3D cultures (organs-on-a-chip) in safety testing and pathophysiology—a t4 report. ALTEX 31, 441–477. https://doi.org/10.14573/altex.1406111.

Andreone, B.J., Lacoste, B., Gu, C., 2015. Neuronal and vascular interactions. Annu. Rev. Neurosci. 38, 25–46.

Bassett, D.S., Gazzaniga, M.S., 2011. Understanding complexity in the human brain. Trends Cognit. Sci. 15 (5), 200–209.

Bellanger, M., Demeneix, B.A., Grandjean, P., et al., 2015. Neurobehavioral deficits, diseases, and associated costs of exposure to endocrine-disrupting chemicals in the European Union. J. Clin. Endocrinol. Metab. 100, 1256–1266. https://doi.org/10.1210/jc.2014-4323.

Bellinger, D.C., 2004. Lead. Pediatrics 113 (4 Suppl.), 1016–1022.

Bhatnagar, S., Taneja, S., 2001. Zinc and cognitive development. Br. J. Nutr. 85 (Suppl. 2), S139–S145. https://doi.org/10.1079/bjn2000306.

Bliss, T.V., Collingridge, G.L., 1993. A synaptic model of memory: long-term potentiation in the hippocampus. Nature 361 (6407), 31–39.

Block, M.L., Calderón-Garcidueñas, L., 2009. Air pollution: mechanisms of neuroinflammation and CNS disease. Trends Neurosci. 32 (9), 506–516. https://doi.org/10.1016/j.tins.2009.05.009.

Boyd, J.L., Lipshitz, N., 2024. Dimensions of consciousness and the moral status of brain organoids. Neuroethics 17, 5. https://doi.org/10.1007/s12152-023-09538-x.

Boyden, E.S., Zhang, F., Bamberg, E., Nagel, G., Deisseroth, K., 2005. Millisecond-timescale, genetically targeted optical control of neural activity. Nat. Neurosci. 8 (9), 1263–1268.

Boyle, J., Yeter, D., Aschner, M., et al., 2021. Estimated IQ points and lifetime earnings lost to early childhood blood lead levels in the United States. Sci. Total Environ. 778, 146307. https://doi.org/10.1016/j.scitotenv.2021.146307.

Buzsáki, G., Draguhn, A., 2004. Neuronal oscillations in cortical networks. Science 304 (5679), 1926–1929.

Cappuccio, G., Khalil, S.M., Osenberg, S., Li, F., Maletic-Savatic, M., 2023. Mass spectrometry imaging as an emerging tool for studying metabolism in human brain organoids. Front. Mol. Biosci. 10, 1181965. https://doi.org/10.3389/fmolb.2023.1181965.

Colgin, L.L., 2016. Rhythms of the hippocampal network. Nat. Rev. Neurosci. 17 (4), 239–249.

Daetwyler, S., Fiolka, R.P., 2023. Light-sheets and smart microscopy, an exciting future is dawning. Commun. Biol. 6 (1), 502. https://doi.org/10.1038/s42003-023-04857-4.

Deco, G., Jirsa, V.K., Robinson, P.A., Breakspear, M., Friston, K., 2008. The dynamic brain: from spiking neurons to neural masses and cortical fields. PLOS Comput. Biol. 4 (8), e1000092. https://doi.org/10.1371/journal.pcbi.1000092.

Deisseroth, K., 2011. Optogenetics. Nat. Methods 8 (1), 26–29.

Di Lullo, E., Kriegstein, A.R., 2017. The use of brain organoids to investigate neural development and disease. Nat. Rev. Neurosci. 18 (10), 573–584. https://doi.org/10.1038/nrn.2017.107.

Dixon, T.A., Muotri, A.R., 2023. Advancing preclinical models of psychiatric disorders with human brain organoid cultures. Mol. Psychiatry 28, 83–95. https://doi.org/10.1038/s41380-022-01708-2.

Eliasmith, C., Stewart, T.C., Choo, X., Bekolay, T., DeWolf, T., Tang, Y., et al., 2012. A large-scale model of the functioning brain. Science 338, 1202–1205. https://doi.org/10.1126/science.1225266.

Eskenazi, B., Marks, A.R., Bradman, A., Harley, K., Barr, D.B., Johnson, C., et al., 2007. Organophosphate pesticide exposure and neurodevelopment in young Mexican-American children. Environ. Health Perspect. 115 (5), 792–798. https://doi.org/10.1289/ehp.9828.

Flanagan, E., Lamport, D., Brennan, L., Burnet, P., Calabrese, V., Cunnane, S.C., et al., 2020. Nutrition and the ageing brain: moving towards clinical applications. Ageing Res. Rev., 101079. https://doi.org/10.1016/j.arr.2020.101079.

Gawryluk, J.R., Palombo, D.J., Curran, J., et al., 2023. Brief diesel exhaust exposure acutely impairs functional brain connectivity in humans: a randomized controlled crossover study. Environ. Health 22, 7. https://doi.org/10.1186/s12940-023-00961-4.

Grandjean, P., Landrigan, P.J., 2014. Neurobehavioural effects of developmental toxicity. Lancet Neurol. 13 (3), 330–338.

Hartung, T., 2023. A call for a human exposome project. ALTEX 40, 4–33. https://doi.org/10.14573/altex.2301061.

Hartung, T., Morales Pantoja, I.E., Smirnova, L., 2024. Brain organoids and Organoid Intelligence (OI) from ethical, legal, and social points of view. Front. Artif. Intell. sec. Organoid Intelligence 6, 1307613. https://doi.org/10.3389/frai.2023.1307613.

Hartung, T., Smirnova, L., Morales Pantoja, I.E., Akwaboah, A., Alam El Din, D.-M., Berlinicke, C.A., et al., 2023. The Baltimore declaration toward the exploration of organoid intelligence. Front. Sci. 1, 1017235. https://doi.org/10.3389/fsci.2023.1017235.

Hartung, T., Stephens, M., Hoffmann, S., 2013. Mechanistic validation. ALTEX 30, 119–130. https://doi.org/10.14573/altex.2013.2.119.

Hartung, T., Smirnova, L., Smirnovaa L., 2024. Brain Organoids as a Translational Model of Human Brain Developmental Neurotoxicity. This book, chapter < #, in press.

Huang, X., Huang, Z., Gao, W., Gao, W., He, R., Li, Y., et al., 2022. Current advances in 3D dynamic cell culture systems. Gels 8 (12), 829. https://doi.org/10.3390/gels8120829.

Hyun, I., Scharf-Deering, J.C., Lunshof, J.E., 2020. Ethical issues related to brain organoid research. Brain Res. 1732, 146653.

Innis, S.M., 2007. Dietary (n-3) fatty acids and brain development. J. Nutr. 137 (4), 855–859. https://doi.org/10.1093/jn/137.4.855. PMID: 17374644.

ISSCR, 2021. Guidelines for stem cell research and clinical translation. Available at: https://www.isscr.org/policy/guidelines-for-stem-cell-research-and-clinical-translation.

Kagan, B.J., Kitchen, A.C., Tran, N.T., Habibollahi, F., Khajehnejad, M., Parker, B.J., et al., 2022. In vitro neurons learn and exhibit sentience when embodied in a simulated game-world. e8. Neuron 110 (23), 3952–3969. https://doi.org/10.1016/j.neuron.2022.09.001.

Kelava, I., Lancaster, M.A., 2016. Stem cell models of human brain development. Cell Stem Cell 18 (6), 736–748. https://doi.org/10.1016/j.stem.2016.05.022.

Kim, S.-h, Chang, M.-Y., 2023. Application of human brain organoids—opportunities and challenges in modeling human brain development and neurodevelopmental diseases. Int. J. Mol. Sci. 24 (15), 12528. https://doi.org/10.3390/ijms241512528.

Kinderlerer, J., 2023. Organoid intelligence: society must engage in the ethics. Available at: https://policylabs.frontiersin.org/content/policy-outlook-julian-kinderlerer-organoid-intelligence-society-must-engage-in-the-ethics (accessed December 2, 2023).

Koo, B., Choi, B., Park, H., Yoon, K.J., 2019. Past, present, and future of brain organoid technology. Mol. Cell 42 (9), 617–627. https://doi.org/10.14348/molcells.2019.0162.
Kriegeskorte, N., 2015. Deep neural networks: a new framework for modeling biological vision and brain information processing. Annu. Rev. Vis. Sci. 1, 417–446.
Lancaster, M.A., Knoblich, J.A., 2014. Generation of cerebral organoids from human pluripotent stem cells. Nat. Protoc. 9 (10), 2329–2340.
Lavazza, A., Massimini, M., 2018. Cerebral organoids: ethical issues and consciousness assessment. J. Med. Ethics 44 (9), 606–610.
Lee, J.H., Sun, W., 2022. Neural organoids, a versatile model for neuroscience. Mol Cells 45 (2), 53–64. https://doi.org/10.14348/molcells.2022.2019.
Leist, M., Hasiwa, M., Daneshian, M., Hartung, T., 2012. Validation and quality control of replacement alternatives—current status and future challenges. Toxicol. Res. 1, 8. https://doi.org/10.1039/C2TX20011B.
Li, M., Yuan, Y., Hou, Z., Hao, S., Jin, L., Wang, B., 2024. Human brain organoid: trends, evolution, and remaining challenges. Neural Regen. Res. 19 (11), 2387–2399. https://doi.org/10.4103/1673-5374.390972.
Lidsky, T.I., Schneider, J.S., 2003. Lead neurotoxicity in children: basic mechanisms and clinical correlates. Brain 126 (Pt 1), 5–19. https://doi.org/10.1093/brain/awg014.
Maenner, M.J., Warren, Z., Williams, A.R., et al., 2023. Prevalence and characteristics of autism spectrum disorder among children aged 8 years—autism and developmental disabilities monitoring network, 11 Sites, United States, 2020. MMWR Surveill. Summ. 72, 1–14. https://doi.org/10.15585/mmwr.ss7202a1.
Magliaro, C., Ahluwalia, A., 2023. To brain or not to brain organoids. Front. Sci. 1, 1148873. https://doi.org/10.3389/fsci.2023.1148873.
Malenka, R.C., Bear, M.F., 2004. LTP and LTD: an embarrassment of riches. Neuron 44 (1), 5–21.
Mansour, A.A., Gonçalves, J.T., Bloyd, C.W., Li, H., Fernandes, S., Quang, D., et al., 2018. An in vivo model of functional and vascularized human brain organoids. Nat. Biotechnol. 36 (5), 432–441.
Marx, U., Akabane, T., Andersson, T.B., Baker, E., Beilmann, M., Beken, S., et al., 2020. Biology-inspired microphysiological systems to advance medicines for patient benefit and animal welfare. ALTEX 37, 364–394. https://doi.org/10.14573/altex.2001241.
Marx, U., Andersson, T.B., Bahinski, A., Beilmann, M., Beken, S., Cassee, F.R., et al., 2016. Biology-inspired microphysiological system approaches to solve the prediction dilemma of substance testing using animals. ALTEX 33, 272–321. https://doi.org/10.14573/altex.1603161.
Miller, G.W., 2023. Organoid intelligence: smarter than the average cell culture. Front. Sci. 1, 1150594. https://doi.org/10.3389/fsci.2023.1150594.
Modafferi, S., Zhong, X., Kleensang, A., Murata, Y., Fagiani, F., Pamies, D., et al., 2021. Gene–environment interactions in developmental neurotoxicity: a case study of synergy between chlorpyrifos and CHD8 knockout in human BrainSpheres. Environ. Health Perspect. 129, 77001. https://doi.org/10.1289/EHP8580.
Morales Pantoja, I.E., Smirnova, L., Muotri, A.R., Wahlin, K.J., Kahn, J., Boyd, L., et al., 2023. First organoid intelligence (OI) workshop to form an OI community(Section Organoid Intelligence). Front. Artif. Intell. 6, 1116870. https://doi.org/10.3389/frai.2023.1116870.
Mulder, L.A., Depla, J.A., Sridhar, A., Wolthers, K., Pajkrt, D., Vieira De Sá, R., 2023. A beginner's guide on the use of brain organoids for neuroscientists: a systematic review. Stem Cell Res. Ther. 14, 87. https://doi.org/10.1186/s13287-023-03302-x.
Murtaza, N., Uy, J., Singh, K.K., 2020. Emerging proteomic approaches to identify the underlying pathophysiology of neurodevelopmental and neurodegenerative disorders. Mol. Autism 11 (1), 27. https://doi.org/10.1186/s13229-020-00334-5.

Obien, M.E.J., Deligkaris, K., Bullmann, T., Bakkum, D.J., Frey, U., 2015. Revealing neuronal function through microelectrode array recordings. Front. Neurosci. 8, 423.
OECD, 2023. Initial recommendations on evaluation of data from the developmental neurotoxicity (DNT) in-vitro testing battery. Ser. Test. Assess. No 377. https://one.oecd.org/document/ENV/CBC/MONO(2023)13/en/pdf.
Paşca, S., 2018. The rise of three-dimensional human brain cultures. Nature 553, 437–445. https://doi.org/10.1038/nature25032.
Pamies, D., Barreras, P., Block, K., Makri, G., Kumar, A., Wiersma, D., et al., 2017. A human brain microphysiological system derived from iPSC to study central nervous system toxicity and disease. ALTEX 34, 362–376. https://doi.org/10.14573/altex.1609122.
Pamies, D., Block, K., Lau, P., Gribaldo, L., Pardo, C., Barreras, P., et al., 2018. Rotenone exerts developmental neurotoxicity in a human brain spheroid model. Toxicol. Appl. Pharmacol. 354, 101–114. https://doi.org/10.1016/j.taap.2018.02.003.
Pasca, S.P., Portmann, T., Voineagu, I., Yazawa, M., Shcheglovitov, A., Pas¸ca, A.M., et al., 2015. Using iPSC-derived neurons to uncover cellular phenotypes associated with Timothy syndrome. Nat. Med. 17 (12), 1657–1662. https://doi.org/10.1038/nm.2576.
Pham, M.T., Pollock, K.M., Rose, M.D., Cary, W.A., Stewart, H.R., Zhou, P., et al., 2018. Generation of human vascularized brain organoids. Neurosurgery 82 (3), 312–325. https://doi.org/10.1097/WNR.0000000000001014.
Qian, X., Nguyen, H.N., Song, M.M., Hadiono, C., Ogden, S.C., Hammack, C., et al., 2018. Brain-region-specific organoids using mini-bioreactors for modeling ZIKV exposure. Cell 165 (5), 1238–1254. https://doi.org/10.1016/j.cell.2016.04.032.
Qian, X., Song, H., Ming, G.L., 2019. Brain organoids: advances, applications and challenges. Development 146 (8), dev166074. https://doi.org/10.1242/dev.166074. PMID: 30992274; PMCID: PMC6503989.
Quadrato, G., Nguyen, T., Macosko, E.Z., Sherwood, J.L., Min Yang, S., Berger, D.R., et al., 2017. Cell diversity and network dynamics in photosensitive human brain organoids. Nature 545 (7652), 48–53. https://doi.org/10.1038/nature22047.
Quirion, R., 2023. Brain organoids: are they for real? Front. Sci. 1, 1148127. https://doi.org/10.3389/fsci.2023.1148127.
Roth, A., Marx, U., Vilén, L., Ewart, L., Griffith, L.G., Hartung, T., et al., MPS-WS Berlin 2019, 2021. Human microphysiological systems for drug development. Science 373, 1304–1306. https://doi.org/10.1126/science.abc3734.
Schantz, S.L., Widholm, J.J., Rice, D.C., 2003. Effects of PCB exposure on neuropsychological function in children. Environ. Health Perspect. 111 (3), 357–576. https://doi.org/10.1289/ehp.5461.
Shen, H., 2018. Core concept: Organoids have opened avenues into investigating numerous diseases. But how well do they mimic the real thing? Proc. Natl. Acad. Sci. U. S. A. 115 (14), 3507–3509. https://doi.org/10.1073/pnas.1803647115.
Shepherd, J., 2018. Ethical (and epistemological) issues regarding consciousness in cerebral organoids. J. Med. Ethics 44 (9), 611–612.
Sillé, F., Hartung, T., 2014. Metabolomics in preclinical drug safety assessment: current status and future trends. Metabolites 14, 98. https://doi.org/10.3390/metabo14020098.
Sloan, S.A., Andersen, J., Paşca, A.M., Birey, F., Pasca, S.P., 2018. Generation and assembly of human brain region-specific three-dimensional cultures. Nat. Protoc. 13 (9), 2062–2085. https://doi.org/10.1038/s41596-018-0032-7.
Smirnova, L., Hartung, T., 2022. Neuronal cultures playing Pong—first steps toward advanced screening and biological computing. Neuron 110, 3855–3856. https://doi.org/10.1016/j.neuron.2022.11.010.

Smirnova, L., Hartung, T., 2024. The promise and potential of brain organoids. Advanced Healthcare Materials, accepted manuscript. Available at: http://dx.doi.org/10.1002/adhm.202302745.

Smirnova, L., Hogberg, H.T., Leist, M., Hartung, T., 2014. Developmental neurotoxicity—challenges in the 21st century and in vitro opportunities. ALTEX 31 (2), 129–156. https://doi.org/10.14573/altex.1403271.

Smirnova, L., Hogberg, H.T., Leist, M., Hartung, T., 2024. Revolutionizing developmental neurotoxicity testing—a journey from animal models to advanced in vitro systems. *ALTEX 2024* 41, 152–178. https://doi.org/10.14573/altex.2403281.

Smirnova, L., Kleinstreuer, N., Corvi, R., Levchenko, A., Fitzpatrick, S.C., Hartung, T., 2018. 3S—systematic, systemic, and systems biology and toxicology. ALTEX (2), 139–162. https://doi.org/10.14573/altex.1804051.

Smirnova, L., Morales Pantoja, I.E., Hartung, T., 2023. Organoid Intelligence (OI)—the ultimate functionality of a brain microphysiological system. ALTEX 40, 191–203. https://doi.org/10.14573/altex.2303261.

Spira, M.E., Hai, A., 2013. Multi-electrode array technologies for neuroscience and cardiology. Nat. Nanotechnol. 8 (2), 83–94.

Stacey, G., 2012. Current developments in cell culture technology. Adv. Exp. Med. Biol. 745, 1–13. https://doi.org/10.1007/978-1-4614-3055-1_1.

Stelzer, E.H.K., Strobl, F., Chang, B.J., et al., 2021. Light sheet fluorescence microscopy. Nat. Rev. Methods Primers 1, 73. https://doi.org/10.1038/s43586-021-00069-4.

Südhof, T.C., 2017. Molecular neuroscience in the 21st century: a personal perspective. Neuron 96 (3), 536–541.

Sun, N., Meng, X., Liu, Y., et al., 2021. Applications of brain organoids in neurodevelopment and neurological diseases. J. Biomed. Sci. 28, 30. https://doi.org/10.1186/s12929-021-00728-4.

Tehranirokh, M., Kouzani, A.Z., Francis, P.S., Kanwar, J.R., 2013. Microfluidic devices for cell cultivation and proliferation. Biomicrofluidics 7 (5), 51502. https://doi.org/10.1063/1.4826935.

Trakoshis, S., Martínez-Cañada, P., Rocchi, F., Canella, C., You, W., Chakrabarti, B., et al., 2020. Intrinsic excitation-inhibition imbalance affects medial prefrontal cortex differently in autistic men versus women. eLife 9, e55684.

Trujillo, C.A., Gao, R., Negraes, P.D., Gu, J., Buchanan, J., Preissl, S., et al., 2019. Complex oscillatory waves emerging from cortical organoids model early human brain network development. Cell Stem Cell 25 (4), 558–569.e7.

Van Vliet, E., Morath, S., Linge, J., Rappsilber, J., Eskes, C., Honegger, P., et al., 2008. A novel in vitro metabolomics approach for neurotoxicity testing, proof of principle for methyl mercury chloride and caffeine. Neurotox 29, 1–12.

Webb, E., Moon, J., Dyrszka, L., Rodriguez, B., Cox, C., Patisaul, H., et al., 2018. Neurodevelopmental and neurological effects of chemicals associated with unconventional oil and natural gas operations and their potential effects on infants and children. Rev. Environ. Health 33 (1), 3–29. https://doi.org/10.1515/reveh-2017-0008.

Xu, R., Boreland, A.J., Li, X., Erickson, C., Jin, M., Atkins, C., et al., 2021. Developing human pluripotent stem cell-based cerebral organoids with a controllable microglia ratio for modeling brain development and pathology. Stem Cell Reports. 16, 1923–1937. https://doi.org/10.1016/j.stemcr.2021.06.011. PMID 34297942.

Yamins, D.L., DiCarlo, J.J., 2016. Using goal-driven deep learning models to understand sensory cortex. Nat. Neurosci. 19 (3), 356–365.

Yang, W., Yuste, R., 2017. In vivo imaging of neural activity. Nat. Methods 14 (4), 349–359.

Zhang, D., Eguchi, N., Okazaki, S., Sora, I., Hishimoto, A., 2023. Telencephalon organoids derived from an individual with ADHD show altered neurodevelopment of early cortical layer structure. Stem Cell Rev. Rep. 19 (5), 1482–1491. https://doi.org/10.1007/s12015-023-10519-z.

Zheng, Y., Zhang, F., Xu, S., Wu, L., 2021. Advances in neural organoid systems and their application in neurotoxicity testing of environmental chemicals. Genes. Environ. 43 (1), 39. https://doi.org/10.1186/s41021-021-00214-1.

Zimmermann, M.B., 2009. Iodine deficiency. Endocr. Rev. 30 (4), 376–408. https://doi.org/10.1210/er.2009-0011.

Zoeller, R.T., 2007. Environmental chemicals targeting thyroid. Hormones 6 (2), 107–114. https://doi.org/10.1093/brain/awg014.

CHAPTER FIVE

Utilization of human stem cells to examine neurotoxic impacts on differentiation

Victoria C. de Leeuw* **and Ellen V.S. Hessel**
National Institute for Public Health and the Environment, Bilthoven, The Netherlands
*Corresponding author. e-mail address: victoria.de.leeuw@rivm.nl

Contents

Advances in Neurotoxicology, Volume 12
ISSN 2468-7480, https://doi.org/10.1016/bs.ant.2024.08.001

Abstract

Cell differentiation is one of the crucial processes for normal brain development. Disruption of differentiation can result in impaired mental and/or physical abilities later in life. Therefore, measuring this endpoint reliably in human-relevant models is essential. Human stem cells have the unique potential to model cell differentiation and study the influence of compounds on proportions between cell populations and on specific cell types, whether that is in altered cellular pathways, morphology or function. In this chapter we discuss the role of differentiation in brain development, disruption of differentiation and its relation to other endpoints and how this can be measured in human stem cells. We discuss the use of embryonic and induced pluripotent stem cells, provide a selection of available readouts for differentiation, list considerations for setting up differentiation protocols and give current examples of in vitro models for differentiation. Lastly, we discuss current and future applications of stem cell-based differentiation models, for example in the context of regulatory testing and for studying gene-environment interactions.

1. The role of cell differentiation in brain development

1.1 Neurodevelopmental disorders

The prevalence of neurodevelopmental disorders (NDDs) is increasing worldwide (Landrigan et al., 2012) and affects around 10% of all births, ranging from minor to severe disabilities that persist throughout life (Astle et al., 2022). The Diagnostic and Statistical Manual of Mental Disorders defines NDDs as a broad category of disorders with onset in the developmental period. Examples are intellectual disability, Autism Spectrum Disorder (ASD), Attention-Deficit/Hyperactivity Disorder, learning and motor disorders (Association, 2013). Many of the NDDs are associated with cognitive impairments, which have great impact on daily life. For instance, a substantial percentage (11%–65%) of school-age children with ASD is reported to have cognitive impairments (Lord et al., 2018).

For preventive purposes, unravelling the causes of NDD and its underlying cognitive problems is urgently needed. Cognitive functioning and other neurobehavior in humans is a result of intricate interactions between brain areas, which is essential to accomplish this human specific complex behavior (Santello et al., 2019). Fully modelling such interactions between brain areas in human with in silico, in vitro or in vivo assays remains an intractable challenge for now. Therefore basic knowledge on the complexity of the developing brain is urgently needed. Mechanistic knowledge and multidisciplinary research is essential to further develop relevant these assays for this purpose.

1.2 Differentiation as a crucial process in early brain development

Brain development starts after conception until the age of around 25 years (Giedd et al., 1999). The adult human brain is divided into hundreds of spatial domains, each comprising tens or hundreds of distinct neuronal, glial, and other cell types. Various stages of development occur in the brain starting with basic processes in early development such as differentiation into neural progenitor cells (NPCs) and finally result in maturation of neurocircuitries, morphology changes and myelination during adolescence (Arain et al., 2013).

By the end of the third week after conception, gastrulation starts and results in three germ layers in the embryo—the ectoderm, mesoderm and endoderm (Stiles & Jernigan, 2010). During the formation of these germ layers, a subset of cells (the later ectoderm) receives signals from migrating cells, which induce their differentiation into NPCs (Moog et al., 2017). These NPCs are capable of producing all of the different cells that make up the central nervous system. Differentiation into the correct cell types in time and space is an essential process. By the end of gastrulation, the NPCs are positioned along the rostral-caudal midline of the upper layer of the three-layered embryo. The NPCs in the most rostral region will give rise to the brain. Through specification, the anterior part will develop into the brain by differentiation, and maturation events that yield to several thousand types of cells.

The adult human brain contains billions of neurons, most of which are produced by mid-gestation. The number of NPCs that is present at the end of gastrulation, that is around three weeks, is far too small to produce all these neurons. Therefore, cell division and proliferation of many NPCs is needed, after which cell differentiation, most importantly neurogenesis and gliogenesis, starts. Neurogenesis is the process by which NPCs differentiate into neurons and gliogenesis is the generation of glia populations from NPCs. These processes begin in early embryogenesis and extend through mid-gestation in most brain areas (Stepien & Wielockx, 2024). Within this period, different populations of neurons arise in different regions of the brain, which forms the basis of proper brain development. As said this complex arrangement of cells guided by differentiation in various brain regions is initially established during the first trimester of development. A recent paper revealed a dissection of spatial, temporal and transcriptional changes that occur in the first trimester of brain development resulting in a fundamental blueprint of the human brain that comprises more than 600 cell types (Braun et al., 2023). This first known comprehensive study of the whole human brain during the first trimester found that although neurons were the most diverse, both pre-astrocytes and oligodendrocyte precursor cells were regionally diverse, and their gene expression suggests region- and cell type–specific supportive functions (Braun et al., 2023). This and other studies highlight the importance of early patterning events and early cell diversification in brain development. Although brain development continues through the age of 25 it seems that differentiation and maturation in the first trimester is crucial for healthy brain development. This knowledge can help to model brain disorders that show region-specific patterns of occurrence or severity for human disorders that affect specific brain cell populations (Braun et al., 2023).

Therefore, models representing parts of human brain development and differentiation in the first trimester and beyond are crucial to study the underlying mechanisms and possible the origin of some brain disorders. This requires the development and/or proper characterization of advanced, human stem cell-based in vitro models and readouts for differentiation. Within these models basic and early differentiation can be modeled including cell-specific differentiation, for example, neuronal (subtype), astrocyte, oligodendrocyte and microglia differentiation or a combination of cell types for example, neuron-astrocyte co-culture or brain region specific differentiation can be modelled (Okabe et al., 1996; Tyzack et al., 2016; Schmidt et al., 2017). Differentiation can also lead to neuronal subtypes of neurons resulting into various neurotransmitter phenotypes, such as gamma-aminobutyric acid (GABA), glutamate, choline, serotonin and dopamine (Borodinsky & Belgacem, 2016). These phenotypes may change during development and adulthood through intrinsic and extrinsic factors (Spitzer, 2012; Spitzer, 2015; Borodinsky & Belgacem, 2016). For example, the role of GABA receptors switches during development from excitatory to inhibitory, which shows that also the function of neuronal molecules can change during development (Spitzer, 2015; Jessell & Sanes, 2000). Furthermore, brain region-specific astrocytes and microglia play important roles in neurophysiology (including structural functions, regulation of metabolism and synapse formation) (Bal-Price et al., 2017). When modelling differentiation in vitro all these aspects could be considered.

1.3 Diseases related to neuronal differentiation defects

As previously stated, neuronal differentiation is a critical process in brain development, and disruptions in this process can have significant effects on NDDs. These disorders typically manifest early in life and are characterized by developmental deficits that impact brain development. Neuronal differentiation is tightly regulated and involves a series of steps including cell proliferation, migration, and organization. Abnormal cell proliferation or apoptosis can result in conditions like microcephaly, where the brain is abnormally small (Sokol & Lahiri, 2023), or megalencephaly, where the brain is abnormally large (Di Donato et al., 2016). Abnormal migration can also lead to disorders such as lissencephaly, characterized by a smooth brain surface due to improper neuronal migration (Kerjan & Gleeson, 2007). But disruption of differentiation can also be associated with a range of rare but also common NDDs, including ASD, schizophrenia, and intellectual disabilities, which can manifest as impairments in cognitive, social, and motor skills (Lu et al., 2022). Understanding the mechanisms of neuronal differentiation and its impact on NDDs is crucial for developing targeted therapies and interventions, but, just as important, to prevent the exposure to chemicals that might affect differentiation of the developing brain.

To sum up, differentiation is an important process throughout brain development and includes various specificities such as the brain region specific composition of brain cells and differentiation in neuronal subtypes. Especially in the first trimester differentiation occurs as a backbone of proper brain development. Later during brain development, although differentiation is still crucial, other processes become more predominant such as synaptogenesis, axon projection and neuronal network formation, maturation of neuronal circuitries and morphology. Still, proper differentiation in early brain development lays the foundation for many of these processes. The focus of this book chapter is on the importance of (disturbances in) differentiation, having crucial effects on later key events, and readouts that can give insight into compound effects on differentiation.

2. Embryonic versus induced pluripotent stem cells as in vitro differentiation models

Both human Embryonic Stem Cells (hESCs) and human induced Pluripotent stem cells (hiPSCs) are virtually infinite sources of cells. Due to self-renewal capacity they can differentiate to cell and tissue types normally originating from any of the three primordial germ layers during embryonic development. In the last 30 years stem cell research took off enormously, which started by using embryonic lines in 1990 and from the late 2000s with iPSCs. In its current definition, stem cells are cells that have the ability to both self-renew and develop in any specialized cell through the process of differentiation. The use of stem cells to model differentiation have gained popularity in developmental neurotoxicity (DNT) research, especially in the last fifteen years (Fritsche et al., 2018). This is potentially be due to improved control over differentiation of human stem cells into NPCs, neurons and glial cells. Here we would like to discuss the stem cell developments the last decades and to elaborate on the advantages and disadvantages of using hESCs and/or hiPSCs to measure neuronal differentiation.

2.1 Human embryonic stem cells

Before hESC, mouse embryonic stem cells (mESC) were predominantly used for studies into (effects on) cell differentiation. While both mESCs and hESCs are obtained from the blastocyst stage, the state of pluripotency of mESC may be more "naïve" than hESC (Nichols, 2011). Maintenance and differentiation of the mESC and hESC is also performed in different kinds of medium with different factors to repress and induce (neuronal) differentiation (Kim et al., 2020). This may not be surprising as rodent and human neural development become progressively different over time; the time for development, the specific events at each point in development and the general structure, cellular composition and function of the brain (Rice & Barone, 2000; Semple et al., 2013; Florio and Huttner, 2014; Oberheim et al., 2006; Masjosthusmann et al., 2018). mESC are valuable study objects, because most compound data is historically derived from mouse or at least rodent experiments, their differentiation and consequences for in vivo is better understood and early differentiation is well conserved across species (Irie & Kuratani, 2011; Kuegler et al., 2010). Although more complicated, human-based neuronal differentiation systems are quickly being improved by a wealth of differentiation protocols that are becoming available and have the obvious advantage in that they represent the ultimate target species: the human.

In the last ten years, the number of culturing and differentiation protocols for these cells greatly expanded and also found its way to the toxicology field. The developments in this field follow at a rapid pace, however historically researchers are more familiar working with rodent cells rather than with human cells, which may be one of the reasons for reproducibility issues (Yoon et al., 2019; Kelava & Lancaster, 2016; Pasca et al., 2015). This issue is becoming less of a problem driven by the fast development of defined culture media and supplements. While for for example, regenerative medicine purposes the aim is to differentiate stem cells into a specific and pure population, basic research and toxicology may also benefit from differentiation protocols that generate mixed cell populations to study the emergence and interaction between these cells. Although commercial (neuronal) differentiation protocols and some NPC lines are becoming more common to use, many researchers, also in the toxicology field, are developing their own custom protocols. Many protocols work with a two-step procedure that allows for selection of cells during differentiation and can vary in length from several weeks to months. Differences can be mainly found in culturing in 2D or three-dimensional (3D) and the kind of cell types that researchers want to achieve. Labs have managed to differentiate stem cells into a mixture of neurons and astrocytes (Pasca et al., 2015; Pistollato et al., 2017; Reubinoff et al., 2000; Carpenter et al., 2001; Chandrasekaran et al., 2016; Lappalainen et al., 2010; De Leeuw et al., 2020), with oligodendrocytes (Pamies et al., 2017; Talens-Visconti et al., 2011; Sandstrom et al., 2017) and microglia that had to be cultured separately and incorporated in relatively simple systems (Schwartz et al., 2015; Abreu et al., 2018) or developed innately in complex cerebral organoids (Ormel et al., 2018; Zhang et al., 2023). The mostly used hESC source is the female H9 (NIH code WA09) cell line.

2.2 Human induced pluripotent stem cells

In 2007, a whole new avenue of research was opened when Shinya Yamanaka proved for the first time that human somatic cells could be reprogrammed back to their pluripotent state, for which he won the Nobel Prize together with John Gurdon (Nobelmedia, 2012). Since the demonstration by Takahashi and Yamanaka (2006) and Takahashi et al. (2007) that mouse and human fibroblasts could be induced into pluripotent stem cells, the application of this technique in human cells holds great promise for many applications for both clinical and research purposes (Takahashi et al., 2007; Takahashi & Yamanaka, 2006).

hiPSC can be generated by taking for example, skin or blood cells from an individual and transfect these cells with pluripotency transcription factors (initially *OCT4*, *SOX2*, *CMYC* and *KLF4* (Takahashi et al., 2007)). As for hESC, hiPSC can be then differentiated into any cell type, with the advantage that no human embryonic material is needed to obtain these cells and the cells can be led back to a specific individual, allowing for personal assessment. Additionally, hiPSC can be generated from various individuals to sample genetic heterogeneity in a population for DNT assessment.

hiPSCs are increasingly used in DNT research (Pistollato et al., 2020; Lieberman et al., 2012; Hofrichter et al., 2017; Kobolak et al., 2020; Zhong et al., 2020). There is some debate about to what extent these cells are programmed back into a truly naïve state as part of the epigenome stays on the cells' genome even after reprogramming, but improved protocols are under development (Perrera & Martello, 2019). Application of hiPSC resolves several restrictions associated with hESC:

- hiPSC are generated from the consenting donors themselves, whereas hESC are obtained from human embryos resulting in different ethical, legal and social perspectives (Moradi et al., 2019);
- hiPSC can be generated from specific individuals with a known life course, for example from patients whose apparent etiology, disease prevention or cure, could benefit from in vitro research with their own cells and genetic makeup;
- hiPSC can be generated from various individuals to sample genetic heterogeneity in a population.

2.3 Challenges related to line to line variability

Experience in hESC and hiPSC lines in the last decades displayed variable capacity to differentiate these cells into specific lineages (Allegrucci & Young, 2007; Ortmann & Vallier, 2017). To work efficiently, the development of universal protocols of differentiation with any stem cell line is challenging. Currently, for many types of differentiation there is no standard protocols available yet, thereby limiting the development of validated assays needed for DNT. To use the models for regulatory DNT, validated protocols that work in a variety of laboratories are crucial. Various general aspects that cause variability in the models such as genetic variation, culture-related effects, experimental variation, genetic stability, epigenetic stability and sex differences need to be considered when using stem cells in differentiation models. These factors are nicely reviewed (Allegrucci & Young, 2007; Ortmann & Vallier, 2017; Waldhorn et al., 2022).

2.3.1 Genetic variation

There is a major advantage of in using stem cells, especially iPSC, in toxicology screening, because each line represents a unique human genetic background. The interaction of compounds tested with the variable alleles that individuals inherit often ensures including different genetic susceptibility to the chemical exposure (Allegrucci & Young, 2007; Ingelman-Sundberg & Rodriguez-Antona, 2005). These genetic effects are not identified in preclinical animal tests, since mainly inbred stains with identical genetic make-up are tested, or in cell-based screens that utilize human, transformed cell lines that were derived from a few individuals. Thus, the provision of human cell types from a wide range of stem cell lines that represent genetic diversity offers a unique opportunity to include genetic variation in the DNT prediction. However, differences in genetic backgrounds will also lead to differences in levels of receptors, transcription factors and growth factors which, in turn could affect the capacity of the stem cells to respond to specific culture conditions. Recently, even mutations in mitochondrial DNA have been raised as a potential additional source of variability in culturing (Allegrucci et al., 2005).

Furthermore, long-term maintenance of stem cells in a laboratory environment may lead to the accumulation of genetic defects in the cells due to normal cell division that includes somatic mutations. Chromosomal abnormalities might also occur, however this is generally not a trend in every hESC culture (Buzzard et al., 2004; Darnfors et al., 2005; Mitalipova et al., 2005), and it is also not clear whether certain cell lines are intrinsically more prone to developing abnormalities or whether their instability is a consequence of certain culture methods (Allegrucci & Young, 2007). This needs further attention and karyotyping of cells is an important control step should be done on a regular basis (ISSCR, 2023).

2.3.2 Culture-related effects, experimental variation

It has been shown that culture conditions might result in line to line variation, the wide range of feeder cells, culture media, additives and passage methods used to derive lines still confounds the interpretation of inter-line differences (Allegrucci et al., 2005). No standardized conditions that are optimal or generically applicable across lines have been established either for their culture or for their differentiation. Although all the available hESC and hiPSC lines derived worldwide share the expression of characteristic pluripotency markers, many differences are emerging between lines that may be more associated with the wide range of culture conditions in current use than the inherent genetic variation of the individual from which stem cells were derived. Several initiatives were undertaken to characterize multiple existing stem cell lines using standardized assay conditions to allow accurate comparison of the data generated (Andrews et al., 2005; Rao & Civin, 2005). Also, for DNT various protocols exist and each laboratory use their own specific protocol which makes it challenging to develop standardized assays that are ready for validation and inter laboratory exchange. Beside preferences based on experiences in protocols between various laboratories this might also be caused by inter-line variations.

2.3.3 Epigenetic stability

Although the stability of the genome of human stem cells have received general attention, little is known about their epigenome. Epigenetic variation between hESCs exist, but the details and their impacts are unknown. A very important example that illustrates this is the effect of X chromosome inactivation. X chromosome inactivation is crucial in human life since this ensures that female embryos express similar levels of X-linked genes to males and therefore is an important mechanism for gene dosage compensation. Discordant reports exist noting variations in epigenetic stability in different hESC lines, in X chromosome inactivation and sometimes this X chromosome inactivation is lost in female stem cell lines which should be taken into consideration when working with female stem cell lines (Allegrucci & Young, 2007).

2.3.4 Sex differences

Biological sex including the sex chromosomes influence development, reproduction, pathogenesis, and medical treatment outcomes. Most research in the last decades, however, focused on males. Modelling sex differences is challenging because autosomal genetic variability may be more dominant than the sex differences and chromosomal versus hormonal effects are hard to separate (Waldhorn et al., 2022). Sex differences in gene expression were identified in neuronal differentiation with hiPSCs, where female hiPSCs more closely resembled the naive pluripotent state than their male counterparts (Waldhorn et al., 2022). Moreover, the model enabled insights of the contributions of X versus Y chromosomes to this phenomenon. More knowledge on sex differences are is needed and stem cells may actually help to investigate the differences.

3. Differentiation as an endpoint for neurotoxicity assessment

3.1 Differentiation alongside other endpoints

By taking differentiation as an endpoint, there is a central assumption to it that cell differentiation is an important feature of (neuronal) development, and that disturbances by compounds will be reflected in molecular and cellular pathways, expression of differentiation markers or neuronal or glial function (see section Considerations for available differentiation routes). While this is obviously true, this is certainly not the only developmental process that may be mimicked in vitro and may be used to study compounds-induced DNT. Other processes in human brain development on a cellular level include cell proliferation, migration, cell-cell interactions, apoptosis, neurite outgrowth, synaptogenesis, and more. These events occur in different (combinations of) cell types and in different developmental time windows, amongst others gastrulation, neurulation, network formation and neural patterning (Hessel et al., 2018). It is important to be aware that there is probably only a very limited part of neurodevelopment that can be mimicked in a single cell-based in vitro assay. That does not mean that it may not be possible to measure several of the aforementioned processes in a single culture. Indeed, there are in vitro tests under review for regulatory implementation in which migration from an embryoid body structure is measured alongside neuronal differentiation and other neurodevelopmental processes (Barenys et al., 2017; Baumann et al., 2016). Measuring different features in one culture system can enhance the mechanistic understanding of compound effects, but the relevance of the parameters should be tested vigorously using compounds with known modes of action to discover which endpoints are sensitive in a specific culture system (Aschner et al., 2017; Blum et al., 2023).

3.2 Available stem cell-based in vitro assays

There are various differentiation assays that have been developed over the years for DNT testing, which each have their own focus on what they are supposed to mimic. A hESC-based version of early differentiation was published with the advantage that not only neuronal differentiation but also other ectodermal lineages can be examined (Tchieu et al., 2017). Analyzing the morphology of neural rosettes analogous to the neural tube has shown a promising method to study genetic defects and the rescue potential of folic acid (Sahakyan et al., 2018; Valensisi et al., 2017) or early neurotoxicants (Lundin et al., 2024; Dreser et al., 2020). Many hESC or hiPSC-based differentiation protocols focus on the part from the NPC stage to neurons or neuron-glia co-cultures. These cultures are grown in 2D structures (Delp et al., 2018; Pistollato et al., 2017; De Leeuw et al., 2020) or a 2.5D structure (Barenys et al., 2017; Baumann et al., 2016), which contain neurons and astrocytes, or neurons and oligodendrocytes (Koch et al., 2022). In these models there is a main focus on imaging-based methods to track differentiation as a measure of DNT. These models are considerably less complex than brain spheres containing neurons, astrocytes (Kobolak et al., 2020; Sirenko et al., 2019) and oligodendrocytes (Pamies et al., 2017; Yla-Outinen et al., 2010; Sandstrom et al., 2017). Microglia, the last major cell population in the brain that is least present in current in vitro cultures, have to be introduced separately in these brain spheres (Schwartz et al., 2015; Abreu et al., 2018), but organotypic cultures mimic neuronal differentiation to such an extent that also microglia can develop spontaneously in the culture (Ormel et al., 2018; Zhang et al., 2023). These models, while more complex and much lengthier, allow for deeper mechanistic insight into the effect of compounds. See the section Considerations for available differentiation routes for guidance on the system of choice.

While there is a lot possible with stem cells, one has to keep in mind that even complex organoid structures may not (yet) be sufficiently complex to study the whole course of neuronal development. Whole non-mammalian organisms such as zebrafish, fruit flies and *Caenorhabditis elegans* can, especially in a toxicological setting where throughput is vital, aid in gaining insight in compound effects on later and more complex neurodevelopmental processes and may bridge cell-based and in vivo toxicology (Collins et al., 2024; Beamish et al., 2023). In conclusion, there are many differentiation models that each have their own advantages and disadvantages, and that can each fill a niche for a specific application in DNT research. Additionally, all these available stem cell assays currently represent only a fraction of the processes that are crucial for early brain development and they can at this moment never predict the complete complexity of the human brain. More research into the human physiology of the developing brain is needed to model this and that might result in more complex in vitro (and in silico) models.

4. Readouts for neuronal differentiation

Differentiation can be measured in a number of ways, assessing in vitro systems on the level of genes, gene expression, protein expression, metabolites, morphology and/or function. The field of toxicology is moving toward a mechanism-based approach, which necessitates readouts that support this. This means that a combination of apical readouts such as morphology, viability and functionality (Schmidt et al., 2017), and mechanistic readouts such as gene expression and other omics is needed. In the following sections, an overview of currently available readouts is provided.

4.1 Mechanistic readouts to discover molecular and cellular pathways

A way to measure differentiation is to look at which mechanisms are regulated in differentiated versus non-differentiated cells, and which processes are disrupted during differentiation by toxic compounds. Toxicogenomics is the field of research where omics technologies (e.g. genome profiling, transcriptomics, proteomics, metabolomics) are applied to study the underlying mechanism of how compound exposure might cause adverse effects on human health and the environment (Nuwaysir et al., 1999). Transcriptomics, the study of gene expression of the whole genome, is currently the most commonly used omics tool in toxicology and is based on the assumption that gene expression of cells is changed when exposed to a toxicant (Robinson & Piersma, 2013; Harrill et al., 2009). Because of its comprehensive and unbiased approach it can aid in the characterization of in vitro models and can explore new biomarkers and gene signatures in these models when they are exposed to compounds (Afshari et al., 2011; Merrick et al., 2015). The technology has become more popular in the past decades due to a decrease in price per sample and major improvements in sequencing and data analysis tools (Liu et al., 2019). There is a rapid increase in the use of RNA sequencing (RNA-Seq), a Next Generation Sequencing (NGS) technology based on massive parallel sequencing of the transcriptome (Hrdlickova et al., 2017; Merrick, 2019). Microarrays used to be the dominant transcriptomics platform for the past twenty years (Duggan et al., 1999), but RNA-Seq is currently the mostly used platform. This technology has comparable performance on endpoint prediction with the advantage for detecting low-abundance genes (Wang et al., 2014; Zhang et al., 2015). Another advantage is that for RNA-seq no prior knowledge of the genome is needed, which is important for the analysis of unknown genes and transcript isoforms (Hrdlickova et al., 2017). Regardless the platform, transcriptomics approaches have proven to be of great use in studies to reveal mechanisms of toxicity and make predictions for sensitive biomarkers that can indicate toxic effects of compounds for human risk assessment (Robinson & Piersma, 2013; Harrill et al., 2009; Liu et al., 2019). Especially in combination with in vitro systems, concentration and time dependent effects of compounds can be studied in a relatively high-throughput manner and can subsequently be linked to other readouts (Pennings et al., 2012; Schulpen et al., 2014; Theunissen et al., 2012; Van Dartel et al., 2009; Chen et al., 2019; Waldmann et al., 2017). Based on NGS data, qPCR markers can be selected to further study of individual genes as a high-throughput and less expensive technology for sensitive detection of gene expression changes in response to compounds.

Transcriptomics provide a wealth of data with at least more than 20,000 protein-coding genes and another 15,000 non-coding genes and splice variants, with new genes being discovered regularly (Merrick, 2019). The analysis of transcriptome data is challenging and guidelines for standardization of omics data have only just being developed (OECD, 2023b; Harrill et al., 2021). While transcriptomic analysis always requires a tailored approach, a standard first step is to analyze which individual genes are differentially regulated using a statistical measure, for example a p-value or false discovery rate value that can optionally be combined with a fold change value of the regulated gene to select for the most regulated genes. A second step may be to analyze in which processes the differentially expressed genes are involved. This can be done for individual genes through databases such as the National Center for Biotechnology Information, GeneCards, Gene Ontology (GO), the Comparative Toxicogenomics Database for compounds reference and the Monarch Initiative and MalaCards for disease reference (Ashburner et al., 2000; Rappaport et al., 2017; Shefchek et al., 2020; Davis et al., 2023; Stelzer et al., 2016; Sayers et al., 2024). Groups of genes can also be analyzed to study their interrelationships. For example, STRING can visualize relationships between proteins and STITCH can do this for proteins and compounds (Szklarczyk et al., 2023). Platforms such as DAVID calculate enrichment of biological pathways, molecular processes, cellular localization and GeneAnalytics additionally calculates cell type, tissue and disease enrichment (Ben-Ari Fuchs et al., 2016; Huang Da et al., 2009; Sherman et al., 2022). KEGG and WikiPathways can map gene or protein expression profile on molecular networks (Kanehisa et al., 2016; Agrawal et al., 2024). Other platforms specialize in prediction of cell types that are present in a cell culture compared to in vivo such as Lifemap Discovery, CoNTexT and TissueEnrich (Ben-Ari Fuchs et al., 2016; Jain & Tuteja, 2019; Stein et al., 2014). A limitation in all these platforms is that they are based on lists of differentially expressed genes and do not use the level of gene expression in their analysis. Gene Set Enrichment Analysis takes into account the whole gene expression profile to determine whether a priori defined gene sets are enriched (Subramanian et al., 2005). To further study upregulated cellular pathways there are both commercial (e.g. Qiagen's Ingenuity Pathway Analysis) and free (e.g. Signaling Pathway Impact Analysis (http://bioinformatics.oxfordjournals.org/cgi/reprint/btn577v1)) software available, which can, as opposed to GO-terms, provide the most likely direction of regulation of a certain pathway.

Single cell RNA sequencing (scRNA-seq) is an upcoming technology that allows to study gene expression of single cells in a cell culture. This is a next level of granularity that can be reached for the characterization of both cell types (see also next section) and processes that are specific for subpopulations of cells. A recent review outlines the advantages and disadvantages of the technology (Tukker & Bowman, 2024). With the rise of this technology and data that comes out of this, it may become possible to predict cell types present in transcriptomics data of a whole cell culture, which has already been shown for blood samples. For more complex cultures this is still under development (Avila Cobos et al., 2018). For studying differentiation in DNT field single cell transcriptomics is a promising technology since the amount of certain cell types can be determined and quantified, which result in a model that can predict how chemicals effect the differentiation of various brain cells (Tukker & Bowman, 2024). It is to be expected that tools alike TissueEnrich and others will be developed on a single cell level, in which scRNA-seq data from in vitro models can be compared to the human developing brain to benchmark the model and further define its biological domain.

4.2 Marker expression to identify cell types

Another central readout for and aspect of neuronal differentiation is the expression of cell type specific markers. While gene expression, and especially whole transcriptome approaches, is a good first lead to discover cell type specific markers, these need to be confirmed on the level of protein expression. Proteins can be measured in multiple manners that are quantitative or semi-quantitative in nature (Avila Cobos et al., 2018). Examples of quantitative readouts are scRNA-Seq, Western blots and proteomics, enzyme-linked immunosorbent assays, flow cytometry and imaging.

As mentioned above, scRNA-Seq is a relatively novel technology that allows to distinguish cell types based on selective markers for that cell type. This has been proven effective for human stem cell cultures that can be benchmarked against in vivo reference samples (La Manno et al., 2016). In a neurotoxicological context, researchers have shown the effect of methyl mercury on the proportions of specific cell types in a stem cell-derived differentiating cortical cell culture (Diana Neely et al., 2021). The power of this approach was recently taken to a next level by a research group making chimeroids of five different iPSC lines that showed individual susceptibility to neurodevelopmental toxicants (valproic acid and ethanol), which affected the proportions of different cell types (Antón-Bolaños et al., 2024). This technology, while based on gene expression, can potentially act as a bridge between unbiased cell marker type identification and follow-up using protein-based read-outs. At the moment, the technology is still very costly and therefore more suited for in-depth mechanistic research than for screening purposes. Still, through the knowledge that is gained by scRNA-seq it is possible for the wider community to learn about new selective cell type markers that can be used to stain these cell populations. This will increase the applicability of the assays for DNT. Other single cell omics technologies, such as single cell proteomics, are still in their infancy (Ahmad & Budnik, 2023).

Flow cytometry is a readout method that allows identifying cell populations and the effect of neurotoxicants on these cell populations by staining cells with selective cell type markers. It is useful for characterization, quantification and proofing reproducibility between batches of differentiated stem cell cultures (Chandrasekaran et al., 2017; Romero et al., 2022). In the context of neurotoxicity experiments with stem cells, there are a limited numbers of studies that have taken this approach using mouse stem cell-derived cultures (Visan et al., 2012; Ogony et al., 2013). One of the reasons it is not used as much may be because of the dissociation step needed to prepare a single cell solution for the flow cytometry processing. This is a challenging step in which the neurites need to disentangle, as far as this is possible, and cells die in this process. Also, only abundant markers can be used, such as transcription factors and components of the cytoskeleton, and an occasional synaptic marker that is also abundantly present in the soma. Lastly, localization information is lost and detection of subtle effects or low-abundance proteins may pose a challenge. These arguments hold also true for Western blot.

Imaging may pose a solution to these aforementioned issues. When using immunocytochemistry the information about localization of proteins in cells can be retained and co-localization can be measured when using a microscope with sufficient resolution and proper pairs of synaptic markers (Verstraelen et al., 2020; Pistollato et al., 2020) (note: these are synapse markers based on mouse primary cells). It is also possible to measure proportions of different cell types using selective markers as a proxy for differentiation (Klose et al., 2022). Quantification is possible but challenging, especially in 3D in vitro systems and dense cell cultures or in tissue. Generation of sufficient samples is needed to measure effects of compounds, which in turn requires powerful software that is able to analyze the large amount of images. There are promising developments in the integration of Artificial Intelligence into current imaging software that may greatly enhance the data quality of image-based technologies. Commonly free platforms to analyze image data are ImageJ and CellProfiler (Mcquin et al., 2018; Rasband, 2012). For large compound screens, high content imaging machines are pivotal (Schmidt et al., 2017; Lickfett et al., 2022), which can be combined with for example, CellPaint for phenotypic profiling (Bray et al., 2016).

4.3 Functional readouts: it looks like a duck, but does it quack like a duck

Ultimately, the central function of well differentiated neuronal cells is to form synapses and to communicate with each other in a neuronal network and for glial cells to support this communication. This section will mainly focus on neuronal function, but some of the readouts can also be applied to neuronal precursors, astrocytes, oligodendrocytes and microglia. As reviewed before by De Groot et al. (2013), intracellular signaling, intercellular imaging and network function are three important aspects of neuronal functioning. Like for the other readouts, measuring this can both aid characterization of the differentiated neurons as well as study the effects of chemical disruption during differentiation on neuronal function.

To measure intracellular signaling of single cells, one can employ for example patch clamping, an electrophysiological approach. Patch clamping allows for the investigation of action potentials and the influence of compound exposure. This technology is very provides very detailed information on the receptor level of a single neuron, but is low throughput and requires highly trained researchers to perform patch clamping. Automated patch clamp systems are available to lessen this burden (Toh et al., 2020). Another single cell technology is calcium imaging, an imaging-based approach. Calcium signaling is a central process in neurons (and astrocytes) with many downstream processes being dependent on this, including neuronal excitability. Following calcium signaling with fluorescent dyes allows for studying spiking behavior and network formation over the course of differentiation or upon manipulation with chemicals (Estévez-Priego et al., 2023; Parmentier et al., 2022). Similar to patch clamping, the spatial resolution of calcium imaging is high, but the temporal resolution and throughput are relatively low. Also, multiple readings of cells are not possible since the dyes wash out and can be toxic to the cells (Parmentier et al., 2023).

Microelectrode array (MEA) is an extracellular electrophysiological technology that overcomes the temporal and some throughput limitations of the aforementioned methods. Moreover, MEA allows multiple readings of the same cell culture over time, which is ideal for following differentiating neurons. A MEA measures the local field potentials in neuronal (and cardiac) cell cultures with a frequency of up to 12.5 kHz. It is possible to directly differentiate human stem cells in a MEA or work from NPCs (Frega et al., 2017; De Leeuw et al., 2020). Individual spikes, bursts, network bursts and many more parameters can be measured to study the effects of compounds on the differentiation of the cell culture. This may result in for example, less electrical activity overall, an altered firing pattern or delay in the development of the culture, which are all indications of alternations in differentiation. MEA, and especially, multi-well MEA, is therefore a suitable screening tool for (developmental) neurotoxicity assessment. However, since the MEA is an integrative measure, further in-depth analysis is needed to study mechanism of compounds. One option for this is a high density MEA with thousands of electrodes per mm^2 (as opposed to around 10 electrodes for a multi-well system) that allows to study single cell function and network connectivity on a much more detailed scale. Both multi-well as high density MEA produce large amounts of data that can be analyzed with proprietary software that come with the MEA hardware and/or open-source scripts (e.g. (Hu et al., 2022)).

By combining in vitro assays and mathematical modelling MEA read-outs are already implemented for the high throughput screening of a possible DNT capacity of compounds (Pistollato et al., 2021; Johnstone et al., 2010). In combination with synaptogenesis related features effects of compounds on brain cell differentiation in vitro and the result on network formation can be investigated with functional readout on the MEA. Although these are relevant developments future optimalisation can result in more in-depth analysis using the MEA technique and in vitro models that better reflect brain development and the underlying mechanisms can be further developed.

Where MEA recordings add to reliable measurement of electrical activity, they lack in distinguishing neuronal subtypes. Measuring proteins and metabolites, in particular neurotransmitters, is therefore another important readout to identify the phenotype of differentiating cells and the effects of compounds. For development, the neurotransmitters glutamate and aspartate are crucial and were shown to be robust markers for DNT testing in differentiating stem cell derived neuronal cells (Cervetto et al., 2023). High-performance liquid chromatography or enzyme-linked immunosorbent assay can be used for this purpose. With the rise of organ-on-chip platforms there is also a push toward the development of miniaturized sensors for easier, real-time, more stable measurement of neurotransmitters, for example, non-enzymatic detection of Glutamate, Acetylcholine, Choline, and Adenosine or for dopamine and serotonin (Shadlaghani et al., 2019;

Rantataro et al., 2023). Broader investigation of small molecules can be done using proteomics or metabolomics, which can be applied in either a targeted or untargeted fashion. It is possible to measure the secretome, that is the proteins expressed that are secreted into the extracellular space, in the medium versus metabolome of the lysate, depending on the research question and the wish to measure multiple times, which is only possible in the secretome.

4.4 Defining the biological domain requires extensive characterization

To for the use of stem cells the applicability domain of in vitro models should be determined to know the cells, region and developmental period it represents. It is clear that due to the reductionist nature of in vitro assays it will require multiple tests to cover the breath of biological space that underlies human brain development. Studying differentiation with models that differentiate, timing is an extra dimension that needs to be taken into account, which makes it even more important to pinpoint what an in vitro test is mimicking at which stage in brain development. This is also crucial in the light of defining the uncertainties of in vitro assays in terms of their relevance and reliability, which is needed for (regulatory) validation of these tests (Paparella et al., 2020). This section described a plethora of readouts that can be used to follow normal and chemically-disrupted neuronal differentiation in vitro. The throughput and costs are being optimized, which makes application for toxicology purposes more attainable. While this section focused on differentiation as endpoint, the described readouts can be used for other endpoints of interest, such as proliferation, apoptosis, migration and neurite outgrowth.

5. Considerations for available differentiation routes

Whether an in vitro assay is deemed useful is highly dependent on whether the complexity of the cell culture is in proportion to the information it needs to provide. It may be that an assay does not sufficiently mimic neuronal differentiation and more complex models may be required, or that the assay turns out to be too complex for providing information on a certain KE or neurodevelopmental process. In this section, an overview and considerations of options for differentiation will be given.

5.1 Cell culture configuration

A first consideration that applies to every in vitro system is the consideration for the configuration of the stem cell culture (Fig. 1). A general consideration for all differentiation platforms to study later neurodevelopment (i.e. from the NPC stage) is whether to do the differentiation of different cell types from a single NPC line or whether to culture cell types separately and merge them together in a later stage of differentiation. The first option can be chosen to replicate the differentiation as it may happen in vivo. The latter can be chosen to accelerate the differentiation process and to have more control over the culture composition. Microglia evolve from the endodermal cell line and therefore often need to be cultured separately before being added to a neuronal culture (Abreu et al., 2018). A related consideration is the choice between monocultures or co-cultures of neurons and glia. For a simple screen or indeed a very detailed study on the mechanisms of one cell type, a monoculture may be the best option. Differentiation is, however, often guided by the interaction between neurons and glia, and compounds may be metabolized by the glia.

An obvious next consideration is whether to grow cells in 2D or 3D. Along the whole differentiation route, stem cells can be grown and studied in 2D configuration. 2D cultures differentiate well in an often robust and relatively cost-efficient manner, are easier to investigate and more suitable for higher throughput (see Table 4 in (Serafini et al., 2024)). There is the option to combine two cultures using a Transwell, for example, when one wants to couple the neuronal differentiation with a blood-brain barrier (Jackson et al., 2019) or placenta cell model. The cells lack, however, the 3D configuration that is closer to their natural environment. For 3D, there are several culture options, which were recently articulated by Pasca et al. (2022). In a 3D spheroid, cells from separate cultures are combined in a 3D culture, which after combination have little self-organizing capacity. In contrast, organoids do have self-organizing capacity as the cells co-differentiate in the culture by interacting with each other. These organoids can either differentiate in an unguided fashion or can be regionalized towards specific brain regions (see section Brain regions). Going even further, regionalized organoids can be fused together creating assembloids, which can be used to study the effects of compounds on for example, cell migration, axon projection and circuit refinement (Pasca et al., 2022). An option in between 2D and 3D is a so-called 2.5D configuration where stem-cell based cells are initially cultured in a sphere and then plated, which

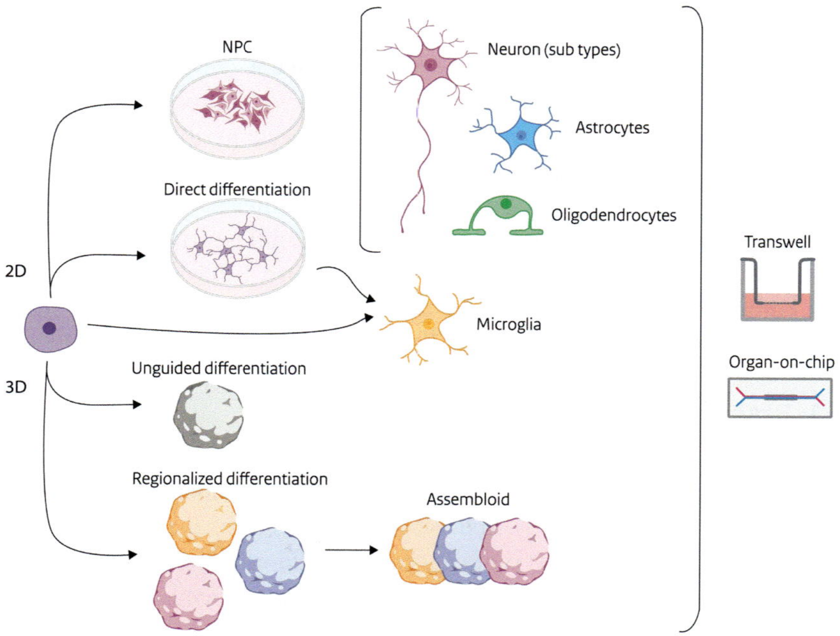

Fig. 1 A schematic overview of considerations for cell culture configurations, starting from hESCs or hiPSCs (purple cell in the middle). Depending on the research question at hand, different configurations and differentiation routes can be chosen. *2D,* two-dimensional; *3D,* three-dimensional; *NPC,* neural progenitor cells. *Adapted from Pasca, S.P., Arlotta, P., Bateup, H.S., Camp, J.G., Cappello, S., Gage, F.H., et al. 2022. A nomenclature consensus for nervous system organoids and assembloids. Nature 609, 907–910.*

causes the cells to grow out of the sphere (e.g. (Baumann et al., 2016)). A final configuration type is an organ-on-chip, which is currently used to a limited extent for DNT (Wang et al., 2018; Koo et al., 2018).

5.2 Cell types

As for cell types, one needs to consider which biological domain should be covered by the assay to mimic the biology for the research question at hand. To study early differentiation, stem cells themselves are obviously more suited. Depending on the configuration, differentiating stem cells can make neural tube like structure (see section Brain regions) or derivatives called neural rosettes (Dreser et al., 2020), which can also be directed towards a more central or peripheral nervous system identity (Lundin et al., 2024). Moving further along the differentiation path towards radial glia, similar assays can be used that are extended in differentiation type. From here, depending on the desired cell types, specific supplements can be added to the media to direct NPCs toward a wide range of neuron subtypes, for example, glutamatergic, GABA-ergic, glycinergic, cholinergic, dopaminergic, serotonergic, motor, (nor)adrenergic, peripheral and enteric neurons, astrocytes, oligodendrocytes, Schwann cells or microglia (Canals et al., 2021; Tao & Zhang, 2016). These supplements sometimes already need to be added before the NPC stage to direct differentiation in the right direction. Protocols for specific differentiations are published in a fast pace and companies are providing more and more standard kits, which add to the standardization and reproducibility of these guided differentiations.

5.3 Brain regions

For some toxicological questions more biological complexity is required, that is not only more specified cell types, but a combination of cell types (either in 2D or 3D) to study toxicological mechanisms that are dependent on interactions between cell types. As for the differentiation of different cell types, the stem cell field is discovering way to culture cells reminiscent of for example, the neural tube (Karzbrun et al., 2021; Xue et al., 2024), hindbrain (Lu et al., 2016), forebrain (Bell et al., 2019), hippocampus (Yu et al., 2014), midbrain (Fiorenzano et al., 2021; Monzel et al., 2017), hindbrain (Lu et al., 2016), cerebellum (Atamian et al., 2024), motor unit (Rimington et al., 2021) and many more. The application of these models is currently quite limited in a toxicological context due to the low throughput of these cultures, which, especially in 3D configurations, require a long time and highly specialized personnel to maintain.

6. Current and future applications of stem cell differentiation models

Stem cell differentiation models can be used for many types of research questions that can range from very mechanistic research to screening purposes. In this final section, we provide two examples where differentiation is being used and can be further leveraged.

6.1 Application of stem cell differentiation models as part of a (regulatory) testing strategy

In regulatory contexts, assessment of neuronal differentiation affected by compounds is one of the endpoints of interest as part of a larger set of tests.

Together with complementary tests, differentiation tests can aid in compound prioritization, hazard identification or risk assessment, depending on the regulatory question at hand and the models used. Neuronal differentiation assays can be used to test for one of the neurodevelopmental processes that have been defined by the Organisation for Economic Co-operation and Development as part of Integrated Approaches to Testing and Assessment for DNT (Sachana et al., 2019). They may represent one or more key events in an adverse outcome pathway (AOP) (Li et al., 2019; Bal-Price et al., 2015). A recent analysis of the AOP landscape that is currently available for DNT showed that many biological processes and adverse outcomes are still missing (Jaylet et al., 2024). Taking a different approach, the DNT-In Vitro Battery (DNT-IVB) contains assays that cover a range of cellular processes that are important for neurodevelopment, including differentiation (OECD, 2023a). However, only general neuronal (Tubulin beta3-positive) cells and oligodendrocytes (sulfatide-positive, detected by the O4 antibody) are currently part of the DNT-IVB. It is recognized that other differentiation paths, such as neuronal subtypes, radial glia and astrocytes are currently missing (OECD, 2023a). Looking specifically at differentiation, there is only one key event (KE) directly related to differentiation available in the AOP-wiki (KE 2229), which means that there is still work to do to better leverage these differentiation assays for test strategy purposes (Jaylet et al., 2024). It should also be mentioned that stem cell-based models can take on more key neurodevelopmental events, as is already the case for some other assays in the DNT-IVB However, they are currently mainly used to study compound effects on relatively basic processes in brain development and have therefore limitations in predicting hazards to human health (Bal-Price et al., 2010). More complex assays, that can be made using stem cell derived cultures, could aid in more human-relevant hazard prediction.

As already discussed, human brain development is very complex and normal behavior and brain functioning is a result of a variety of interactions between neuronal networks and brain regions. Advancements in stem cell models including iPSCs and 3D cell culture systems, such as organoids, have brought a new level of complexity and human relevance to in vitro models, potentially offering more efficient and relevance for DNT. Assays that focus on the differentiation of specific brain areas and the neuronal subtypes and interaction of these cells are therefore needed to better represent the complexity of the developing brain and to better predict DNT capacity of compounds. Using iPSCs with various genetic backgrounds can have more insight into the mechanisms and future inter-species variation can be included by investigating gene-environment interactions in iPSC based in vitro assays as discussed in the section GxE interaction. Again, stem cells and the development of technical tools could provide an important addition to filling these data gaps in future. Despite these all these advancements and developments, capturing the full complexity of brain development in vitro remains a significant scientific challenge.

An interesting question is in which direction stem cell-based in vitro assays may evolve to contribute to a testing strategy. Will relatively simple models that mimic simple key events suffice or do we need to build complex 3D or organotypic cultures for the toxicological questions at hand? This is partly dependent on how fast different fields in toxicology and other disciplines will develop. In the future, (parts of) human physiology may be fully digitalized in what is called the Virtual Human (Piersma et al., 2019). These kind computational tools may require limited input from high-throughput in vitro assays and all the other steps toward human-relevant results may be calculated in silico (DeWoskin et al., 2024; Barham et al., 2024). On the other hand, the (developing) brain is still a poorly understood organ, which hampers capturing this organ in silico. Developing more complex in vitro models is therefore probably the more likely direction taken in the short term. It will also ultimately depend on which depth of information is regarded to be sufficient according to the toxicological community to make safe predictions for human toxicology.

6.2 Future applications of stem cell differentiation models in GxE DNT research

Causes of neuro (developmental) disorders are mostly unknown and multifactorial with interactions between genetic background and environmental factors, amongst others chemical exposure (Persico & Bourgeron, 2006; Weintraub, 2011). Because of the multifactorial nature of the causes of NDDs studies on gene-environment interaction are needed to unravel how exposure chemicals can cause NDDs (Persico & Bourgeron, 2006; Weintraub, 2011). Gene-environment-interaction in DNT research is an area already discussed in 2010 at the International Neurotoxicology Association meeting (INA-12), and although very relevant it is currently not very well investigated in the DNT field. iPSC models with various genetic backgrounds can be used to study the interaction of genetic background with chemical exposure (Finkelstein et al., 2010). There are some examples to study gene-environment interactions in 2D and 3D models in DNT field recently nicely reviewed (Tukker et al., 2021; Suciu et al., 2023; Smirnova & Hartung, 2024; Modafferi et al., 2021). For example, a study showed a potential synergy between mutation in chromodomain helicase DNA binding protein 8 (*CHD8*), a high-risk autism gene, and environmental exposure to an organophosphate pesticide (chlorpyrifos) in an iPSC-derived human 3D brain model (Modafferi et al., 2021). They found that expression of CHD8 protein was significantly lower in CHD8 heterozygous knockout BrainSpheres compared to brainspheres without a mutation. Exposure to chlorpyrifos further reduced CHD8 protein levels further and affected neurite outgrowth. Another recent example combined five iPSC lines into one organoid where the individual lines clustered among their own cells in neural rosettes (Anton-Bolanos et al., 2024). Exposure to valproic acid or ethanol revealed individual susceptibility to these compounds. This platform also exemplifies a strategy to increase throughput to test interindividual variability by combining cell lines in one cell culture.

Tukker et al. showed recently that monogenic high penetrant mutations are ideal for toxicant challenge studies with a wide variety of environmental stressors (Tukker et al., 2021). These models promote mechanistic understandings of gene-environment interactions and biological pathways relevant to both Mendelian and related sporadic complex disease outcomes by creating a sensitized background for relevant environmental risk factors.

This complexity needs further attention in toxicology and one opportunity to do that are with iPSCs with gene mutations or a variety in genetic background. iPSCs from various genetic backgrounds or with very specific genetic mutations can be used to study this. The most efficient way to make specific genetic modifications is by using an RNA-protein complex called CRISPR-Cas. Using a guide RNA-oligomer, the Cas-endonuclease can be used to create double strand DNA breaks at a specific place in the genome, allowing genes to be removed or added at those locations (Mali et al., 2013). Using modified Cas-proteins that introduce single strand breaks ("nicks") in the genome, very precise genetic modifications can be made in a process termed "prime editing" (Chen & Liu, 2023).

Introduction of relevant genetic mutations facilitates the identification and validation of appropriate endpoints in the stem cell differentiation models and enables susceptibility investigations of specific sensitive groups affected by such mutations.

7. Concluding remarks

In this chapter we have discussed that human stem cells, both hESC and hiPSC, can be used to model cell differentiation, an important process in early brain development. Differentiation is a complex process and includes a variety of (neuronal) cell specifications that result in the formation of neuronal circuitries and brain region specific maturation. Researchers increasingly master the differentiation of stem cells to more complex and specialized cell cultures. There is still much to develop and standardize, and the field working hard on getting a better grip on differentiation to make specialized 2D and 3D stem cell models and further develop and apply readouts, including molecular, morphological and functional readouts, to model the complexity and underlying mechanisms of the developing brain. In the light of toxicological applications, protocols need to be robust and harmonized, throughput should be increased and costs should go down for widespread use. While already applied in testing strategies for regulatory purposes, there is a great potential for stem cells to be applied more widely in this field. For that, more complex stem cell models for differentiation should be standardized and added to the current testing strategies for DNT. In the near future, we envision that differentiating iPSCs can be leveraged for gene-environment interaction studies by including a variety of genetic backgrounds to investigate individual susceptibility to chemical exposure. To conclude, studying differentiation in human stem cells has great potential for human-relevant and animal-free for DNT assessment.

Acknowledgements

This research is funded by the Dutch Ministry of Health, Welfare and Sports. We would like to thank Julia Meerman for a critical review of the manuscript and Marcia Oldenburger for making the figure for this chapter.

References

Abreu, C.M., Gama, L., Krasemann, S., Chesnut, M., Odwin-Dacosta, S., Hogberg, H.T., et al., 2018. Microglia increase inflammatory responses in iPSC-derived human brainspheres. Front. Microbiol. 9, 2766.

Afshari, C.A., Hamadeh, H.K., Bushel, P.R., 2011. The evolution of bioinformatics in toxicology: advancing toxicogenomics. Toxicol. Sci. 120 (Suppl. 1), S225–237.

Agrawal, A., Balci, H., Hanspers, K., Coort, S.L., Martens, M., Slenter, D.N., et al., 2024. WikiPathways 2024: next generation pathway database. Nucleic Acids Res. 52, D679–D689.

Ahmad, R., Budnik, B., 2023. A review of the current state of single-cell proteomics and future perspective. Anal. Bioanal. Chem. 415, 6889–6899.

Allegrucci, C., Denning, C.N., Burridge, P., Steele, W., Sinclair, K.D., Young, L.E., 2005. Human embryonic stem cells as a model for nutritional programming: an evaluation. Reprod. Toxicol. 20, 353–367.

Allegrucci, C., Young, L.E., 2007. Differences between human embryonic stem cell lines. Hum. Reprod. Update 13, 103–120.

Andrews, P.W., Benvenisty, N., Mckay, R., Pera, M.F., Rossant, J., Semb, H., et al., 2005. The international stem cell initiative: toward benchmarks for human embryonic stem cell research. Nat. Biotechnol. 23, 795–797.

Anton-Bolanos, N., Faravelli, I., Faits, T., Andreadis, S., Kastli, R., Trattaro, S., et al., 2024. Brain Chimeroids reveal individual susceptibility to neurotoxic triggers. Nature 631, 142–149.

Antón-Bolaños, N., Faravelli, I., Faits, T., Andreadis, S., Kastli, R., Trattaro, S., et al., 2024. Brain Chimeroids reveal individual susceptibility to neurotoxic triggers. Nature 631, 142–149.

Arain, M., Haque, M., Johal, L., Mathur, P., Nel, W., Rais, A., et al., 2013. Maturation of the adolescent brain. Neuropsychiatr. Dis. Treat. 9, 449–461.

Aschner, M., Ceccatelli, S., Daneshian, M., Fritsche, E., Hasiwa, N., Hartung, T., et al., 2017. Reference compounds for alternative test methods to indicate developmental neurotoxicity (DNT) potential of chemicals: example lists and criteria for their selection and use. Altex 34, 49–74.

Ashburner, M., Ball, C.A., Blake, J.A., Botstein, D., Butler, H., Cherry, J.M., et al., 2000. Gene ontology: tool for the unification of biology. The gene ontology consortium. Nat. Genet. 25, 25–29.

Association, A.P. 2013. Diagnostic and Statistical Manual of Mental Disorders, fifth ed. (DSM-5). American Psychiatric Association, 5th edition.

Astle, D.E., Holmes, J., Kievit, R., Gathercole, S.E., 2022. Annual research review: the transdiagnostic revolution in neurodevelopmental disorders. J. Child. Psychol. Psychiatry 63, 397–417.

Atamian, A., Birtele, M., Hosseini, N., Nguyen, T., Seth, A., Del Dosso, A., et al., 2024. Human cerebellar organoids with functional Purkinje cells. Cell Stem Cell 31 (39–51), e6.

Avila Cobos, F., Vandesompele, J., Mestdagh, P., De Preter, K., 2018. Computational deconvolution of transcriptomics data from mixed cell populations. Bioinformatics 34, 1969–1979.

Bal-Price, A., Crofton, K.M., Leist, M., Allen, S., Arand, M., Buetler, T., et al., 2015. International STakeholder NETwork (ISTNET): creating a developmental neurotoxicity (DNT) testing road map for regulatory purposes. Arch. Toxicol. 89, 269–287.

Bal-Price, A.K., Hogberg, H.T., Buzanska, L., Lenas, P., Van Vliet, E., Hartung, T., 2010. In vitro developmental neurotoxicity (DNT) testing: relevant models and endpoints. Neurotoxicology 31, 545–554.

Bal-Price, A., Lein, P.J., Keil, K.P., Sethi, S., Shafer, T., Barenys, M., et al., 2017. Developing and applying the adverse outcome pathway concept for understanding and predicting neurotoxicity. Neurotoxicology 59, 240–255.

Barenys, M., Gassmann, K., Baksmeier, C., Heinz, S., Reverte, I., Schmuck, M., et al., 2017. Epigallocatechin gallate (EGCG) inhibits adhesion and migration of neural progenitor cells in vitro. Arch. Toxicol. 91, 827–837.

Barham, K., Spencer, R., Baker, N.C., Knudsen, T.B., 2024. Engineering a computable epiblast for in silico modeling of developmental toxicity. Reprod. Toxicol. 128, 108625.

Baumann, J., Gassmann, K., Masjosthusmann, S., Deboer, D., Bendt, F., Giersiefer, S., et al., 2016. Comparative human and rat neurospheres reveal species differences in chemical effects on neurodevelopmental key events. Arch. Toxicol. 90, 1415–1427.

Beamish, C.R., Dey, A., Sarkar, S., Rand, M.D., 2023. Chapter Eight - Perspectives for advancing neurotoxicity studies with Drosophila. In: Batista Rocha, J., Aschner, M., Costa, L.G. (Eds.), Advances in Neurotoxicology. Academic Press.

Bell, S., Hettige, N.C., Silveira, H., Peng, H., Wu, H., Jefri, M., et al., 2019. Differentiation of human induced pluripotent stem cells (iPSCs) into an effective model of forebrain neural progenitor cells and mature neurons. Bio. Protoc 9, e3188.

Ben-Ari Fuchs, S., Lieder, I., Stelzer, G., Mazor, Y., Buzhor, E., Kaplan, S., et al., 2016. GeneAnalytics: an integrative gene set analysis tool for next generation sequencing, RNAseq and microarray data. OMICS 20, 139–151.

Blum, J., Masjosthusmann, S., Bartmann, K., Bendt, F., Dolde, X., Dönmez, A., et al., 2023. Establishment of a human cell-based in vitro battery to assess developmental neurotoxicity hazard of chemicals. Chemosphere 311, 137035.

Borodinsky, L.N., Belgacem, Y.H., 2016. Crosstalk among electrical activity, trophic factors and morphogenetic proteins in the regulation of neurotransmitter phenotype specification. J. Chem. Neuroanat. 73, 3–8.

Braun, E., Danan-Gotthold, M., Borm, L.E., Lee, K.W., Vinsland, E., Lönnerberg, P., et al., 2023. Comprehensive cell atlas of the first-trimester developing human brain. Science 382, eadf1226.

Bray, M.A., Singh, S., Han, H., Davis, C.T., Borgeson, B., Hartland, C., et al., 2016. Cell painting, a high-content image-based assay for morphological profiling using multiplexed fluorescent dyes. Nat. Protoc. 11, 1757–1774.

Buzzard, J.J., Gough, N.M., Crook, J.M., Colman, A., 2004. Karyotype of human ES cells during extended culture. Nat. Biotechnol. 22, 381–2; author reply 382.

Canals, I., Quist, E., Ahlenius, H., 2021. Transcription factor-based strategies to generate neural cell types from human pluripotent stem cells. Cell Reprogram. 23, 206–220.

Carpenter, M.K., Inokuma, M.S., Denham, J., Mujtaba, T., Chiu, C.P., Rao, M.S., 2001. Enrichment of neurons and neural precursors from human embryonic stem cells. Exp. Neurol. 172, 383–397.

Cervetto, C., Pistollato, F., Amato, S., Mendoza-De Gyves, E., Bal-Price, A., Maura, G., et al., 2023. Assessment of neurotransmitter release in human iPSC-derived neuronal/glial cells: a missing in vitro assay for regulatory developmental neurotoxicity testing. Reprod. Toxicol. 117, 108358.

Chandrasekaran, A., Avci, H.X., Leist, M., Kobolak, J., Dinnyes, A., 2016. Astrocyte differentiation of human pluripotent stem cells: new tools for neurological disorder research. Front. Cell Neurosci. 10, 215.

Chandrasekaran, A., Avci, H.X., Ochalek, A., Rösingh, L.N., Molnár, K., László, L., et al., 2017. Comparison of 2D and 3D neural induction methods for the generation of neural progenitor cells from human induced pluripotent stem cells. Stem Cell Res. 25, 139–151.

Chen, P.J., Liu, D.R., 2023. Prime editing for precise and highly versatile genome manipulation. Nat. Rev. Genet. 24, 161–177.

Chen, H., Seifikar, H., Larocque, N., Kim, Y., Khatib, I., Fernandez, C.J., et al., 2019. Using a multi-stage hESC model to characterize BDE-47 toxicity during neurogenesis. Toxicol. Sci. 171, 221–234.

Collins, E.S., Hessel, E.V.S., Hughes, S., 2024. How neurobehavior and brain development in alternative whole-organism models can contribute to prediction of developmental neurotoxicity. Neurotoxicology 102, 48–57.

Darnfors, C., Flodin, A., Andersson, K., Caisander, G., Lindqvist, J., Hyllner, J., et al., 2005. High-resolution analysis of the subtelomeric regions of human embryonic stem cells. Stem Cell 23, 483–488.

Davis, A.P., Wiegers, T.C., Johnson, R.J., Sciaky, D., Wiegers, J., Mattingly, C.J., 2023. Comparative toxicogenomics database (CTD): update 2023. Nucleic Acids Res. 51, D1257–D1262.

De Groot, M.W., Westerink, R.H., Dingemans, M.M., 2013. Don't judge a neuron only by its cover: neuronal function in in vitro developmental neurotoxicity testing. Toxicol. Sci. 132, 1–7.

De Leeuw, V.C., Van Oostrom, C.T.M., Westerink, R.H.S., Piersma, A.H., Heusinkveld, H.J., Hessel, E.V.S., 2020. An efficient neuron-astrocyte differentiation protocol from human embryonic stem cell-derived neural progenitors to assess chemical-induced developmental neurotoxicity. Reprod. Toxicol. 98, 107–116.

Delp, J., Gutbier, S., Klima, S., Hoelting, L., Pinto-Gil, K., Hsieh, J.H., et al., 2018. A high-throughput approach to identify specific neurotoxicants/developmental toxicants in human neuronal cell function assays. ALTEX 35, 235–253.

DeWoskin, R.S., Knudsen, T.B., Shah, I., 2024. Virtual models (aka: in silico or computational models). Encyclopedia of Toxicology 9, 779–793.

Di Donato, N., Jean, Y.Y., Maga, A.M., Krewson, B.D., Shupp, A.B., Avrutsky, M.I., et al., 2016. Mutations in CRADD result in reduced Caspase-2-mediated neuronal apoptosis and cause megalencephaly with a rare lissencephaly variant. Am. J. Hum. Genet. 99, 1117–1129.

Diana Neely, M., Xie, S., Prince, L.M., Kim, H., Tukker, A.M., Aschner, M., et al., 2021. Single cell RNA sequencing detects persistent cell type- and methylmercury exposure paradigm-specific effects in a human cortical neurodevelopmental model. Food Chem. Toxicol. 154, 112288.

Dreser, N., Madjar, K., Holzer, A.K., Kapitza, M., Scholz, C., Kranaster, P., et al., 2020. Development of a neural rosette formation assay (RoFA) to identify neurodevelopmental toxicants and to characterize their transcriptome disturbances. Arch. Toxicol. 94, 151–171.

Duggan, D.J., Bittner, M., Chen, Y., Meltzer, P., Trent, J.M., 1999. Expression profiling using cDNA microarrays. Nat. Genet. 21, 10–14.

Estévez-Priego, E., Moreno-Fina, M., Monni, E., Kokaia, Z., Soriano, J., Tornero, D., 2023. Long-term calcium imaging reveals functional development in hiPSC-derived cultures comparable to human but not rat primary cultures. Stem Cell Rep. 18, 205–219.

Finkelstein, Y., Fox, D.A., Aschner, M., Boyes, W.K., 2010. Gene-environment interactions in neurotoxicology: the 12th biennial meeting of the International Neurotoxicology Association. Neurotoxicology 31, 543–544.

Fiorenzano, A., Sozzi, E., Birtele, M., Kajtez, J., Giacomoni, J., Nilsson, F., et al., 2021. Single-cell transcriptomics captures features of human midbrain development and dopamine neuron diversity in brain organoids. Nat. Commun. 12, 7302.

Florio, M., Huttner, W.B., 2014. Neural progenitors, neurogenesis and the evolution of the neocortex. Development 141 (11), 2182–2194.

Frega, M., Van Gestel, S.H., Linda, K., Van Der Raadt, J., Keller, J., Van Rhijn, J.R., et al., 2017. Rapid neuronal differentiation of induced pluripotent stem cells for measuring network activity on micro-electrode arrays. J. Vis. Exp.

Fritsche, E., Barenys, M., Klose, J., Masjosthusmann, S., Nimtz, L., Schmuck, M., et al., 2018. Current availability of stem cell-based in vitro methods for developmental neurotoxicity (DNT) testing. Toxicol. Sci. 165, 21–30.

Giedd, J.N., Blumenthal, J., Jeffries, N.O., Castellanos, F.X., Liu, H., Zijdenbos, A., et al., 1999. Brain development during childhood and adolescence: a longitudinal MRI study. Nat. Neurosci. 2, 861–863.

Harrill, A.H., Ross, P.K., Gatti, D.M., Threadgill, D.W., Rusyn, I., 2009. Population-based discovery of toxicogenomics biomarkers for hepatotoxicity using a laboratory strain diversity panel. Toxicol. Sci. 110, 235–243.

Harrill, J.A., Viant, M.R., Yauk, C.L., Sachana, M., Gant, T.W., Auerbach, S.S., et al., 2021. Progress towards an OECD reporting framework for transcriptomics and metabolomics in regulatory toxicology. Regul. Toxicol. Pharmacol. 125, 105020.

Hessel, E.V.S., Staal, Y.C.M., Piersma, A.H., 2018. Design and validation of an ontology-driven animal-free testing strategy for developmental neurotoxicity testing. Toxicol. Appl. Pharmacol. 354, 136–152.

Hofrichter, M., Nimtz, L., Tigges, J., Kabiri, Y., Schröter, F., Royer-Pokora, B., et al., 2017. Comparative performance analysis of human iPSC-derived and primary neural progenitor cells (NPC) grown as neurospheres in vitro. Stem Cell Res. 25, 72–82.

Hrdlickova, R., Toloue, M., Tian, B., 2017. RNA-Seq methods for transcriptome analysis. Wiley Interdiscip. Rev. RNA 8.

Hu, M., Frega, M., Tolner, E.A., Van Den Maagdenberg, A., Frimat, J.P., Le Feber, J., 2022. MEA-ToolBox: an open source toolbox for standardized analysis of multi-electrode array data. Neuroinformatics 20, 1077–1092.

Huang Da, W., Sherman, B.T., Lempicki, R.A., 2009. Systematic and integrative analysis of large gene lists using DAVID bioinformatics resources. Nat. Protoc. 4, 44–57.

Ingelman-Sundberg, M., Rodriguez-Antona, C., 2005. Pharmacogenetics of drug-metabolizing enzymes: implications for a safer and more effective drug therapy. Philos. Trans. R. Soc. Lond. B Biol. Sci. 360, 1563–1570.

Irie, N., Kuratani, S., 2011. Comparative transcriptome analysis reveals vertebrate phylotypic period during organogenesis. Nat. Commun. 2, 248.

ISSCR, 2023. Standards for Human Stem Cell Use in Research. https://doi.org/10.1016/j.stemcr.2023.08.003.

Jackson, S., Meeks, C., Vezina, A., Robey, R.W., Tanner, K., Gottesman, M.M., 2019. Model systems for studying the blood-brain barrier: applications and challenges. Biomaterials 214, 119217.

Jain, A., Tuteja, G., 2019. TissueEnrich: tissue-specific gene enrichment analysis. Bioinformatics 35, 1966–1967.

Jaylet, T., Coustillet, T., Smith, N.M., Viviani, B., Lindeman, B., Vergauwen, L., et al., 2024. Comprehensive mapping of the AOP-Wiki database: identifying biological and disease gaps. Front. Toxicol. 6, 1285768.

Jessell, T.M., Sanes, J.R., 2000. Development. The decade of the developing brain. Curr. Opin. Neurobiol. 10, 599–611.

Johnstone, A.F., Gross, G.W., Weiss, D.G., Schroeder, O.H., Gramowski, A., Shafer, T.J., 2010. Microelectrode arrays: a physiologically based neurotoxicity testing platform for the 21st century. Neurotoxicology 31, 331–350.

Kanehisa, M., Sato, Y., Kawashima, M., Furumichi, M., Tanabe, M., 2016. KEGG as a reference resource for gene and protein annotation. Nucleic Acids Res. 44, D457–D462.

Karzbrun, E., Khankhel, A.H., Megale, H.C., Glasauer, S.M.K., Wyle, Y., Britton, G., et al., 2021. Human neural tube morphogenesis in vitro by geometric constraints. Nature 599, 268–272.

Kelava, I., Lancaster, M.A., 2016. Dishing out mini-brains: current progress and future prospects in brain organoid research. Dev. Biol. 420, 199–209.

Kerjan, G., Gleeson, J.G., 2007. Genetic mechanisms underlying abnormal neuronal migration in classical lissencephaly. Trends Genet. 23, 623–630.

Kim, J., Koo, B.K., Knoblich, J.A., 2020. Human organoids: model systems for human biology and medicine. Nat. Rev. Mol. Cell Biol. 21, 571–584.

Klose, J., Pahl, M., Bartmann, K., Bendt, F., Blum, J., Dolde, X., et al., 2022. Neurodevelopmental toxicity assessment of flame retardants using a human DNT in vitro testing battery. Cell Biol. Toxicol. 38, 781–807.

Kobolak, J., Teglasi, A., Bellak, T., Janstova, Z., Molnar, K., Zana, M., et al., 2020. Human induced pluripotent stem cell-derived 3D-neurospheres are suitable for neurotoxicity screening. Cells 9.

Koch, K., Bartmann, K., Hartmann, J., Kapr, J., Klose, J., Kuchovska, E., et al., 2022. Scientific validation of human neurosphere assays for developmental neurotoxicity evaluation. Front. Toxicol. 4, 816370.

Koo, Y., Hawkins, B.T., Yun, Y., 2018. Three-dimensional (3D) tetra-culture brain on chip platform for organophosphate toxicity screening. Sci. Rep. 8, 2841.

Kuegler, P.B., Zimmer, B., Waldmann, T., Baudis, B., Ilmjärv, S., Hescheler, J., et al., 2010. Markers of murine embryonic and neural stem cells, neurons and astrocytes: reference points for developmental neurotoxicity testing. Altex 27, 17–42.

La Manno, G., Gyllborg, D., Codeluppi, S., Nishimura, K., Salto, C., Zeisel, A., et al., 2016. Molecular diversity of midbrain development in mouse, human, and stem cells. Cell 167, 566–580.e19.

Landrigan, P.J., Lambertini, L., Birnbaum, L.S., 2012. A research strategy to discover the environmental causes of autism and neurodevelopmental disabilities. Environ. Health Perspect. 120, a258–a260.

Lappalainen, R.S., Salomaki, M., Yla-Outinen, L., Heikkila, T.J., Hyttinen, J.A., Pihlajamaki, H., et al., 2010. Similarly derived and cultured hESC lines show variation in their developmental potential towards neuronal cells in long-term culture. Regen. Med. 5, 749–762.

Li, J., Settivari, R., Lebaron, M.J., Marty, M.S., 2019. An industry perspective: a streamlined screening strategy using alternative models for chemical assessment of developmental neurotoxicity. Neurotoxicology 73, 17–30.

Lickfett, S., Menacho, C., Zink, A., Telugu, N.S., Beller, M., Diecke, S., et al., 2022. High-content analysis of neuronal morphology in human iPSC-derived neurons. STAR. Protoc. 3, 101567.

Lieberman, R., Levine, E.S., Kranzler, H.R., Abreu, C., Covault, J., 2012. Pilot study of iPS-derived neural cells to examine biologic effects of alcohol on human neurons in vitro. Alcohol. Clin. Exp. Res. 36, 1678–1687.

Liu, Z., Huang, R., Roberts, R., Tong, W., 2019. Toxicogenomics: a 2020 vision. Trends Pharmacol. Sci. 40, 92–103.

Lord, C., Elsabbagh, M., Baird, G., Veenstra-Vanderweele, J., 2018. Autism spectrum disorder. Lancet 392, 508–520.

Lu, X., Yang, J., Xiang, Y., 2022. Modeling human neurodevelopmental diseases with brain organoids. Cell Regen. 11, 1.

Lu, J., Zhong, X., Liu, H., Hao, L., Huang, C.T., Sherafat, M.A., et al., 2016. Generation of serotonin neurons from human pluripotent stem cells. Nat. Biotechnol. 34, 89–94.

Lundin, B.F., Knight, G.T., Fedorchak, N.J., Krucki, K., Iyer, N., Maher, J.E., et al., 2024. RosetteArray(R) platform for quantitative high-throughput screening of human neurodevelopmental risk. bioRxiv.

Mali, P., Yang, L., Esvelt, K.M., Aach, J., Guell, M., Dicarlo, J.E., et al., 2013. RNA-guided human genome engineering via Cas9. Science 339, 823–826.

Masjosthusmann, S., Becker, D., Petzuch, B., Klose, J., Siebert, C., Deenen, R., et al., 2018. A transcriptome comparison of time-matched developing human, mouse and rat neural progenitor cells reveals human uniqueness. Toxicol. Appl. Pharmacol. 354, 40–55.

Mcquin, C., Goodman, A., Chernyshev, V., Kamentsky, L., Cimini, B.A., Karhohs, K.W., et al., 2018. CellProfiler 3.0: next-generation image processing for biology. PLoS Biol. 16, e2005970.

Merrick, B.A., 2019. Next generation sequencing data for use in risk assessment. Curr. Opin. Toxicol. 18, 18–26.

Merrick, B.A., Paules, R.S., Tice, R.R., 2015. Intersection of toxicogenomics and high throughput screening in the Tox21 program: an NIEHS perspective. Int. J. Biotechnol. 14, 7–27.

Mitalipova, M.M., Rao, R.R., Hoyer, D.M., Johnson, J.A., Meisner, L.F., Jones, K.L., et al., 2005. Preserving the genetic integrity of human embryonic stem cells. Nat. Biotechnol. 23, 19–20.

Modafferi, S., Zhong, X., Kleensang, A., Murata, Y., Fagiani, F., Pamies, D., et al., 2021. Gene-environment interactions in developmental neurotoxicity: a case study of synergy between chlorpyrifos and chd8 knockout in human brainspheres. Env. Health Perspect. 129, 77001.
Monzel, A.S., Smits, L.M., Hemmer, K., Hachi, S., Moreno, E.L., Van Wuellen, T., et al., 2017. Derivation of human midbrain-specific organoids from neuroepithelial stem cells. Stem Cell Rep. 8, 1144–1154.
Moog, N.K., Entringer, S., Heim, C., Wadhwa, P.D., Kathmann, N., Buss, C., 2017. Influence of maternal thyroid hormones during gestation on fetal brain development. Neuroscience 342, 68–100.
Moradi, S., Mahdizadeh, H., Saric, T., Kim, J., Harati, J., Shahsavarani, H., et al., 2019. Research and therapy with induced pluripotent stem cells (iPSCs): social, legal, and ethical considerations. Stem Cell Res. Ther. 10, 341.
Nichols, J.S.A., 2011. The origin and identity of embryonic stem cells. Development 3–8.
Nobelmedia 2012. The Nobel Prize in Physiology or Medicine. Press Release. https://www.nobelprize.org/prizes/medicine/2012/press-release/.
Nuwaysir, E.F., Bittner, M., Trent, J., Barrett, J.C., Afshari, C.A., 1999. Microarrays and toxicology: the advent of toxicogenomics. Mol. Carcinog. 24, 153–159.
Oberheim, N.A., Wang, X., Goldman, S., Nedergaard, M., 2006. Astrocytic complexity distinguishes the human brain. Trends Neurosci. 29, 547–553.
OECD. 2023a. Initial Recommendations on Evaluation of Data from the Developmental Neurotoxicity (DNT) In-Vitro Testing Battery. OECD Publishing, Paris.
OECD. 2023b. OECD Omics Reporting Framework (OORF): Guidance on Reporting Elements for the Regulatory Use of Omics Data from Laboratory-based Toxicology Studies. OECD Publishing, Paris.
Ogony, J.W., Malahias, E., Vadigepalli, R., Anni, H., 2013. Ethanol alters the balance of Sox2, Oct4, and Nanog expression in distinct subpopulations during differentiation of embryonic stem cells. Stem Cell Dev. 22, 2196–2210.
Okabe, S., Forsberg-Nilsson, K., Spiro, A.C., Segal, M., Mckay, R.D., 1996. Development of neuronal precursor cells and functional postmitotic neurons from embryonic stem cells in vitro. Mech. Dev. 59, 89–102.
Ormel, P.R., Vieira De Sa, R., Van Bodegraven, E.J., Karst, H., Harschnitz, O., Sneeboer, M.A.M., et al., 2018. Microglia innately develop within cerebral organoids. Nat. Commun. 9, 4167.
Ortmann, D., Vallier, L., 2017. Variability of human pluripotent stem cell lines. Curr. Opin. Genet. Dev. 46, 179–185.
Pamies, D., Barreras, P., Block, K., Makri, G., Kumar, A., Wiersma, D., et al., 2017. A human brain microphysiological system derived from induced pluripotent stem cells to study neurological diseases and toxicity. ALTEX 34, 362–376.
Paparella, M., Bennekou, S.H., Bal-Price, A., 2020. An analysis of the limitations and uncertainties of in vivo developmental neurotoxicity testing and assessment to identify the potential for alternative approaches. Reprod. Toxicol. 96, 327–336.
Parmentier, T., James, F.M.K., Hewitson, E., Bailey, C., Werry, N., Sheridan, S.D., et al., 2022. Human cerebral spheroids undergo 4-aminopyridine-induced, activity associated changes in cellular composition and microrna expression. Sci. Rep. 12, 9143.
Parmentier, T., Lamarre, J., Lalonde, J., 2023. Evaluation of neurotoxicity with human pluripotent stem cell-derived cerebral organoids. Curr. Protoc. 3, e744.
Pasca, S.P., Arlotta, P., Bateup, H.S., Camp, J.G., Cappello, S., Gage, F.H., et al., 2022. A nomenclature consensus for nervous system organoids and assembloids. Nature 609, 907–910.
Pasca, A.M., Sloan, S.A., Clarke, L.E., Tian, Y., Makinson, C.D., Huber, N., et al., 2015. Functional cortical neurons and astrocytes from human pluripotent stem cells in 3D culture. Nat. Methods 12, 671–678.

Pennings, J.L., Theunissen, P.T., Piersma, A.H., 2012. An optimized gene set for transcriptomics based neurodevelopmental toxicity prediction in the neural embryonic stem cell test. Toxicology 300, 158–167.

Perrera, V., Martello, G., 2019. How does reprogramming to pluripotency affect genomic imprinting? Front. Cell Dev. Biol. 7, 76.

Persico, A.M., Bourgeron, T., 2006. Searching for ways out of the autism maze: genetic, epigenetic and environmental clues. Trends Neurosci. 29, 349–358.

Piersma, A.H., van Benthem, J., Ezendam, J., Staal, Y.C.M., Kienhuis, A.S., 2019. The virtual human in chemical safety assessment. Curr. Opin. Toxicol. 15, 26–32.

Pistollato, F., Canovas-Jorda, D., Zagoura, D., Price, A., 2017. Protocol for the differentiation of human induced pluripotent stem cells into mixed cultures of neurons and glia for neurotoxicity testing. J. Vis. Exp.

Pistollato, F., Carpi, D., Mendoza-De Gyves, E., Paini, A., Bopp, S.K., Worth, A., et al., 2021. Combining in vitro assays and mathematical modelling to study developmental neurotoxicity induced by chemical mixtures. Reprod. Toxicol. 105, 101–119.

Pistollato, F., De Gyves, E.M., Carpi, D., Bopp, S.K., Nunes, C., Worth, A., et al., 2020. Assessment of developmental neurotoxicity induced by chemical mixtures using an adverse outcome pathway concept. Env. Health 19, 23.

Rantataro, S., Parkkinen, I., Airavaara, M., Laurila, T., 2023. Real-time selective detection of dopamine and serotonin at nanomolar concentration from complex in vitro systems. Biosens. Bioelectron. 241, 115579.

Rao, M.S., Civin, C.I., 2005. Translational research: toward better characterization of human embryonic stem cell lines. Stem Cell 23, 1453.

Rappaport, N., Twik, M., Plaschkes, I., Nudel, R., Iny Stein, T., Levitt, J., et al., 2017. MalaCards: an amalgamated human disease compendium with diverse clinical and genetic annotation and structured search. Nucleic Acids Res. 45, D877–D887.

Rasband: Schindelin, J., Arganda-Carreras, I., Frise, E., Kaynig, V., Longair, M., Pietzsch, T., et al., 2012. Fiji: an open-source platform for biological-image analysis. *Nat Methods.* 9, 676–682.

Reubinoff, B.E., Pera, M.F., Fong, C.Y., Trounson, A., Bongso, A., 2000. Embryonic stem cell lines from human blastocysts: somatic differentiation in vitro. Nat. Biotechnol. 18, 399–404.

Rice, D., Barone JR, S., 2000. Critical periods of vulnerability for the developing nervous system: evidence from humans and animal models. Env. Health Perspect. 108 (Suppl 3), 511–533.

Rimington, R.P., Fleming, J.W., Capel, A.J., Wheeler, P.C., Lewis, M.P., 2021. Bioengineered model of the human motor unit with physiologically functional neuromuscular junctions. Sci. Rep. 11, 11695.

Robinson, J.F., Piersma, A.H., 2013. Toxicogenomic approaches in developmental toxicology testing. Methods Mol. Biol. 947, 451–473.

Romero, J.C., Berlinicke, C., Chow, S., Duan, Y., Wang, Y., Chamling, X., et al., 2022. Oligodendrogenesis and myelination tracing in a CRISPR/Cas9-engineered brain microphysiological system. Front. Cell Neurosci. 16, 1094291.

Sachana, M., Bal-Price, A., Crofton, K.M., Bennekou, S.H., Shafer, T.J., Behl, M., et al., 2019. International regulatory and scientific effort for improved developmental neurotoxicity testing. Toxicol. Sci. 167, 45–57.

Sahakyan, V., Duelen, R., Tam, W.L., Roberts, S.J., Grosemans, H., Berckmans, P., et al., 2018. Folic acid exposure rescues spina bifida aperta phenotypes in human induced pluripotent stem cell model. Sci. Rep. 8, 2942.

Sandstrom, J., Eggermann, E., Charvet, I., Roux, A., Toni, N., Greggio, C., et al., 2017. Development and characterization of a human embryonic stem cell-derived 3D neural tissue model for neurotoxicity testing. Toxicol. Vitro 38, 124–135.

Santello, M., Toni, N., Volterra, A., 2019. Astrocyte function from information processing to cognition and cognitive impairment. Nat. Neurosci. 22, 154–166.

Sayers, E.W., Beck, J., Bolton, E.E., Brister, J.R., Chan, J., Comeau, D.C., et al., 2024. Database resources of the National Center for Biotechnology Information. Nucleic Acids Res. 52, D33–D43.

Schmidt, B.Z., Lehmann, M., Gutbier, S., Nembo, E., Noel, S., Smirnova, L., et al., 2017. In vitro acute and developmental neurotoxicity screening: an overview of cellular platforms and high-throughput technical possibilities. Arch. Toxicol. 91, 1–33.

Schulpen, S.H., Pennings, J.L., Tonk, E.C., Piersma, A.H., 2014. A statistical approach towards the derivation of predictive gene sets for potency ranking of chemicals in the mouse embryonic stem cell test. Toxicol. Lett. 225, 342–349.

Schwartz, M.P., Hou, Z., Propson, N.E., Zhang, J., Engstrom, C.J., Santos Costa, V., et al., 2015. Human pluripotent stem cell-derived neural constructs for predicting neural toxicity. Proc. Natl Acad. Sci. U S A 112, 12516–12521.

Semple, B.D., Blomgren, K., Gimlin, K., Ferriero, D.M., Noble-Haeusslein, L.J., 2013. Brain development in rodents and humans: identifying benchmarks of maturation and vulnerability to injury across species. Prog. Neurobiol. 106-107, 1–16.

Serafini, M.M., Sepehri, S., Midali, M., Stinckens, M., Biesiekierska, M., Wolniakowska, A., et al., 2024. Recent advances and current challenges of new approach methodologies in developmental and adult neurotoxicity testing. Arch. Toxicol. 98, 1271–1295.

Shadlaghani, A., Farzaneh, M., Kinser, D., Reid, R.C., 2019. Direct electrochemical detection of glutamate, acetylcholine, choline, and adenosine using non-enzymatic electrodes. Sensors (Basel) 19 (3), 447.

Shefchek, K.A., Harris, N.L., Gargano, M., Matentzoglu, N., Unni, D., Brush, M., et al., 2020. The monarch initiative in 2019: an integrative data and analytic platform connecting phenotypes to genotypes across species. Nucleic Acids Res. 48, D704–D715.

Sherman, B.T., Hao, M., Qiu, J., Jiao, X., Baseler, M.W., Lane, H.C., et al., 2022. DAVID: a web server for functional enrichment analysis and functional annotation of gene lists (2021 update). Nucleic Acids Res 50. pp. W216–W221.

Sirenko, O., Parham, F., Dea, S., Sodhi, N., Biesmans, S., Mora-Castilla, S., et al., 2019. Functional and mechanistic neurotoxicity profiling using human iPSC-derived neural 3D cultures. Toxicol. Sci. 167, 58–76.

Smirnova, L., Hartung, T., 2024. The promise and potential of brain organoids. Adv. Healthc. Mater., e2302745. https://doi.org/10.1002/adhm.202302745.

Sokol, D.K., Lahiri, D.K., 2023. Neurodevelopmental disorders and microcephaly: how apoptosis, the cell cycle, tau and amyloid-β precursor protein APPly. Front. Mol. Neurosci. 16, 1201723.

Spitzer, N.C., 2012. Activity-dependent neurotransmitter respecification. Nat. Rev. Neurosci. 13, 94–106.

Spitzer, N.C., 2015. DEVELOPMENTAL NEUROSCIENCE. Neurotransmitter-tailored dendritic trees. Science 350, 510–511.

Stein, J.L., De La Torre-Ubieta, L., Tian, Y., Parikshak, N.N., Hernandez, I.A., Marchetto, M.C., et al., 2014. A quantitative framework to evaluate modeling of cortical development by neural stem cells. Neuron 83, 69–86.

Stelzer, G., Rosen, N., Plaschkes, I., Zimmerman, S., Twik, M., Fishilevich, S., et al., 2016. The genecards suite: from gene data mining to disease genome sequence analyses. Curr. Protoc. Bioinforma. 54, 1 30 1–1 30 33.

Stepien, B.K., Wielockx, B., 2024. From vessels to neurons-the role of hypoxia pathway proteins in embryonic neurogenesis. Cells 13.

Stiles, J., Jernigan, T.L., 2010. The basics of brain development. Neuropsychol. Rev. 20, 327–348.

Subramanian, A., Tamayo, P., Mootha, V.K., Mukherjee, S., Ebert, B.L., Gillette, M.A., et al., 2005. Gene set enrichment analysis: a knowledge-based approach for interpreting genome-wide expression profiles. Proc. Natl Acad. Sci. U S A 102, 15545–15550.

Suciu, I., Pamies, D., Peruzzo, R., Wirtz, P.H., Smirnova, L., Pallocca, G., et al., 2023. G × E interactions as a basis for toxicological uncertainty. Arch. Toxicol. 97, 2035–2049.

Szklarczyk, D., Kirsch, R., Koutrouli, M., Nastou, K., Mehryary, F., Hachilif, R., et al., 2023. The STRING database in 2023: protein-protein association networks and functional enrichment analyses for any sequenced genome of interest. Nucleic Acids Res. 51, D638–D646.

Takahashi, K., Tanabe, K., Ohnuki, M., Narita, M., Ichisaka, T., Tomoda, K., et al., 2007. Induction of pluripotent stem cells from adult human fibroblasts by defined factors. Cell 131, 861–872.

Takahashi, K., Yamanaka, S., 2006. Induction of pluripotent stem cells from mouse embryonic and adult fibroblast cultures by defined factors. Cell 126, 663–676.

Talens-Visconti, R., Sanchez-Vera, I., Kostic, J., Perez-Arago, M.A., Erceg, S., Stojkovic, M., et al., 2011. Neural differentiation from human embryonic stem cells as a tool to study early brain development and the neuroteratogenic effects of ethanol. Stem Cell Dev. 20, 327–339.

Tao, Y., Zhang, S.C., 2016. Neural subtype specification from human pluripotent stem cells. Cell Stem Cell 19, 573–586.

Tchieu, J., Zimmer, B., Fattahi, F., Amin, S., Zeltner, N., Chen, S., et al., 2017. A modular platform for differentiation of human PSCs into all major ectodermal lineages. Cell Stem Cell 21 (399-410), e7.

Theunissen, P.T., Robinson, J.F., Pennings, J.L., De Jong, E., Claessen, S.M., Kleinjans, J.C., et al., 2012. Transcriptomic concentration-response evaluation of valproic acid, cyproconazole, and hexaconazole in the neural embryonic stem cell test (ESTn). Toxicol. Sci. 125, 430–438.

Toh, M.F., Brooks, J.M., Strassmaier, T., Haedo, R.J., Puryear, C.B., Roth, B.L., et al., 2020. Application of high-throughput automated patch-clamp electrophysiology to study voltage-gated ion channel function in primary cortical cultures. SLAS Discov. 25, 447–457.

Tukker, A.M., Bowman, A.B., 2024. Application of single cell gene expression technologies to neurotoxicology. Curr. Opin. Toxicol. 37.

Tukker, A.M., Royal, C.D., Bowman, A.B., Mcallister, K.A., 2021. The impact of environmental factors on monogenic mendelian diseases. Toxicol. Sci. 181, 3–12.

Tyzack, G., Lakatos, A., Patani, R., 2016. Human stem cell-derived astrocytes: specification and relevance for neurological disorders. Curr. Stem Cell Rep. 2, 236–247.

Valensisi, C., Andrus, C., Buckberry, S., Doni Jayavelu, N., Lund, R.J., Lister, R., et al., 2017. Epigenomic landscapes of hESC-derived neural rosettes: modeling neural tube formation and diseases. Cell Rep. 20, 1448–1462.

Van Dartel, D.A., Pennings, J.L., Hendriksen, P.J., Van Schooten, F.J., Piersma, A.H., 2009. Early gene expression changes during embryonic stem cell differentiation into cardiomyocytes and their modulation by monobutyl phthalate. Reprod. Toxicol. 27, 93–102.

Verstraelen, P., Garcia-Diaz Barriga, G., Verschuuren, M., Asselbergh, B., Nuydens, R., Larsen, P.H., et al., 2020. Systematic quantification of synapses in primary neuronal culture. iScience 23, 101542.

Visan, A., Hayess, K., Sittner, D., Pohl, E.E., Riebeling, C., Slawik, B., et al., 2012. Neural differentiation of mouse embryonic stem cells as a tool to assess developmental neurotoxicity in vitro. Neurotoxicology 33, 1135–1146.

Waldhorn, I., Turetsky, T., Steiner, D., Gil, Y., Benyamini, H., Gropp, M., et al., 2022. Modeling sex differences in humans using isogenic induced pluripotent stem cells. Stem Cell Rep. 17, 2732–2744.

Waldmann, T., Grinberg, M., König, A., Rempel, E., Schildknecht, S., Henry, M., et al., 2017. Stem cell transcriptome responses and corresponding biomarkers that indicate the transition from adaptive responses to cytotoxicity. Chem. Res. Toxicol. 30, 905–922.

Wang, C., Gong, B., Bushel, P.R., Thierry-Mieg, J., Thierry-Mieg, D., Xu, J., et al., 2014. The concordance between RNA-seq and microarray data depends on chemical treatment and transcript abundance. Nat. Biotechnol. 32, 926–932.

Wang, Y., Wang, L., Zhu, Y., Qin, J., 2018. Human brain organoid-on-a-chip to model prenatal nicotine exposure. Lab. Chip 18, 851–860.

Weintraub, K., 2011. The prevalence puzzle: Autism counts. Nature 479, 22–24.

Xue, X., Kim, Y.S., Ponce-Arias, A.I., O'laughlin, R., Yan, R.Z., Kobayashi, N., et al., 2024. A patterned human neural tube model using microfluidic gradients. Nature 628, 391–399.

Yla-Outinen, L., Heikkila, J., Skottman, H., Suuronen, R., Aanismaa, R., Narkilahti, S., 2010. Human cell-based micro electrode array platform for studying neurotoxicity. Front. Neuroeng. 3.

Yoon, S.J., Elahi, L.S., Pasca, A.M., Marton, R.M., Gordon, A., Revah, O., et al., 2019. Reliability of human cortical organoid generation. Nat. Methods 16, 75–78.

Yu, D.X., Di Giorgio, F.P., Yao, J., Marchetto, M.C., Brennand, K., Wright, R., et al., 2014. Modeling hippocampal neurogenesis using human pluripotent stem cells. Stem Cell Rep. 2, 295–310.

Zhang, W., Jiang, J., Xu, Z., Yan, H., Tang, B., Liu, C., et al., 2023. Microglia-containing human brain organoids for the study of brain development and pathology. Mol. Psychiatry 28, 96–107.

Zhang, W., Yu, Y., Hertwig, F., Thierry-Mieg, J., Zhang, W., Thierry-Mieg, D., et al., 2015. Comparison of RNA-seq and microarray-based models for clinical endpoint prediction. Genome Biol. 16, 133.

Zhong, X., Harris, G., Smirnova, L., Zufferey, V., Sá, R., Baldino Russo, F., et al., 2020. Antidepressant paroxetine exerts developmental neurotoxicity in an iPSC-derived 3D human brain model. Front. Cell Neurosci. 14, 25.

CHAPTER SIX

The role of stem cells in the study and treatment of neurodegenerative diseases with environmental etiology

Ribhav Mishra* **and Aaron B. Bowman**
School of Health Sciences, Purdue University, West Lafayette, IN, United States
*Corresponding author. e-mail address: mishr233@purdue.edu

Contents

Abstract

Implications of environmental toxins and other toxic substances on human brain health has shown the importance of resident brain stem cells. Various findings have indicated that toxicants/toxins can affect different cellular mechanisms in brain-related stem cells like proliferation, differentiation, DNA damage repair, and apoptosis. The effect of neurotoxicity can result in deleterious consequences on the progeny of these stem cells like neurons, glia and other supporting brain cells and the effects may finally manifest in more severe forms as neurodegenerative or

Advances in Neurotoxicology, Volume 12
ISSN 2468-7480, https://doi.org/10.1016/bs.ant.2024.07.001

neurodevelopmental diseases and disorders. In addition, human stem cells can model brain disease by generating patient-derived models of the neuronal systems. This chapter aims to address how environmental toxicants are contributing to disease risk by directly altering the biology of neural stem cells and their differentiated progeny. Thus, neural stem cells contribute to a variety of neurological diseases and disorders like Alzheimer Disease, Parkinson Disease, Amyotrophic lateral sclerosis, and others. The chapter analyses the affected molecular pathways in stem cell derived neurons and neural stem cells themselves, due to environmental toxin/toxicant exposures and the role of such mechanisms in the etiology of neurodegenerative diseases (NDDs). Furthermore, we evaluate the value of stem cell-based neuro-glial models to address experimental questions related to neurodegenerative phenotypes and the use of stem cells and their derivatives as therapeutic tools, and the challenges faced in using such approaches. The chapter concludes by examining the crucial role stem cells and stem-cell based models play in understanding the importance of gene-environment interactions on the development of neurodegenerative disorders and applying stem cell therapy can be a potential future effective treatment strategy to treat environmental toxicants generated NDDs.

1. Introduction

1.1 An overview of neurodegenerative diseases (NDDs)

Neurodegenerative diseases (NDDs) are a group of diseases in which brain functions are progressively lost over time mainly due to degeneration of the brain anatomy and brain cells (Amanullah et al., 2017; Andreone et al., 2020). One of the very important brain cells that are found to be affected by the NDDs are the neurons of the brain. The loss of the neurons in the brain can lead to various debilitating conditions like declining memory, decreasing cognition, and loss of motor skills (Lamptey et al., 2022; Jellinger, 2010). Such deficits in brain function are also observed in various types of NDDs like Alzheimer Disease (AD), Parkinson Disease (PD), and Amyotrophic lateral sclerosis (ALS). Although in different NDDs brain regions affected can be different, many of these NDDs share some commonly affected pathological and cellular mechanisms like protein aggregation, oxidative stress, endoplasmic stress, neuroinflammation, and apoptosis (Mishra et al., 2020; Bowman et al., 2011).

NDDs are a significant burden on the global population with several etiologies being identified as either familial (e.g., genetic) or sporadic (e.g., environmental) (Huang et al., 2023; Feigin et al., 2020). While, genetic factors are major contributors to the development of NDDs, emerging evidence suggests that toxic environmental interactions with humans also play significant roles in the generation of such NDDs (Ayeni et al., 2022; Ruffini et al., 2020). Exposure of an environmental neurotoxicant/neurotoxin to a human population can be categorized as (a) acute: In acute environmental contamination, shortly after neurotoxicant exposure, days to weeks, neurological symptoms appear however such phenotypes may or may not be caused due to cell loss. However, in other (b) chronic exposures; the exposure would occur from weeks to years and sometimes in decades and would result in changes like cellular dysfunctions, cellular loss, and such kind of neurotoxicant exposures are found to be likely cause of NDDs (Nabi and Tabassum, 2022; Genuis and Kelln, 2015; Modgil et al., 2014).

Although environmental contaminants are found to induce NDD-like effects, it is interesting to note that humans in their life span get exposed to a variety of environmental chemicals but only a few of such toxin/toxicants have clearly been shown to be capable of inducing NDDs. Many such toxicants are found to have common characteristics and affect brain cells by crossing the blood-brain barrier, disrupting cellular transport, inducing cell apoptosis, impacting cell organelle functions, and getting metabolized to harmful products inside cells (Cresto et al., 2023; Barrios-Arpi et al., 2022; Harischandra et al., 2019). Furthermore, single neurotoxicants have been implicated in the induction of more than one type of NDD, potentially due to multiple different mechanisms of action of the neurotoxicants, suggesting a single neurotoxicant can induce different kinds of neuropathological changes in different contexts or in different individuals (Gorell et al., 1999; Eid et al., 2016).

1.2 Different classes of environmental toxicants

Environmental toxins/toxicants are toxic materials that are either present naturally or are anthropogenic in nature. There are varieties of environmental toxins like heavy metals, persistent organic pollutants, pesticides, air pollutants, endocrine disruptors, industrial chemicals, and radioactive contaminants. In the text below we have described a few of the common environmental contaminants (heavy metals, persistent organic pollutants and pesticides) that are known to have a significant impact on human brain health and are commonly present in our environment and people are more prone to exposure to them and more likely to have effects on their brains due to these exposures.

Heavy metals are metallic elements that are of high density and of high molecular weight. Heavy metals are naturally present and are found in the earth's crust. However, human exposure to heavy metals is elevated due to different kinds of human activities like mining, smelting, industrial work, agricultural, and domestic use of heavy metals (He et al., 2005). Heavy metals can cause environmental contamination by different exposure routes like corrosion, leaching, soil erosion, etc. (Nriagu, 1989). Natural processes like volcanic eruption and weather changes can also lead to heavy metal leaks into the environment resulting in heavy metal pollution (Fergusson, 1990; Bradl, 2005). Industrial refineries, coal burning, fuel burning, and nuclear power stations are significant heavy metal contributors to the environment (Arruti et al., 2010; Sträter et al., 2010). Some heavy metals like cobalt, copper, chromium, iron, magnesium, manganese, molybdenum, nickel, selenium, and zinc are needed in small quantities for essential physiological functions of the human body and a deficiency in their levels would lead to pathological conditions (Who, 2020; Witkowska et al., 2021; Singh et al., 2011). However, overexposure to these heavy metals is found to cause different types of metabolic, neurodevelopmental, neurodegenerative, and cancer- associated diseases (Yavuz et al., 2018; Chowdhury et al., 2018; Xu et al., 2021; Chen et al., 2016; Kim et al., 2015).

Persistent organic pollutants (POPs) are lipophilic chemicals that bioaccumulate in human and animal tissues and these pollutants are released in the environment by industrial activity (Pumarega et al., 2016; Tian et al., 2020). Some examples of such POPs are chemicals like polychlorinated dibenzo-p-dioxins (PCDDs), organochlorine pesticides, and polychlorinated biphenyls (Lee et al., 2018b). These POPs are internationally agreed to be released in a limited amount in the environment but the same are not monitored for environmental exposure (Kumari et al., 2021). Exposure to POPs have been linked with different kinds of pathologies like metabolic, neurotoxicity, and cancer (Lee et al., 2018a; Pessah et al., 2019; Zhang et al., 2021). Some POPs are intentionally produced like DDT, Chlordane, Dieldrin, etc, whereas some are by-products of industrial products like Hexachlorobenzene, and PCBs. POPs tend to resist environmental degradation and hence they tend to travel longer distances by air, water, soil (Göktaş and Macleod, 2016).

Pesticides are the chemicals that protect crops/food from pests, and are often neurotoxicants to at least the pests themselves, and due to evolutionary conservation often to people and other mammals even if to a lesser extent. The application of pesticides is observed to increase crop yield and help in the management of foods suitable for human consumption (Nicolopoulou-Stamati et al., 2016). However, it is observed that some of these chemicals tend to be of serious concern for human health as they can lead to pathological conditions of inflammation, dermatological diseases, metabolic syndromes, and neurological disorders (Gangemi et al., 2016).

There are still many environmental pollutants or toxins that affect human health in general but describing every pollutant is out of scope for this chapter and there are excellent reviews available to understand the effect of such chemicals on human health (Fuller et al., 2022; González et al., 2019; Aravindan et al., 2024; Shrivastav et al., 2024). In the upcoming section of this chapter, our focus will be more on finding the implications of the effects of environmental toxins on different aspects of stem cells in brain including its growth, differentiation, and how such effects lead to different types of pathological alterations in stem cells derived neurons and can cause NDDs.

1.3 Neuropathological effects of environmental contaminants on stem cell biology and its importance in NDDs

Stem cells have the ability for unlimited self-renewal, as well as they have the potential to produce a multitude of different types of differentiated mature cells (Zakrzewski et al., 2019). Neural stem cells (NSCs) are brain stem cells and their differentiation leads to different brain cells such as neurons, and glia (Ding et al., 2020). Stem cells impacted by environmental toxins can have deficits in developing into cells of the nervous system and hence can lead to pathological symptoms of neurodevelopment and neurodegeneration. Many reports suggest that different environmental toxicants e.g. Heavy metals, organic pollutants, pesticides, industrial chemicals, etc. affect the stem cells proliferative and regenerative potential leading to neurodegenerative and neurodevelopmental defects (Tiwari et al., 2015; Li et al., 2013; Senut et al., 2014).

A plausible explanation of the impact of environmental factors on stem cells, including NSCs, is that the variety of development potential, and the plethora of cellular processes in these cells make them prone to impacts by environmental exposures and these effects are then magnified by the breadth of developmental trajectories of these cells. One study found that the Bisphenol-A (BPA), an endocrine disruptor molecular present in the coating of packaged foods affects the Wnt/β-catenin signaling pathways in NSCs. BPA decreases the levels of cellular β-catenin and p-GSK-3β in NSCs, which causes reduced NSCs proliferation, and hence produces defective neuronal differentiation in the hippocampus and subventricular zone (Tiwari et al., 2015). Another finding demonstrated that exposure to the pesticides Paraquat and Manabe in combination can lead to a decreased proliferation of NSCs due to decreased expression of cell cycle genes such as cyclin D1, Rb1, P19, and cyclin D2 (Colle et al., 2018). In another publication, nanoceria a diesel additive, commonly used in industrial settings was found to hinder neuronal differentiation by inhibiting the expression of β3 tubulin in NSCs showing the differentiation inhibiting potential of environmental toxicants (Gliga et al., 2017). All these findings suggest that NSCs are highly sensitive to environmental toxicants as these toxins affects the critical mechanisms of NSCs differentiation and proliferation. Finally, the effect of such toxins on NSCs differentiation and proliferation may manifest as NSCs-derived neuronal abnormality leading to neurological disorders.

Along with the effect on stem cell proliferation and differentiation, environmental toxins/toxicants can induce abnormalities in the functions of stem cell derived progeny in the brain. In one study performed in our own lab, we found the environmentally relevant concentrations of MeHg can cause abnormalities in the differentiation of the glutamatergic (GLUergic) neurons derived from the iPSCs. The study provided evidence that MeHg-exposed stem cells differentiated in GLUergic neurons have an abnormality in the expression of neuronal markers like MAP2, DCX and TBR1 suggesting abnormal neurons generation (Prince et al., 2021). In NDDs like PD, and AD, significant loss in the population of neurons, glia, and in other supporting cells of brain occurs, however, in a healthy environment the same reduction in the brain cells may be replenished with the help of NSCs (Deture and Dickson, 2019; Mangalmurti and Lukens, 2022; Cai et al., 2017). However, as explained further by these authors, due to the neurotoxic influence of environmental toxins on stem cell differentiation and proliferation capacity, such protective effects of the stem cells in NDDs are lost (Addae et al., 2013; Colle et al., 2018).

Similarly, it has been shown in numerous studies that exposure of stem cells, including NSCs to toxic chemicals can cause functional and other characteristic changes in the stem cell-derived neurons of the brain like loss in neuronal synapse formation, neuronal axonal transport, retraction of neuronal dendrites, changes in neuronal firing and thus contributing to either neurodevelopmental or neurodegenerative conditions. A recent study shows the cyanobacteria, and microscopic algae produce a neurotoxicant β-N-methylamino-L-alanine (BMAA) which can act as a risk factor for ALS or PD pathology. The study proposed, exposure of hippocampal NSCs to the BMAA toxin negatively affects the neuronal synapse formation, differentiation, survival, and methylation (Pierozan et al., 2020).

More work in a separate study shows silver nanoparticle exposed stem cells when differentiated into neurons have lower levels of important synaptic proteins like PSD95 and synaptophysin and are prone to degeneration (Repar et al., 2018). Furthermore, another study proposed that exposure to the mixture of five chemicals i.e. (diethylphosphate, octachlorodipropyl ether, cotinine, selenium, and mercury) can cause a significant reduction in the size of the stem cell based model, neurospheres (cell aggregates of NSCs and progenitor cells), and lead to altered gene expression and a decrease in neural differentiation (Da Silva Siqueira et al., 2021). In the same study, it was found that the same chemical mixture can modify the DNA methylation of the genome and hence the work showed evidence of the epigenetic (additional inheritable changes to genome) effects of environmental toxins in the stem cell biology (Arai and Nishino, 2023). Collectively, all these cited studies have significantly contributed to current understanding of the harmful effects of environmental toxicants on the stem cells and their neuro-glial derivates key cellular functions. Hence, a loss in activity of stem cells and their progeny cells due to such effects may lead to pathological conditions of NDDs and other brain diseases.

Furthermore, many studies have now provided sufficient evidence in support of how the exposure of stem cells and NSCs to environmentally relevant neurotoxicants can cause different neuronal and glial cell cellular pathway anomalies like elevated reactive oxygen species production, protein aggregation, cellular transport mechanisms, and mitochondrial dysfunction. In one finding, tetrabromobisphenol A (TBBPA), a toxic substance commonly found in plastics can induce apoptosis in NSCs and the same condition generate oxidative stress and mitophagy at the in vitro level in primary mouse neuronal culture (Cho et al., 2020). Another study performed by the same group found the toxic effects of TBBPA in mouse hippocampus and it decreased the NSCs population in hippocampus leading to impairment in memory formation (Kim et al., 2017). Moreover, in another study, a knockout of Glyoxalase I in induced pluripotent stem cell (iPSC) derived neurons makes them susceptible to carbonyl stress from the environment. The carbonyl exposure resulted in mitochondrial impairment and generation of oxidative species in the stem cells and the same leads to cell death (Hara et al., 2021). Hence, the above results from different research groups have shown how environmental pollutants can cause dysregulation in brain stem cells and their neuronal progeny and thereby contribute to the etiology of NDD and other brain health issues. Finding solutions to these problems may require additional efforts from the scientists including devising new targets in such affected mechanisms that can be used for alleviating the toxic effects of these environmental contaminants.

Several of the altered stem cell mechanisms noted above share features with known irregularities of cellular pathways identified in different kinds of NDDs like AD, PD, and ALS (Mishra et al., 2021). Hence, it strongly gives evidence of the importance of stem cells, particularly NSCs and their biology in contributing to NDDs when they are exposed to relevant environmental risk factors. Many of the alterations to functional and cellular pathways within NSC and their derived progeny are observed to occur at the transcriptomic, proteomics, or epigenetic levels (Waldmann et al., 2017). In Fig. 1, we provide a schematic overview of the use of patient derived stem cells (adult stem cells, ESCs and iPSCs) in modeling the progression of NDDs. Moreover, in Fig. 1, we have also illustrated different stages of differentiation for the stem cells which finally leads to generation of neuro-glial networks in vitro and how the same neuronal/neuro-glial cultures can be used for various applications like drug screening, disease modeling and exploring novel pathways to better understand the pathology of different neurodegenerative disorders. A detailed explanation of the effects of various types of environmental factors in generating different NDDs with emphasis on NDDs, including AD, PD, and ALS, and due to harmful impacts on NSCs and other stem cells is discussed in the next section of our text.

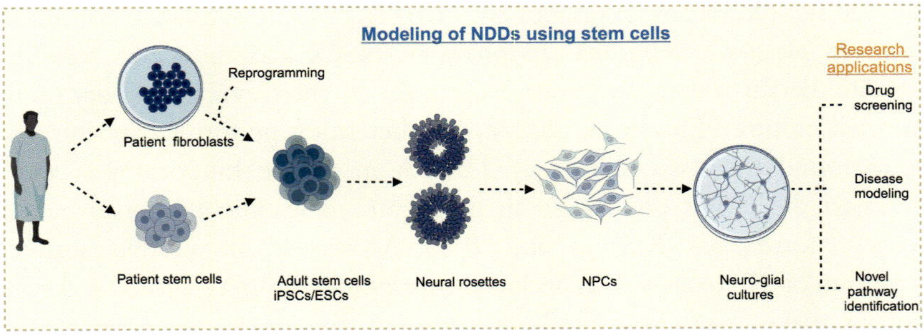

Fig. 1 Studying neurodegenerative diseases with patient-derived stem cells: This figure illustrates how patient derived stem cells (e.g. adult stem cells, ESCs, or iPSCs) can be differentiated to neuro-glial networks. During the differentiation process these cells progress through different stages including neural rosette formation and neural progenitor cells (NPCs), etc. Finally, the differentiated neuronal and glial cells can be used for drug screening, disease modeling or identifying novel pathways in NDDs. *Figure created with BioRender.com.*

2. Effect of environmental toxins/toxicants on NSCs that may contribute to the etiology of different types of NDDs

Neurotoxicants are observed to be contributing factors in the development of different kinds of NDDs. Different reports have indicated that these environmental factors impact NSCs in the lifespan of a healthy human starting from early embryological development and leading to neurological conditions which finally manifest in the form of NDDs. We describe next the mechanisms and association of specific neurodegenerative conditions with the role of NSCs and other brain stem cells:

2.1 Alzheimer disease (AD)

AD is the most common form of dementia and the disease is identified as a slowly progressing neurodegeneration. The anatomical studies performed in AD patients have shown an accumulation of amyloid-beta (Aβ) peptide in the brain with characteristic development of plaques and tangles in the brain (Silva et al., 2019). In AD pathology, a substantial loss of neurons in the brain is observed (Breijyeh and Karaman, 2020). Currently, there are more than 50 million AD patients in the whole-world with the number predicted to increase 3 times by 2050 (Breijyeh and Karaman, 2020). Findings we discuss here next have indicated the critical role of stem cells and the impact of environmental toxins on stem cells on AD progression.

The importance of stem cell-based models in understanding the development of AD due to neurotoxicant exposures is exemplified in the following study. Pb exposure to iPSC-derived GLUergic neurons can lead to the development of AD-like symptoms in these stem-cell derived cortical neurons (Xie et al., 2023). The authors demonstrated that Pb affects Ca^{2+} ion homeostasis in the iPSC-derived neurons and leads to the variety of different phenotypes resembling AD-like neuropathobiology. The work also showed a low concentration of Pb exposure to early progenitor cells derived from human iPSCs for 48 h mimics an early developmental exposure effect of Pb. Also observed were a more anisotropic nuclear shape after developmental Pb exposure in immature neurons and the same persists in mature neurons. Early Pb exposure also affected the changes to Ca^{2+} ion responsive genes in the neurons as identified by RNA-seq analysis. Moreover, developmental Pb exposure decreases the level of pre-synaptic protein Synapsin1 and HOMER1 which indicate lowering of synapse formation between neurons exposed to early Pb. Furthermore, an association of developmental Pb exposure and AD phenotype was present and an early Pb exposure can lead to increased phosphorylation of Tau protein at Thr181 in mature neurons (Day 45 of differentiation) which is the most abundant Tau phosphorylation observed in middle-stage AD. Moreover, an increase in the level of pThr181-Tau was reported using different techniques like western blotting, FRET. Also the authors noted an elevated level of Aβ42 peptide in the solution whereas Aβ40 remains unchanged (Xie et al., 2023). Overall, this work provides insight into the effect of developmental Pb exposures on precursor cells and the same is present in mature neurons as aberrant cellular functions such as Ca^{2+} ions dyshomeostasis, loss in synaptic protein and synapse, Tau protein hyperphosphorylation, and an increase in Aβ42 level. All these phenotypes in mature neurons shares common features with AD pathology and hence the study provides a good link on how early neurotoxin/neurotoxicant exposure of stem cells can develop into a serious NDD-like AD-like effect in later stages of life of an individual (Guan et al., 2021).

Pb exposure is also observed to cause harmful effects on the AD-associated gene apolipoprotein E (ApoE). ApoE4 is a human isoform of the ApoE gene and is associated strongly with AD risk, and is not present in mice but genetically engineered mice have been generated that express this isoform of ApoE. Pb was found to cause cognitive decline in ApoE4- knock in (KI) mice. ApoE is expressed in adult NPCs in the dentate gyrus (DG) and in vivo studies using ApoE4-KI mice found that ApoE4 alters adult NSCs born neuron survival and maturation in an age- and sex-dependent fashion. Pb was observed to impair the maturation, dendritic development and differentiation of adult NSCs born neurons in the DG of hippocampus leading to cognitive decline, spatial working memory deficit, decrease of spontaneous alteration and a contextual fear memory impairment. Interestingly the study observed some of the above phenotypes like impaired contextual fear memory, reduced spontaneous alterations specific to ApoE4 KI female mice indicating a higher sensitivity of Pb exposure to female mice than to male suggesting a sex-based gene-environment interaction in inducing different phenotypes (Engstrom et al., 2017). The study provides an example of the effect of Pb on adult NSCs and how Pb impacts adult NSCs derived neuronal maturation, differentiation, dendrite formation and increase late-stage AD risk.

Similar to the above cited work, other environmental toxicant exposures are also found to have gene-environment interaction in ApoE4-KI mice model suggesting susceptibility of ApoE4 allele to different environmental toxins. In one work, performed on humanized ApoE4-KI mice, authors found that neurotoxicant, Cadmium (Cd), may cause elevated loss of neurogenesis in the hippocampus and hence can generate AD-like symptoms and pathobiology. This work demonstrated the sensitivity of ApoE4 allele to Cd, where an exposure to Cd in ApoE4-KI mice has resulted in loss of adult NSCs in hippocampus resulting in a decreased neurogenesis and neuronal differentiation. The authors' also concluded that due to Cd neurotoxicity APOE4-KI mice performed poorly in T-maze test, object location test later in its life span. The authors' further concluded that due to gene by environment (GxE) interaction of Cd with ApoE4-KI mice there is a loss of NSCs based neurogenesis and differentiation in hippocampus and that may contribute significantly to poor performance of the mice in different memory and recognition-based tests (Zhang et al., 2020; Chung et al., 2014). Hence all the above work clearly demonstrates how GxE interactions between different toxicants and gene variants, like ApoE4, can produce severe neurological defects in the hippocampus. The work contributes on defining the role of GxE interactions in individuals carrying the ApoE4 allele exposed to different environmental toxicants and how differences in susceptibility to developing dementia like AD in later stages of life can be linked to both exposures and genetics.

A growing literature of work on environmental toxicant effects on understanding dementia related phenotypes using stem cells have used iPSC-based model systems. In a separate study conducted in Gulf War US veterans it was found that the environmental toxin organophosphate can have deleterious effects on the iPSC-derived GLUergic neurons. This work explored the effects of stress hormone cortisol and an environmental toxin diisopropyl fluorophosphate (DFP) on the iPSC-derived mature GLUergic neurons of Gulf war veterans and their control and the work suggests that the DFP+cortisol causes increased level of total and phosphorylated tau, abnormal mitochondrial dynamics and reduced neuronal activity in gulf war veteran iPSC derived GLUergic neurons suggesting they are more susceptible to develop late-stage AD (Yates et al., 2021). Furthermore, the work explains how an early exposure of neurotoxicant DFP is toxic to the mental health of the veterans even if the exposure is stopped suggesting a persistent effect of neurotoxicants which can later manifests as NDD form.

Environmental exposure to metals like Mn are also observed to induce AD-like phenotypes, though has more typically been linked to inducing PD-like symptoms (Tong et al., 2014). Mn is an essential trace element and plays critical roles as a co-factor for enzymes and is required for brain development; yet excessive Mn exposure is associated with neurotoxicity.

Mn is found to be accumulated in the hippocampus of the brain, and hence can affect cognition and memory formation ability of the brain; a key symptom of AD (Martins et al., 2019). In one study, Mn-exposed brain regions from mice were transplanted with NSCs in the hippocampus. The work showed that the Mn exposure causes impairment of neurogenesis in brain whereas transplantation of NSC in the Mn exposed mice improves their spatial cognition behavior (Shu et al., 2021). During the study, authors first isolated NSCs from the mice at postnatal day 0 and maintained them in in vitro condition and after establishing the multipotent nature of the NSCs they stereotactically transplanted the cells in the lateral ventricles or hippocampus of Mn exposed mice. Although, the study did not definitely observe a direct Mn effect on stem cell and the subsequent development to AD, the worked provided evidence via stem-cell transplantation of the possible role of stem cells in increasing neurogenesis and improving cognitive behavioral phenotypes in Mn over-exposed mice. The study found that transplanted stem cells migrated to the lateral ventricles or to the dentate gyrus of the hippocampus to potentially contribute via neurogenesis in adults in these specific regions.

Hence from all of the studies we discussed above on the effects of environmental toxins on NSCs and etiology of dementia or other features of AD, we can emphasize that neurotoxicant exposures of NSCs can alter their proliferation, differentiation and maturation and cause a defective progeny of stem cell derived neurons in specific brain regions like the hippocampus (a brain region affected in AD). Additionally, it was observed that environmental toxicant exposures of stem cell (e.g. iPSCs) derived cortical neurons can generate AD-like phenotypes such as tau aggregation, tau hyperphosphorylation, Ca^{2+}ions dyshomeostasis, synaptic plasticity dysregulation, and neuronal dysfunctions. Nonetheless more work needs to be carried out to understand the molecular mechanisms in NSCs that are affected by such toxicants and how these effects may at times lead to AD pathogenesis.

2.2 Parkinson disease (PD)

PD is a chronic movement disorder that is caused mainly by loss of dopaminergic (DAergic) neurons in the basal ganglion area of the brain leading to disruption in the DAergic transmission by the neurons. The relationship between environmental toxicants leading to PD is not a recent discovery but it goes back to time when a toxin-induced PD mice model was first developed for research use. It was found that a well-known environmental toxicant 1-methyl-4-phenyl-1,2,3,6-tetrahydropyridine (MPTP) produced as a side-product in the synthesis of 1-methyl-4-phenyl-4-propionoxypiperidine (an opioid) induces Parkinson similar symptoms in the users of the opioid (Langston et al., 1983). Later, the mechanism linked with MPTP was explored and the study explained MPTP can cross the blood-brain barrier and get accumulated in the brain where in astrocytes it is metabolized to MPP+ and enter the DAergic neurons to produce PD symptoms (Javitch et al., 1985). Afterwards, other toxins were also found responsible to cause PD such as neurotoxin 6-OHDA (6-hydroxydopamine), rotenone (a pesticide), paraquat (herbicide), maneb (herbicide) which indicated the role of such environmental toxins in inducing PD-like conditions (Simola et al., 2007; Betarbet et al., 2000; Saravanan et al., 2005; Costello et al., 2009).

Mn has also been shown to be a risk-factor for producing PD-like symptoms and contribute to the development of parkinsonian features. Overexposure to Mn can result in a condition known as Manganism. The condition shares features with PD including bradykinesia and rigidity. Excessive Mn exposure is found to be accumulated in basal ganglia of the brain specifically in different regions globus pallidus, subthalamic nucleus, and striatum, those which are involved in the control of motor and non-motor functions (Bouabid et al., 2014; Dietz et al., 2001; Klos et al., 2006). However contrasting reports also exist, suggesting Mn-induced Manganism differs from PD as it accumulates in pallidium and striatum, rather than other regions of the basal ganglia such as the substantia nigra which are commonly degenerated in PD (Olanow, 2004).

The basic mechanism(s) by which Mn may induce the PD-like symptoms is not fully understood but several reports show that Mn can alter NSC and neuronal biology and hence lead to changes in the neurons which degenerate in the pathology of PD. In one work, Mn impact on mice ventricular NSCs were observed and even a moderate Mn concentration (100–300 μM) caused a disruption in differentiation of NSCs as the treatment reduced the neurite length in NSCs differentiation. The study further found a decrease in the cytoskeletal f-actin polymerization when the NSCs were exposed to very high Mn levels (800 μM; a level well beyond what has been observed in the human brain following high exposures). Furthermore, authors observed, a decrease in DCX (a progenitor marker for migration) and GFAP level (glial cell marker) suggesting Mn treatment can reduce the migration of stem cells (Parsons-White and Spitzer, 2018). NSCs exposed to Mn are also prone to apoptosis by the mitochondrial mediated pathway and cytosol release of cytochrome C, and activation of caspase-3 (Tamm et al., 2008).

In our own lab, we have observed some effects of elevated Mn exposure on the iPSC-derived DAergic neurons. It was previously presumed that since Mn inhibits mitochondrial respiratory systems similar to rotenone (a PD neurotoxicant) and that both Mn and rotenone would share common mechanism of inducing oxidative stress in DAergic neurons if their exposure dose and time was accounted for. However, we observed a distinct oxidative stress profile for both these neurotoxicants in DAergic system using 3 different oxidative stress read-outs i.e., DCF assay, isoprostanes generation and glutathione (GSH-a antioxidant) production. Mn induced a concentration (0.5 μM–200 μM) and time dependent (0–240 min) increase in the intracellular reactive oxygen nitrogen species (RONS) production in the DAergic neurons but the same effect was absent in rotenone exposure (0.1 μM−10 μM and 0–240 min). RONS elevation was reduced in presence of rasagiline (a monoaminase inhibitor in PD therapy). Mn (200 μM) exposure for 5-hr also initially reduced glutathione (an antioxidant), but the level return to the normal after 24-hr however the DAergic neurons had still lower GSH level in rotenone (1 and 10 μM) treatment. In another oxidative stress read-out of isoprostane levels (formed due to lipid peroxidation) we found that their levels did not change after Mn (200 μM) treatment in any exposure duration yet their levels were elevated by rotenone exposure (1 and 10 μM). We also observed a small (5%) change in reduction of DAergic neurons neurites length and neuronal marker tyrosine hydroxylase (TH) in Mn treatment showing small variations in DAergic neurons morphology due to the Mn exposure. Our work provided evidence that although both Mn and rotenone have been found to inhibit the mitochondrial respiratory complexes in cells, their mechanism of action for inducing oxidative stress differs in DAergic neurons. Mn primarily impacted oxidative balance in DAergic neurons by inducing RONS production, while rotenone primarily induced the isoprostane marker of lipid peroxidation. Both altered glutathione levels. Hence our findings bolster effects of Mn and rotenone neurotoxicity in DAergic neurons and thus study of effects of environmental factors on DAergic neurons may provide key insight into the understanding of environmental factors that lead to elevated risk of PD-like pathogenesis as well as insight in their mechanism of action in inducing the PD similar pathologies (Neely et al., 2017).

Furthermore, our own lab has performed more work to understand Mn exposure effects on DAergic neurons. In one such study, we differentiated human iPSCs obtained from control and Huntington disease (HD) patients to different lineages of cortical, striatal, and midbrain NPCs at Day 11 and then later to the respective lineage specific neurons (Joshi et al., 2019). We observed that the NPCs in the HD phenotypes were more sensitive to Mn toxic insult (viability) for cortical and striatal lineages but not for midbrain. However, we observed that at Day 25 differentiation mid-brain DAergic neurons from HD patients were more sensitive to Mn than control subject neurons, while no difference between HD and control was seen for the maturing cortical and striatal neurons (Joshi et al., 2019). Hence our study is really crucial to understand that an early exposure to environmental toxin like Mn can make neurons especially DAergic neurons (pathologically lost in PD) more vulnerable to later Mn insult and thus potentially contributing to the development of a parkinsonian phenotype.

The importance of stem cell models in pre-clinical studies of PD can be appreciated from the use of stem cell-based iPSC platforms for screening of environmental toxicants for PD risk. A very recent study has used engineered and fluorescent reporter labeled iPSC-derived DAergic neurons from PD patient to screen for pesticides responsible for causing the disease. The study used an iPSC based cellular system to screen for 39 pesticides that can alone or in specific combination cause PD-like effects in the model system. Toxicity studies for these selected pesticides were performed in the reporter iPSC line using a live imaging cell viability assay. The report of the work showed 10 of pesticides were highly toxic to DAergic neurons, with altered viability at 30 μM and the report also found new pesticides like folpet, endothall, endosulfan, and naled with toxic effects (Paul et al., 2023). PD is one of the NDDs which has seen an extensive application of stem cell therapy and the same is described more in our section on stem cell-based therapy for NDDs. Nonetheless a more complete understanding of the progression and etiology of PD and how environmental toxicants induced stem cell pathologies leading to PD is needed, specifically more research on finding the key cellular pathways affected due to environmental toxin exposures leading to the selective sensitivity of DAergic neurons.

2.3 Amyotrophic lateral sclerosis (ALS)

ALS is a disease which was initially identified as motor neuron disease but increasing findings have indicated it to be a multi-system neurological disorder. Clinically, ALS patients has symptoms of muscle wasting and muscle weakness (Zarei et al., 2015). The disease is highly progressive in nature and due to respiratory muscle wasting in the disease, the median survival of the patients is mostly observed to be 2–5 years in duration (Masrori and Van Damme, 2020). At the gene level, more than 70% of familial ALS is characterized by mutations of at least four different genes including superoxide dismutase 1 (SOD1), TAR DNA Binding Protein-43 (TDP-43), c9ORF72, and FUS, whereas in sporadic ALS several new genes and their contribution to disease development are continuing to be identified on an almost routine basis. However, at molecular level, various cellular pathways are found to be affected in ALS including pathways related to cellular protein aggregation, neuronal inflammation, endoplasmic reticulum stress, and defective RNA processing (Chakraborty and Diwan, 2022).

New findings are being published on the role of environmental factors in the development and risk of ALS using stem cell-based models. In one such work, authors have found evidence that dioxin-based polychlorinated biphenyls and polycyclic aromatic hydrocarbons environmental toxicants induce aryl hydrocarbon receptor (AHR- a transcription factor) and the same factors are shown to be risk factor for ALS. The study found that the environmental toxicants can induce the expression of AHR and the same lead to a higher expression of soluble and insoluble TDP-43 in ALS patients (G298S mutation) iPSCs differentiated motor neurons (MN). The study proved that due to AHR activation (transcription factor) there is slower TDP-43 degradation and can lead to its aggregation in the MN and hence develop into the pathology of ALS. The study further showed that AHR has its binding site in the TDP-43 gene TAR-DNA binding protein 43 (TARDBP) transcript and an agonism of AHR induces TARDBP transcript however the same doesn't significantly translate into a higher protein expression of TDP-43 and the mechanistic reason for TDP-43 aggregation is due to its slower degradation (Ash et al., 2017). Thus, the study provided a different perspective on the effects of the environmental toxicant on TDP-43 both at protein and gene level and concluded that environmental toxicant effects on TDP-43 pathology may be due to effects on TDP-43 catabolism.

In the pathology of ALS, primary motor neurons degenerate. Thus, understanding the effects of environmental toxicants on the health of motor neurons will give directions for exploring the same in the pathology of ALS. One study explored the impact of MeHg on iPSC-derived motor neurons. The work reported that MeHg in an acute exposure paradigm (0.1–0.5 μM for 1 h) followed by 24 h of recovery period can alter intracellular Ca^{2+} ion concentration in motor neurons by modulating the level of α-amino-3-hydroxy-5-methyl-4-isoxazolepropionic acid receptor (AMPAR) in motor neurons. Further, the work also showed that MeHg can cause death in motor neurons by inducing a large influx of Ca^{2+} ions. Motor neurons lack buffering capacity for large Ca^{2+} ions influx as they don't have calbindin protein and hence a high influx causes excitotoxicity. iPSC-based modeling helped the authors to characterize in depth the role of AMPAR in MeHg induced neurotoxicity in ALS which was not previously established. Both these findings have important implications in elucidating the role of MeHg in the pathology of motor neurons and the same can be understood to explore more effects of MeHg in ALS where motor neurons are primarily affected (Colón-Rodríguez et al., 2020).

A separate study performed between 2011 and 2014 at a referral center collected blood samples from ALS patients (156 cases) and healthy controls (128 controls) and screened this blood to detect a panel of 122 organic pollutants. The study found a direct correlation between ALS and blood levels of organochlorine pesticides (OCP), polychlorinated biphenyls, and brominated flame retardants. The work showed high blood levels of OCP cis-chlordane were associated with ALS patients and interestingly in healthy controls the organochlorine pesticide Dacthal was found to be enriched (Su et al., 2016). The above study identified potential risk factors associated with ALS. In other work, the mechanistic action of the *cis*-chlordane in the context of ALS was explored. In this work, selective neurotoxicity by cell-type (i.e. motor neurons) was observed for cis-chlordane (an OCP). A finding that is highly relevant in defining the role of environmental neurotoxicants in inducing ALS. Specifically, the authors differentiated human ESCs to motor neurons and then checked the toxicity of cis-chlordane on the motor neurons and generated a dose-response curve for the same. The authors then found that the EC_{50} of cis-chlordane was 12 μM in motor neurons whereas other cells like cortical neurons and fibroblasts had higher EC_{50} values indicating that the motor neurons were more sensitive to the OCP treatment. The authors identified that a potential mechanism of action of cis-chlordane was due to its ability to severely inhibit the neuronal action potentials of iPSC-derived motor neurons as well as reducing motor neuron viability (Kulick et al., 2022). All the above work provided evidence that different environmental toxins can cause deleterious molecular changes in stem cell derived motor neurons such as impaired catabolism, Ca^{2+} ion dyshomeostasis, altered action potential propagation; all consistent with known pathological features of ALS. The findings on environmental factors and their role in development of ALS by negatively altering the function and survival of NSCs and their motor neurons progeny are limited and more studies can help in finding suitable environmental conditions for ALS patients to increase their survival chances. More work in this area should focus on finding pathogenic mechanisms in stem cell derived model systems and revealing novel therapeutic targets of the motor neurons and other cells that contribute to ALS, each of which can be generated by stem cell-based modeling.

2.4 Other neurodegenerative diseases

Iron overload in the cells can also cause deleterious consequences for the cell. In one study done in the case of Friedrich ataxia (FA) pathology it was observed that excess Iron (Fe) in FA patient iPSC derived cardiomyocytes cells suffers from different defects like impairment of Calcium homeostasis, more reactive oxygen species production (Verheijen et al., 2022; Lee et al., 2014). Another NDD which our lab has studied and contributed using stem cells in context of effect of an environmental toxicant is HD. Our research on HD patient derived stem cells supported work from cell line and in vivo models that the HD risk genotype was associated with a resistance to Mn neurotoxicity – due to a decrease in net transport of Mn in HD cells. Using stem cell models we found that Mn dependent p53 phosphorylation is modulated via the ataxia telangiectasia mutated kinase (ATM) pathway and that Mn-dependent activation of p53 phosphorylation is regulated by ATM and is diminished in HD NPCs (Tidball et al., 2014). Our findings implicate Mn biology in the disease processes occurring in HD patients.

The next segment of this chapter explores how stem cells-based approaches may provide therapeutic benefit for many NDDs. We highlight recent studies in both pre-clinical and clinical trial-based approaches of stem cell therapy.

3. Stem cell therapy: a promising approach to treat environmental contaminants induced NDDs

Stem cell therapy is a developing approach that seeks to treat NDDs, thus the same strategy could be utilized to alleviate the symptoms of the neurological effects caused by environmental toxicants (Hoang et al., 2022; Dantuma et al., 2010). Stem cell therapy has certain advantages over the traditional chemotherapeutic applications of the drugs to treat NDDs as the stem cells have the unique ability to regenerate damaged tissues, regulate inflammatory responses and, hence promote neuronal repair and survival. Different kinds of stem cells such as embryonic stem cells (ESCs), iPSCs, NSCs and mesenchymal stem cells (MSCs) are investigated for their therapeutic applications in preclinical and clinical settings to treat NDDs (Hoang et al., 2022).

ESCs are obtained from early-stage embryos and have the potential to differentiate into all types of cells in an adult and are thus known as pluripotent. iPSCs are also pluripotent in nature but are generated in lab by reprogramming of somatic cells and they allow to study a particular disease or mechanism using a patient-specific approach. In contrast to pluripotent, multipotent cells are stem cells which will give rise to a specific lineage of cell population e.g. Mesenchymal and NSCs. MSCs are multipotent stem cells of stromal origins (coming from connective tissues or supporting tissues) which can be isolated from different tissues like bone marrow, embryonic tissue, adipose tissue, cord blood, and perinatal tissue and they can be differentiated into neuronal cells using defined conditions. They are of significant potential value in NDDs and other human disease conditions including orthopedic pathological conditions, and they are used in regenerative medicine such as transplantation and in immunosuppression and gene therapy (Mattei and Delle Monache, 2024). In comparison to past ESC- and iPSC-based transplantation methods, MSCs have shown specific efficacious advantages like decreased immune rejection, relative ease of availability, as they can be obtained from the same patient (compared to ethical challenges associated with ESCs), fewer concerns about genomic instability (like observed in iPSCs) and hence they hold great promise in treating NDDs in both a cost-effective and safe manner (Gopalarethinam et al., 2023). However, owing to the advantages of being "self", as well as being able to generate functional replacement cells and tissues, iPSC-based transplantation approaches are expected to gain favor once technical challenges have been resolved.

Typically, NSCs and other stem cells to treat NDDs are of neuronal origin or lineage and hence have potential to show fewer side effects and are also less prone to transplantation rejection in NDDs treatment. NSC derived from the same patients would also have ideal gene and protein expression and after transplant would have less chances of adverse immune reactions compared to use of other stem cells. Additionally, NSCs for transplantation in NDDs can secrete trophic factors for neuronal growth, they can integrate better with the host environment and thus elevate improved neuronal plasticity, and may have increased transplantation efficiency. Hence, NSCs are an ideal candidate for transplantation in NDDs. Along with the MSCs, NSC are most extensively applied for transplantation studies in different pre-clinical and clinical settings. NSCs for transplantation to treat NDDs can be obtained from different sources such as fetal neural tissue, MSCs themselves, as well as differentiation from ESCs and iPSCs (Fan et al., 2023). NSCs have the capability to differentiate in either neurons or glial cell population. NSCs like other stem cells also have the property of self-division and multipotency. At first, NSCs also known as neuroepithelial cells differentiate into radial glial cells which then divide and form neural progenitor cells (NPCs). Both NPCs and NSCs are the proliferating cells in the CNS and have the capability to form neuron by neurogenesis and glial cells (astrocytes, oligodendrocytes) by gliogenesis. Research in the field of stem cell therapy in the treatment of NDDs is focused on the use of NPCs and NSCs to regenerate the cells lost in the brain due to their death in NDDs. Different growth factors: extrinsic and intrinsic are known to assist in the process of neuronal and glial cells development in early stages.

3.1 Applications and future directions

In preclinical studies, stem cell therapy has shown promising results in mitigating neurodegeneration induced by environmental toxins. It is observed that stem cell transplantation in animals exposed to environmental toxins shows improved neurogenesis, enhanced neuronal survival, and ameliorate cognition and motor functions (Shu et al., 2021; Barati et al., 2022). To give some examples, in chronically Mn exposed mice, hippocampal or ventricular transplantation of NSCs improved spatial learning and memory function indicating the importance of stem cells as a therapeutic agent in reducing the Mn induced effects on learning and memory. The authors proposed in their work that transplantation of stem cells may increase neurogenesis in adult mice hippocampus as they observed stem cells migrating to regions such as ventricles and hippocampus involved in neurogenesis (Shu et al., 2021). Additionally, one more study implicated the therapeutic effects of MSCs in a copper chelator cuprizone mice model. Cuprizone is environmentally toxic and cuprizone exposed mice has symptoms of reactive astrocytes and demyelination. MSCs were injected into the right lateral ventricle of cuprizone mice and after two weeks of transplantation, authors observed a decrease in number of reactive A1 astrocytes, increase in number of oligodendrocytes and in remyelination, and a decrease of proinflammatory factors and improved behavioral deficits (rotarod test) in the cuprizone mice model (Barati et al., 2022). These findings demonstrate applications of stem cells in treatment of environmentally induced neurological deficits. The neuroprotective actions of the transplanted stem cells are thought to be by either replenishing the number of lost neurons in the brain by neurogenesis and/or by providing the necessary neurotrophic or nutritional factors (e.g. in re-myelination) needed for maintenance of neuronal function in the exposed brain.

Advantages of stem cell therapy in treating NDDs over conventional therapy have been suggested by various reports. A study based on treating neurotoxin 6-OHDA induced PD model with stem cells has used 5-Bromo-2-deoxyuridine- (BrdU) labeled MSCs for transplantation. BrdU labeling is used to identify the proliferation of MSCs and their differentiation into neuronal lineages. The findings of the work indicated improved rotational test behavior of the PD rats and an increase in the number of neurons stained positive by DAergic markers in the affected PD striatum of the rat after MSC administration. Hence the work implicated the importance of MSCs in preventing the loss of DAergic neurons in PD pathology and thus alleviating, at least partially, PD symptoms (Chen et al., 2017). It remains unknown if MSCs can become new DAergic neurons, versus prevent further declines or induce regeneration of DAergic neurons from resident neural stem cells, though the later seems more likely. Similarly, a separate study showed a combined effect of stem cell and gene therapy in combating PD effects. The study showed a transplantation of LIM homeobox transcription factor 1 (LMX1A) and Neurturin (NTN) transfected MSCs along with adenovirus containing NTN and tyrosine hydroxylase (TH) in the substantia niagra and striatum of MPTP lesioned monkey have higher expression of DAergic neurons and have improved PD behavioral tests results. The transfection of LMX1A and NTN in MSCs would drive their differentiation towards DAergic neuronal fate and thus replenish (at least partially) the loss of DAergic neurons associated with PD (Zhou et al., 2013).

The potential for therapeutic efficacy of stem cells has not only been shown by its transplantation effects but reports suggesting that stem cells secrete extracellular vesicles (like exosomes) which can also have a protective effect against toxicant induced neurodegeneration. Exosomes are spherical extracellular vesicles secreted by cells for cell-to-cell communication and contain different types of biomolecules (DNA, RNA, proteins etc.) and factors which have neuroprotective roles (Kalluri and Lebleu, 2020). One study demonstrated that exosomes derived from MSCs when used to treat cadmium exposed mice improved the total antioxidant status, elevated growth factors i.e., BDNF and NGF expression level to ameliorate the oxidative stress and inflammation (Zaazaa et al., 2022). In a similar type of work, MSC derived exosomes were found to be neuroprotective against the Aβ oligomers (AβO) toxicity. The study found that the MSCs derived exosomes can reduce the toxicity of AβO by decreasing the loss of post-synaptic density 95 and hence maintains the synapse integrity in the neuronal culture. Since, AβO have been implicated in the pathogenesis of AD, this finding holds significance in the treatment of AD (De Godoy et al., 2018). Another finding also demonstrates that NSCs derived exosomes have a neuroprotective effect against the 6-OHDA toxin (causes death of dopaminergic neurons in the substantia nigra) induced PD in both at in vitro and in vivo systems by reducing the intracellular ROS production, decreasing inflammatory markers and downregulating apoptotic pathways. In this study authors found exosomes to carry higher level of miRNAs like miRNAs hsa-mir-183-5p, and hsa-mir-17-5p which showed neuroprotective effects by preserving neurite length (Lee et al., 2022). Stem cell derived exosome applications in treating environmental toxicant induced NDDs was further bolstered by a study where use of human umbilical cord MSC exosomes was successful in blocking the 6-OHDA induced neurodegeneration in SH-SY5Y cells by inhibiting apoptosis and inducing autophagy (Chen et al., 2020).

Stem cell therapy was also found to be potentially effective in treating sporadic NDDs of unknown etiology like multiple system atrophy (MSA). A report showed that the MSA-like phenotypes induced by polyamines and cholesterol exposure respond to treatment with the help of MSC transplantation. Beneficial effects of MSC transplantation further included a reduction of behavioral disorder, cytoprotection of DAergic neurons, anti-inflammatory effects and a decrease in polyamine and cholesterol level in the MSA mice (Park et al., 2020). The use of MSCs in treating the environmental toxicant produced brain damage is also validated by the use of MSCs intravenously injected in Lipopolysaccharide (LPS) exposed mice. The study found the beneficial effects of MSCs injection on improved memory impairment, reduction in Aβ(1−42) accumulation, improving expression of acetyl-choline receptors in brain and brain mitochondria (Lykhmus et al., 2019).

Future applications of stem cells in treating environmental toxin related NDDs may involve using alternative strategies like inducing the resident NSCs populations to expand and undergo neuroprotective changes. In one example, allopregnanolone promoted survival and proliferation of endogenous stem cells, and the same showed efficacy in improving learning and memory tasks in the 3xTgAD (transgenic AD mice with 3 mutations) mice (Singh et al., 2012). Furthermore, a different study explored how different growth factors such as granulocyte colony stimulating factor (GCSF), and stromal-cell derived factor-1 α can stimulate the recruitment of bone marrow-derived hematopoietic progenitor cells in the brain of an Alzheimer mice model and helps in overall memory function improvement but had no effect on Aβ deposition (Shin et al., 2011).

Similar to the above alternative approaches, another study used the small molecule Phenserine (a cholinesterase inhibitor) to induce NSC differentiation in an in vitro and in vivo AD condition and has observed improvement in AD-like phenotypes such as a decrease in amyloid precursor protein (APP) levels and reduction in reactive astrocytes by checking levels of glial fibrillary acidic protein. Phenserine did not change expression levels of the APP gene suggesting the molecule modulated APP levels post-transcriptionally (Marutle et al., 2007). The work provided support for the conclusion that improving the differentiation potential of NSCs and other resident brain stem cells can be an effective means of treating NDDs and hence should be investigated more in the future.

An exciting application of stem cell therapy is the use of patient derived somatic cells in the generation of iPSCs and how the same cells can be differentiated to produce neurons which have future implications for transplantation and treatment of NDDs. The field is still developing but it has shown some tentatively promising results in treatment of NDDs especially PD in different pre-clinical settings. In one study performed in MPTP treated monkey, healthy humans and PD patients iPSCs and their derived dopaminergic neurons were used for transplantation. The progenitor cells used for transplantation were rich in DAergic progenitor marker Forkhead box protein (FOXA2) and injected into the striatal putamen of MPTP treated mice. To prevent immune-rejection and for immunosuppression, FK506 was administered to the monkeys. Monkeys were analyzed based on the neurological scale and no symptoms of bradykinesia was observed. The study also proved efficacious relative to a standard L-Dopa therapy used for PD. MRI based experimental and immunohistological analysis revealed no evidence for side-effects like tumor formation that have been observed in other stem-cell therapies (Kikuchi et al., 2017). The study provided the rationale for a subsequent (2018) clinical trial of stem cells transplantation done in PD patients (JMA-IIA00384, UMIN000033564). However, there is no update on clinical trials as the same were suspended due to risks that remain to be satisfactorily addressed (Brianna et al., 2022).

In another study done by the same above group, human clinical grade iPSCs were used to derive DAergic neurons and the same were then transplanted into the lesioned striatum of the 6-OHDA treated mice. This work also showed the iPSC derived DAergic neurons generated had no continuing population of progenitor cells, had no detected genetic aberrations or cancer related genes and were successful in eliminating PD symptoms in mice based on methamphetamine rotation tests (Doi et al., 2020). Similarly, in another work, human derived NSCs made from patient derived peripheral blood mononuclear cells were induced to differentiate into DAergic neurons and the DAergic precursors were transplanted into striatum of a 6-OHDA PD mouse model. Examination of these animals showed an increase in signal intensity of TH+ cells indicating that the precursor cells successfully integrated into the host environment, that they had formed DAergic neurons and thus were the basis for the detected and improved PD phenotypes as measured by a behavioral test of apomorphine induced rotations (Yuan et al., 2018). In another similar work, PD patient derived iPSCs were transplanted into a 6-OHDA PD mouse model with significant alleviation in PD related phenotypes like improved behavior test using rotarod and higher number of dopaminergic neurons by staining with DAergic marker of TH indicating the importance of iPSC in PD treatment (Song et al., 2020). However, generating iPSCs from an academic lab for therapeutic purpose is an expensive affair and will likely require more financial support from the government and non-government sources to be a viable therapy.

Stem cell based therapeutic efficacy can be improved with the help of different gene editing technologies coming into the foray like CRISPR-Cas9. As CRISPR-Cas9 based editing of the stem cells can help to change the gene expression of the stem cells affected by toxicants inducing neurodegeneration (Heman-Ackah et al., 2016; Ye et al., 2021). Editing of the gene can also become useful as a transplantation therapy in different kind of NDDs; especially in cases where the patient carries a disease risk allele. Beyond applications of stem cells for NDD treatment, many barriers exist in using stem cells for treatment of NDDs and few of such problems are listed in the next section of our book chapter.

3.2 Challenges in using stem cell approach

Despite the potential of stem cells as a therapy for NDDs, many challenges exist for the stem cell application as a clinical tool in NDDs. Many of these difficulties include challenges with cell delivery methods such as inaccessibility of targets in the brain, safety and efficacy, ethical concerns of using human ESCs and legal regulatory inspections (Scopetti et al., 2020; Alipour et al., 2019). One more factor likely to hinder progress in stem cell therapy of NDDs is the source of stem cells used for therapy. It has been observed that stem cell derived from varying sources may behave quite differently in various modalities of therapy. For example, MSCs can be isolated from various sources like bone marrow and adipose tissue but findings suggests that they have different profiles of secreted cytokine proteins and also have different proliferative capacities and these qualities can play a vital role in the behavior of the cells post-transplantation in NDDs (Lee et al., 2004; Ahmadian Kia et al., 2011). Hence, different secretion and proliferative profiles of MSCs can affect their transplantation and treatment efficiency in the NDDs.

One of the approaches used in early 2000s in clinics to treat PD was to apply fetal cells transplantation as a stem cell resource. Three independent studies performed as prospective, double-blind, randomized, placebo-controlled found no positive effect of fetal-cells transplantation on PD patients. In addition, patients in two of the studies developed dyskinesis which progressed despite a reduction in the treatment (Goodarzi et al., 2015; Lindvall and Kokaia, 2009; Politis et al., 2010). Moreover, additional work is required to understand the complex interplay present

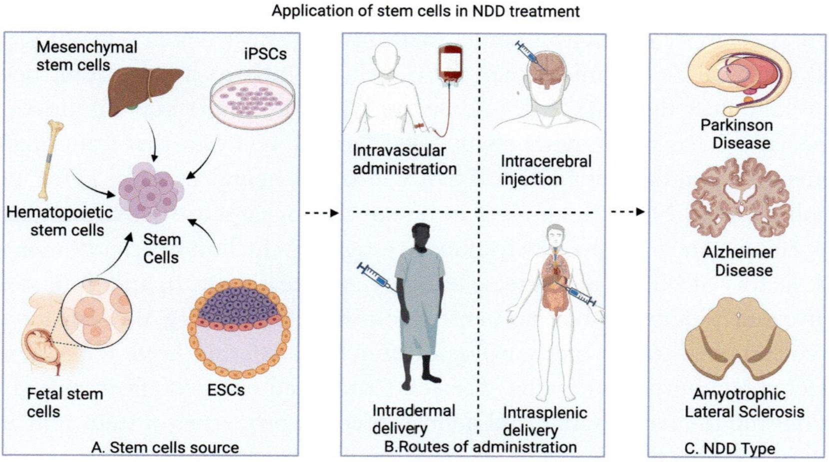

Fig. 2 Therapeutic potential of stem cells: (A) Stem cells can be isolated from different sources such as hematopoietic, mesenchymal, iPSCs, fetal, ESCs, etc. and the same cells can be delivered (B) to the patients using various approaches like intravascular, intracerebral, intradermal and intrasplenic. (C) The platform of stem cells-based therapy can be a boon for neurodegenerative disorders patients of PD, AD and ALS. *Figure created with BioRender.com.*

between environmental contaminants, cellular pathways and NDDs. Longitudinal studies are required to elucidate the effect of chronic low dose exposure to multiple pollutants on neurological health and thus find potential therapeutic targets. Similar to challenges faced in using stem cells for PD, difficulties lie in also using stem cells in other NDDs like ALS. MSCs were used in few pre-clinical and also in some clinical studies to treat ALS, but the major challenge in ALS reside in using MSCs include source of stem cells, clinical results, safety, dose optimization, route of administration, patient selection (Mazzini et al., 2003; Cudkowicz et al., 2022; Najafi et al., 2023). In Fig. 2, we illustrate potential applications of different stem cells sources and various routes of their administration and how the same can be used to treat varieties of neurodegenerative diseases such as AD, PD and ALS.

4. Conclusion

In conclusion, NDDs induced by environmental neurotoxicants/ neurotoxins poses a significant public health challenge, necessitating innovative therapeutic approaches. Environmental toxins effects on NSCs and other brain-derived stem cells need to studied in depth in respect to how the variety of environmental toxicants can affect which stage of stem cell neuronal development (e.g. from earliest NSCs to mature adult neurons) and how the same leads to different types of changes in brain function especially in relation with stem cells derived neurons that can finally generate symptoms of NDDs. Stem cell-based neural models, given the implication of NSCs themselves in disease and as targets of environmental risk factors, are an important tool in investigating the link of environmental risk factors of NDDs and disease etiology and treatment. In addition, stem cell-based therapies hold immense promise in mitigating the impact of environmental toxins on neurological function, offering hope for patients affected by neurodegenerative diseases. Continued research efforts aimed at harnessing the regenerative and neuroprotective properties of stem cells are crucial for advancing towards effective treatments and improving outcomes in this burgeoning field of neurology.

Acknowledgments

This work was supported in part by grants from the National Institutes of Health (USA), NIEHS R01 ES010563 and R01 ES07331 and R01 ES031401 and NIA R01 AG080917.

References

Addae, C., Cheng, H., Martinez-Ceballos, E., 2013. Effect of the environmental pollutant hexachlorobenzene (HCB) on the neuronal differentiation of mouse embryonic stem cells. Int. J. Environ. Res. Public Health 10, 5244–5256.

Ahmadian Kia, N., Bahrami, A.R., Ebrahimi, M., Matin, M.M., Neshati, Z., Almohaddesin, M.R., et al., 2011. Comparative analysis of chemokine receptor's expression in mesenchymal stem cells derived from human bone marrow and adipose tissue. J. Mol. Neurosci. 44, 178–185.

Alipour, M., Nabavi, S.M., Arab, L., Vosough, M., Pakdaman, H., Ehsani, E., et al., 2019. Stem cell therapy in Alzheimer's disease: possible benefits and limiting drawbacks. Mol. Biol. Rep. 46, 1425–1446.

Amanullah, A., Upadhyay, A., Joshi, V., Mishra, R., Jana, N.R., Mishra, A., 2017. Progressing neurobiological strategies against proteostasis failure: challenges in neurodegeneration. Prog. Neurobiol. 159, 1–38.

Andreone, B.J., Larhammar, M., Lewcock, J.W., 2020. Cell death and neurodegeneration. Cold Spring Harb. Perspect. Biol. 12.

Arai, Y., Nishino, K., 2023. Epigenetic mutagen-like environmental chemicals alter neural differentiation of human induced pluripotent stem cells. J. Toxicol. Sci. 48, 571–583.

Aravindan, A., Newell, M.E., Halden, R.U., 2024. Literature review and meta-analysis of environmental toxins associated with increased risk of Parkinson's disease. Sci. Total Environ. 931, 172838.

Arruti, A., Fernández-Olmo, I., irabien, A., 2010. Evaluation of the contribution of local sources to trace metals levels in urban PM2.5 and PM10 in the Cantabria region (Northern Spain). J. Environ. Monit. 12, 1451–1458.

Ash, P.E.A., Stanford, E.A., Al Abdulatif, A., Ramirez-Cardenas, A., Ballance, H.I., Boudeau, S., et al., 2017. Dioxins and related environmental contaminants increase TDP-43 levels. Mol. Neurodegener. 12, 35.

Ayeni, E.A., Aldossary, A.M., Ayejoto, D.A., Gbadegesin, L.A., Alshehri, A.A., Alfassam, H.A., et al., 2022. Neurodegenerative diseases: implications of environmental and climatic influences on neurotransmitters and neuronal hormones activities. Int. J. Environ. Res. Public Health 19.

Barati, S., Kashani, I.R., Tahmasebi, F., 2022. The effects of mesenchymal stem cells transplantation on A1 neurotoxic reactive astrocyte and demyelination in the cuprizone model. J. Mol. Histol. 53, 333–346.

Barrios-Arpi, L., Arias, Y., Lopez-Torres, B., Ramos-Gonzalez, M., Ticli, G., Prosperi, E., et al., 2022. In vitro neurotoxicity of flumethrin pyrethroid on SH-SY5Y neuroblastoma cells: apoptosis associated with oxidative stress. Toxics 10, 131.

Betarbet, R., Sherer, T.B., Mackenzie, G., Garcia-Osuna, M., Panov, A.V., Greenamyre, J.T., 2000. Chronic systemic pesticide exposure reproduces features of Parkinson's disease. Nat. Neurosci. 3, 1301–1306.

Bouabid, S., Delaville, C., De Deurwaerdère, P., Lakhdar-Ghazal, N., Benazzouz, A., 2014. Manganese-induced atypical parkinsonism is associated with altered basal ganglia activity and changes in tissue levels of monoamines in the rat. PLoS One 9, e98952.

Bowman, A.B., Kwakye, G.F., Herrero Hernández, E., Aschner, M., 2011. Role of manganese in neurodegenerative diseases. J. Trace Elem. Med. Biol. 25, 191–203.

Bradl, H., 2005. Heavy Metals in the Environment: Origin, Interaction and Remediation. Elsevier.

Breijyeh, Z., Karaman, R., 2020. Comprehensive review on Alzheimer's disease: causes and treatment. Molecules 25.

Brianna, Ling, A.P.K., Wong, Y.P., 2022. Applying stem cell therapy in intractable diseases: a narrative review of decades of progress and challenges. Stem Cell Investig. 9, 4.

Cai, Z., Wan, C.-Q., Liu, Z., 2017. Astrocyte and Alzheimer's disease. J. Neurol. 264, 2068–2074.
Chakraborty, A., Diwan, A., 2022. Biomarkers and molecular mechanisms of Amyotrophic Lateral Sclerosis. AIMS Neurosci. 9, 423–443.
Chen, D., Fu, W., Zhuang, W., Lv, C., Li, F., Wang, X., 2017. Therapeutic effects of intranigral transplantation of mesenchymal stem cells in rat models of Parkinson's disease. J. Neurosci. Res. 95, 907–917.
Chen, H.-X., Liang, F.-C., Gu, P., Xu, B.-L., Xu, H.-J., Wang, W.-T., et al., 2020. Exosomes derived from mesenchymal stem cells repair a Parkinson's disease model by inducing autophagy. Cell Death Dis. 11, 288.
Chen, P., Miah, M.R., Aschner, M., 2016. Metals and neurodegeneration. F1000Research 5.
Cho, J.-H., Lee, S., Jeon, H., Kim, A.H., Lee, W., Lee, Y., et al., 2020. Tetrabromobisphenol A-induced apoptosis in neural stem cells through oxidative stress and mitochondrial dysfunction. Neurotox. Res. 38, 74–85.
Chowdhury, R., Ramond, A., O'keeffe, L.M., Shahzad, S., Kunutsor, S.K., Muka, T., et al., 2018. Environmental toxic metal contaminants and risk of cardiovascular disease: systematic review and meta-analysis. BMJ 362.
Chung, S.J., Kim, M.-J., Kim, Y.J., Kim, J., You, S., Jang, E.H., et al., 2014. CR1, ABCA7, and APOE genes affect the features of cognitive impairment in Alzheimer's disease. J. Neurological Sci. 339, 91–96.
Colle, D., Farina, M., Ceccatelli, S., Raciti, M., 2018. Paraquat and maneb exposure alters rat neural stem cell proliferation by inducing oxidative stress: new insights on pesticide-induced neurodevelopmental toxicity. Neurotox. Res. 34, 820–833.
Colón-Rodríguez, A., Colón-Carrión, N.M., Atchison, W.D., 2020. AMPA receptor contribution to methylmercury-mediated alteration of intracellular Ca(2+) concentration in human induced pluripotent stem cell motor neurons. Neurotoxicology 81, 116–126.
Costello, S., Cockburn, M., Bronstein, J., Zhang, X., Ritz, B., 2009. Parkinson's disease and residential exposure to maneb and paraquat from agricultural applications in the central valley of California. Am. J. Epidemiol. 169, 919–926.
Cresto, N., Forner-Piquer, I., Baig, A., Chatterjee, M., Perroy, J., Goracci, J., et al., 2023. Pesticides at brain borders: Impact on the blood-brain barrier, neuroinflammation, and neurological risk trajectories. Chemosphere 324, 138251.
Cudkowicz, M.E., Lindborg, S.R., Goyal, N.A., Miller, R.G., Burford, M.J., Berry, J.D., et al., 2022. A randomized placebo-controlled phase 3 study of mesenchymal stem cells induced to secrete high levels of neurotrophic factors in amyotrophic lateral sclerosis. Muscle Nerve 65, 291–302.
Da Silva Siqueira, L., majolo, F., da Silva, A.P.B., da Costa, J.C., marinowic, D.R., 2021. Neurospheres: a potential in vitro model for the study of central nervous system disorders. Mol. Biol. Rep. 48, 3649–3663.
Dantuma, E., Merchant, S., Sugaya, K., 2010. Stem cells for the treatment of neurodegenerative diseases. Stem Cell Res. Ther. 1, 37.
De Godoy, M.A., saraiva, L.M., de Carvalho, L.R.P., vasconcelos-Dos-Santos, A., beiral, H.J.V., ramos, A.B., et al., 2018. Mesenchymal stem cells and cell-derived extracellular vesicles protect hippocampal neurons from oxidative stress and synapse damage induced by amyloid-β; oligomers. J. Biol. Chem. 293, 1957–1975.
Deture, M.A., Dickson, D.W., 2019. The neuropathological diagnosis of Alzheimer's disease. Mol. Neurodegener. 14, 32.
Dietz, M., Ihrig, A., Wrazidlo, W., Bader, M., Jansen, O., Triebig, G., 2001. Results of magnetic resonance imaging in long-term manganese dioxide-exposed workers. Environ. Res. 85, 37–40.

Ding, W.Y., Huang, J., Wang, H., 2020. Waking up quiescent neural stem cells: molecular mechanisms and implications in neurodevelopmental disorders. PLoS Genet. 16, e1008653.

Doi, D., Magotani, H., Kikuchi, T., Ikeda, M., Hiramatsu, S., Yoshida, K., et al., 2020. Pre-clinical study of induced pluripotent stem cell-derived dopaminergic progenitor cells for Parkinson's disease. Nat. Commun. 11, 3369.

Eid, A., Bihaqi, S.W., Renehan, W.E., Zawia, N.H., 2016. Developmental lead exposure and lifespan alterations in epigenetic regulators and their correspondence to biomarkers of Alzheimer's disease. Alzheimer's & Dementia: Diagn. Assess. Dis. Monit. 2, 123–131.

Engstrom, A.K., Snyder, J.M., Maeda, N., Xia, Z., 2017. Gene-environment interaction between lead and Apolipoprotein E4 causes cognitive behavior deficits in mice. Mol. Neurodegener. 12, 14.

Fan, Y., Goh, E.L.K., Chan, J.K.Y., 2023. Neural cells for neurodegenerative diseases in clinical trials. Stem Cell Transl. Med. 12, 510–526.

Feigin, V.L., Vos, T., Nichols, E., Owolabi, M.O., Carroll, W.M., Dichgans, M., et al., 2020. The global burden of neurological disorders: translating evidence into policy. Lancet Neurol. 19, 255–265.

Fergusson, J.E., 1990. The Heavy Elements: Chemistry, Environmental Impact and Health Effects. Pergamon Press, Oxford, pp. 85–547.

Fuller, R., Landrigan, P.J., Balakrishnan, K., Bathan, G., Bose-O'reilly, S., Brauer, M., et al., 2022. Pollution and health: a progress update. Lancet Planet. Health 6, e535–e547.

Gangemi, S., Miozzi, E., Teodoro, M., Briguglio, G., De Luca, A., Alibrando, C., et al., 2016. Occupational exposure to pesticides as a possible risk factor for the development of chronic diseases in humans (review). Mol. Med. Rep. 14, 4475–4488.

Genuis, S.J., Kelln, K.L., 2015. Toxicant exposure and bioaccumulation: a common and potentially reversible cause of cognitive dysfunction and dementia. Behavioural Neurol. 2015, 620143.

Gliga, A.R., Edoff, K., Caputo, F., Källman, T., Blom, H., Karlsson, H.L., et al., 2017. Cerium oxide nanoparticles inhibit differentiation of neural stem cells. Sci. Rep. 7, 9284.

Göktaş, R.K., Macleod, M., 2016. Remoteness from sources of persistent organic pollutants in the multi-media global environment. Environ. Pollut. 217, 33–41.

González, N., Marquès, M., Nadal, M., Domingo, J.L., 2019. Occurrence of environmental pollutants in foodstuffs: a review of organic vs. conventional food. Food Chem. Toxicol. 125, 370–375.

Goodarzi, P., Aghayan, H.R., Larijani, B., Soleimani, M., Dehpour, A.R., Sahebjam, M., et al., 2015. Stem cell-based approach for the treatment of Parkinson's disease. Med. J. Islam. Repub. Iran. 29, 168.

Gopalarethinam, J., Nair, A.P., Iyer, M., Vellingiri, B., Subramaniam, M.D., 2023. Advantages of mesenchymal stem cell over the other stem cells. Acta Histochemica 125, 152041.

Gorell, J.M., Johnson, C.C., Rybicki, B.A., Peterson, E.L., Kortsha, G.X., Brown, G.G., et al., 1999. Occupational exposure to manganese, copper, lead, iron, mercury and zinc and the risk of Parkinson's disease. Neurotoxicology 20, 239–247.

Guan, P.P., Cao, L.L., Yang, Y., Wang, P., 2021. Calcium ions aggravate Alzheimer's disease through the aberrant activation of neuronal networks, leading to synaptic and cognitive deficits. Front. Mol. Neurosci. 14, 757515.

Hara, T., Toyoshima, M., Hisano, Y., Balan, S., Iwayama, Y., Aono, H., et al., 2021. Glyoxalase I disruption and external carbonyl stress impair mitochondrial function in human induced pluripotent stem cells and derived neurons. Transl. Psychiatry 11, 275.

Harischandra, D.S., Ghaisas, S., Zenitsky, G., Jin, H., Kanthasamy, A., Anantharam, V., et al., 2019. Manganese-induced neurotoxicity: new insights into the triad of protein misfolding, mitochondrial impairment, and neuroinflammation. Front. Neurosci. 13.

He, Z.L., Yang, X.E., Stoffella, P.J., 2005. Trace elements in agroecosystems and impacts on the environment. J. Trace Elem. Med. Biol. 19, 125–140.

Heman-Ackah, S.M., Bassett, A.R., Wood, M.J.A., 2016. Precision modulation of neurodegenerative disease-related gene expression in human iPSC-derived neurons. Sci. Rep. 6, 28420.

Hoang, D.M., Pham, P.T., Bach, T.Q., Ngo, A.T.L., Nguyen, Q.T., Phan, T.T.K., et al., 2022. Stem cell-based therapy for human diseases. Signal. Transduct. Target. Ther. 7, 272.

Huang, Y., Li, Y., Pan, H., Han, L., 2023. Global, regional, and national burden of neurological disorders in 204 countries and territories worldwide. J. Global Health 13, 04160.

Javitch, J.A., D'amato, R.J., Strittmatter, S.M., Snyder, S.H., 1985. Parkinsonism-inducing neurotoxin, N-methyl-4-phenyl-1,2,3,6 -tetrahydropyridine: uptake of the metabolite N-methyl-4-phenylpyridine by dopamine neurons explains selective toxicity. Proc. Natl. Acad. Sci. U. S. A. 82, 2173–2177.

Jellinger, K.A., 2010. Basic mechanisms of neurodegeneration: a critical update. J. Cell Mol. Med. 14, 457–487.

Joshi, P., Bodnya, C., Ilieva, I., Neely, M.D., Aschner, M., Bowman, A.B., 2019. Huntington's disease associated resistance to Mn neurotoxicity is neurodevelopmental stage and neuronal lineage dependent. Neurotoxicology 75, 148–157.

Kalluri, R., Lebleu, V.S., 2020. The biology, function, and biomedical applications of exosomes. Science 367.

Kikuchi, T., Morizane, A., Doi, D., Magotani, H., Onoe, H., Hayashi, T., et al., 2017. Human iPS cell-derived dopaminergic neurons function in a primate Parkinson's disease model. Nature 548, 592–596.

Kim, A.H., Chun, H.J., Lee, S., Kim, H.S., Lee, J., 2017. High dose tetrabromobisphenol A impairs hippocampal neurogenesis and memory retention. Food Chem. Toxicol. 106, 223–231.

Kim, H.S., Kim, Y.J., Seo, Y.R., 2015. An overview of carcinogenic heavy metal: molecular toxicity mechanism and prevention. J. Cancer Prev. 20, 232.

Klos, K., Chandler, M., Kumar, N., Ahlskog, J., Josephs, K., 2006. Neuropsychological profiles of manganese neurotoxicity. Eur. J. Neurol. 13, 1139–1141.

Kulick, D., Moon, E., Riffe, R.M., Teicher, G., Deursen, V.A.N., Berson, S., et al., 2022. Amyotrophic lateral sclerosis-associated persistent organic pollutant cis-chlordane causes GABAA-independent toxicity to motor neurons, providing evidence toward an environmental component of sporadic amyotrophic lateral sclerosis. ACS Chem. Neurosci. 13, 3567–3577.

Kumari, K., Swamy, S., Singh, A., 2021. Global Monitoring Plan on Persistent Organic Pollutants (POPs). Persistent Organic Pollutants. CRC Press.

Lamptey, R.N.L., Chaulagain, B., Trivedi, R., Gothwal, A., Layek, B., Singh, J., 2022. A review of the common neurodegenerative disorders: current therapeutic approaches and the potential role of nanotherapeutics. Int. J. Mol. Sci. 23.

Langston, J.W., Ballard, P., Tetrud, J.W., Irwin, I., 1983. Chronic Parkinsonism in humans due to a product of meperidine-analog synthesis. Science 219, 979–980.

Lee, E.J., Choi, Y., Lee, H.J., Hwang, D.W., Lee, D.S., 2022. Human neural stem cell-derived extracellular vesicles protect against Parkinson's disease pathologies. J. Nanobiotechnol. 20, 198.

Lee, R.H., Kim, B., Choi, I., Kim, H., Choi, H.S., Suh, K., et al., 2004. Characterization and expression analysis of mesenchymal stem cells from human bone marrow and adipose tissue. Cell Physiol. Biochem. 14, 311–324.

Lee, Y.-M., Jacobs Jr, D.R., Lee, D.-H., 2018a. Persistent organic pollutants and type 2 diabetes: a critical review of review articles. Front. Endocrinol. 9, 416951.

Lee, Y.-M., Kim, S.-A., Choi, G.-S., Park, S.-Y., Jeon, S.W., Lee, H.S., et al., 2018b. Association of colorectal polyps and cancer with low-dose persistent organic pollutants: a case-control study. PLoS One 13, e0208546.

Lee, Y.K., Ho, P.W., Schick, R., Lau, Y.M., Lai, W.H., Zhou, T., et al., 2014. Modeling of Friedreich ataxia-related iron overloading cardiomyopathy using patient-specific-induced pluripotent stem cells. Pflug. Arch. 466, 1831–1844.

Li, T., Wang, W., Pan, Y.-W., Xu, L., Xia, Z., 2013. A hydroxylated metabolite of flame-retardant PBDE-47 decreases the survival, proliferation, and neuronal differentiation of primary cultured adult neural stem cells and interferes with signaling of ERK5 MAP kinase and neurotrophin 3. Toxicol. Sci. 134, 111–124.

Lindvall, O., Kokaia, Z., 2009. Prospects of stem cell therapy for replacing dopamine neurons in Parkinson's disease. Trends Pharmacol. Sci. 30, 260–267.

Lykhmus, O., Koval, L., Voytenko, L., Uspenska, K., Komisarenko, S., Deryabina, O., et al., 2019. Intravenously injected mesenchymal stem cells penetrate the brain and treat inflammation-induced brain damage and memory impairment in mice. Front. Pharmacol. 10, 355.

Mangalmurti, A., Lukens, J.R., 2022. How neurons die in Alzheimer's disease: implications for neuroinflammation. Curr. Opin. Neurobiol. 75, 102575.

Martins, A.C. JR., Morcillo, P., Ijomone, O.M., Venkataramani, V., Harrison, F.E., Lee, E., et al., 2019. New insights on the role of manganese in Alzheimer's disease and Parkinson's disease. Int. J. Environ. Res. Public Health 16.

Marutle, A., Ohmitsu, M., Nilbratt, M., Greig, N.H., Nordberg, A., Sugaya, K., 2007. Modulation of human neural stem cell differentiation in Alzheimer (APP23) transgenic mice by phenserine. Proc. Natl. Acad. Sci. 104, 12506–12511.

Masrori, P., Van Damme, P., 2020. Amyotrophic lateral sclerosis: a clinical review. Eur. J. Neurol. 27, 1918–1929.

Mattei, V., Delle Monache, S., 2024. Mesenchymal stem cells and their role in neurodegenerative diseases. Cells 13.

Mazzini, L., Fagioli, F., Boccaletti, R., Mareschi, K., Oliveri, G., Olivieri, C., et al., 2003. Stem cell therapy in amyotrophic lateral sclerosis: a methodological approach in humans. Amyotroph. Lateral Scler. Other Mot. Neuron Disord. 4, 158–161.

Mishra, R., Amanullah, A., Upadhyay, A., Dhiman, R., Dubey, A.R., Singh, S., et al., 2020. Ubiquitin ligase LRSAM1 suppresses neurodegenerative diseases linked aberrant proteins induced cell death. Int. J. Biochem. Cell Biol. 120, 105697.

Mishra, R., Bansal, A., Mishra, A., 2021. LISTERIN E3 ubiquitin ligase and ribosome-associated quality control (RQC) mechanism. Mol. Neurobiol. 58, 6593–6609.

Modgil, S., Lahiri, D.K., Sharma, V.L., Anand, A., 2014. Role of early life exposure and environment on neurodegeneration: implications on brain disorders. Transl. Neurodegener. 3, 9.

Nabi, M., Tabassum, N., 2022. Role of environmental toxicants on neurodegenerative disorders. Front. Toxicol. 4, 837579.

Najafi, S., Najafi, P., Kaffash Farkhad, N., Hosseini Torshizi, G., Assaran Darban, R., Boroumand, A.R., et al., 2023. Mesenchymal stem cell therapy in amyotrophic lateral sclerosis (ALS) patients: a comprehensive review of disease information and future perspectives. Iran. J. Basic. Med. Sci. 26, 872–881.

Neely, M.D., Davison, C.A., Aschner, M., Bowman, A.B., 2017. From the cover: manganese and rotenone-induced oxidative stress signatures differ in iPSC-derived human dopamine neurons. Toxicol. Sci. 159, 366–379.

Nicolopoulou-Stamati, P., Maipas, S., Kotampasi, C., Stamatis, P., Hens, L., 2016. Chemical pesticides and human health: the urgent need for a new concept in agriculture. Front. Public Health 4, 148.

Nriagu, J.O., 1989. A global assessment of natural sources of atmospheric trace metals. Nature 338, 47–49.

Olanow, C.W., 2004. Manganese-induced parkinsonism and Parkinson's disease. Ann. N. Y. Acad. Sci. 1012, 209–223.

Park, K.-R., Hwang, C.J., Yun, H.-M., Yeo, I.J., Choi, D.-Y., Park, P.-H., et al., 2020. Prevention of multiple system atrophy using human bone marrow-derived mesenchymal stem cells by reducing polyamine and cholesterol-induced neural damages. Stem Cell Res. Ther. 11, 63.

Parsons-White, A.B., Spitzer, N., 2018. Environmentally relevant manganese overexposure alters neural cell morphology and differentiation in vitro. Toxicol. Vitro 50, 22–28.

Paul, K.C., Krolewski, R.C., Lucumi Moreno, E., Blank, J., Holton, K.M., Ahfeldt, T., et al., 2023. A pesticide and iPSC dopaminergic neuron screen identifies and classifies Parkinson-relevant pesticides. Nat. Commun. 14, 2803.

Pessah, I.N., Lein, P.J., Seegal, R.F., Sagiv, S.K., 2019. Neurotoxicity of polychlorinated biphenyls and related organohalogens. Acta Neuropathol. 138, 363–387.

Pierozan, P., Cattani, D., Karlsson, O., 2020. Hippocampal neural stem cells are more susceptible to the neurotoxin BMAA than primary neurons: effects on apoptosis, cellular differentiation, neurite outgrowth, and DNA methylation. Cell Death Dis. 11, 910.

Politis, M., Wu, K., Loane, C., Quinn, N.P., Brooks, D.J., Rehncrona, S., et al., 2010. Serotonergic neurons mediate dyskinesia side effects in Parkinson's patients with neural transplants. Sci. Transl. Med. 2, 38–946.

Prince, L.M., Neely, M.D., Warren, E.B., Thomas, M.G., Henley, M.R., Smith, K.K., et al., 2021. Environmentally relevant developmental methylmercury exposures alter neuronal differentiation in a human-induced pluripotent stem cell model. Food Chem. Toxicol. 152, 112178.

Pumarega, J., Gasull, M., Lee, D.-H., López, T., Porta, M., 2016. Number of persistent organic pollutants detected at high concentrations in blood samples of the United States population. PLoS One 11, e0160432.

Repar, N., Li, H., Aguilar, J.S., Li, Q.Q., Drobne, D., Hong, Y., 2018. Silver nanoparticles induce neurotoxicity in a human embryonic stem cell-derived neuron and astrocyte network. Nanotoxicology 12, 104–116.

Ruffini, N., Klingenberg, S., Schweiger, S., Gerber, S., 2020. Common factors in neurodegeneration: a meta-study revealing shared patterns on a multi-omics scale. Cells 9.

Saravanan, K.S., Sindhu, K.M., Mohanakumar, K.P., 2005. Acute intranigral infusion of rotenone in rats causes progressive biochemical lesions in the striatum similar to Parkinson's disease. Brain Res. 1049, 147–155.

Scopetti, M., Santurro, A., Gatto, V., La Russa, R., Manetti, F., D'errico, S., et al., 2020. Mesenchymal stem cells in neurodegenerative diseases: opinion review on ethical dilemmas. World J. Stem Cell 12, 168–177.

Senut, M.-C., Sen, A., Cingolani, P., Shaik, A., Land, S.J., Ruden, D.M., 2014. Lead exposure disrupts global DNA methylation in human embryonic stem cells and alters their neuronal differentiation. Toxicol. Sci. 139, 142–161.

Shin, J.W., Lee, J.K., Lee, J.E., Min, W.K., Schuchman, E.H., Jin, H.K., et al., 2011. Combined effects of hematopoietic progenitor cell mobilization from bone marrow by granulocyte colony stimulating factor and AMD3100 and chemotaxis into the brain using stromal cell-derived factor-1α in an Alzheimer's disease mouse model. Stem Cell 29, 1075–1089.

Shrivastav, A., Swetanshu, Singh, P., 2024. The impact of environmental toxins on cardiovascular diseases. Curr. Probl. Cardiol. 49, 102120.

Shu, H., Guo, Z., Chen, X., Qi, S., Xiong, X., Xia, S., et al., 2021. Intracerebral transplantation of neural stem cells restores manganese-induced cognitive deficits in mice. Aging Dis. 12, 371–385.

Silva, M.V.F., Loures, C.D.M.G., Alves, L.C.V., De Souza, L.C., Borges, K.B.G., Carvalho, M.D.G., 2019. Alzheimer's disease: risk factors and potentially protective measures. J. Biomed. Sci. 26, 33.

Simola, N., Morelli, M., Carta, A.R., 2007. The 6-hydroxydopamine model of Parkinson's disease. Neurotox. Res. 11, 151–167.

Singh, C., Liu, L., Wang, J.M., Irwin, R.W., Yao, J., Chen, S., et al., 2012. Allopregnanolone restores hippocampal-dependent learning and memory and neural progenitor survival in aging 3xTgAD and nonTg mice. Neurobiol. Aging 33, 1493–1506.

Singh, R., Gautam, N., Mishra, A., Gupta, R., 2011. Heavy metals and living systems: an overview. Indian J. Pharmacol. 43, 246–253.

Song, B., Cha, Y., Ko, S., Jeon, J., Lee, N., Seo, H., et al., 2020. Human autologous iPSC–derived dopaminergic progenitors restore motor function in Parkinson's disease models. J. Clin. Investig. 130, 904–920.

Sträter, E., Westbeld, A., Klemm, O., 2010. Pollution in coastal fog at Alto Patache, Northern Chile. Environ. Sci. Pollut. Res. Int. 17, 1563–1573.

Su, F.-C., Goutman, S.A., Chernyak, S., Mukherjee, B., Callaghan, B.C., Batterman, S., et al., 2016. Association of environmental toxins with amyotrophic lateral sclerosis. JAMA Neurol. 73, 803–811.

Tamm, C., Sabri, F., Ceccatelli, S., 2008. Mitochondrial-mediated apoptosis in neural stem cells exposed to manganese. Toxicol. Sci. 101, 310–320.

Tian, Y., Gui, W., Rimal, B., Koo, I., Smith, P.B., Nichols, R.G., et al., 2020. Metabolic impact of persistent organic pollutants on gut microbiota. Gut Microbes 12, 1848209.

Tidball, A.M., Bryan, M.R., Uhouse, M.A., Kumar, K.K., Aboud, A.A., Feist, J.E., et al., 2014. A novel manganese-dependent ATM-p53 signaling pathway is selectively impaired in patient-based neuroprogenitor and murine striatal models of Huntington's disease. Hum. Mol. Genet. 24, 1929–1944.

Tiwari, S.K., Agarwal, S., Seth, B., Yadav, A., Ray, R.S., Mishra, V.N., et al., 2015. Inhibitory effects of bisphenol-A on neural stem cells proliferation and differentiation in the rat brain are dependent on Wnt/β-catenin pathway. Mol. Neurobiol. 52, 1735–1757.

Tong, Y., Yang, H., Tian, X., Wang, H., Zhou, T., Zhang, S., et al., 2014. High manganese, a risk for Alzheimer's disease: high manganese induces amyloid-β related cognitive impairment. J. Alzheimers Dis. 42, 865–878.

Verheijen, M.C.T., Krauskopf, J., Caiment, F., Nazaruk, M., Wen, Q.F., Van Herwijnen, M.H.M., et al., 2022. iPSC-derived cortical neurons to study sporadic Alzheimer disease: a transcriptome comparison with post-mortem brain samples. Toxicol. Lett. 356, 89–99.

Waldmann, T., Grinberg, M., König, A., Rempel, E., Schildknecht, S., Henry, M., et al., 2017. Stem cell transcriptome responses and corresponding biomarkers that indicate the transition from adaptive responses to cytotoxicity. Chem. Res. Toxicol. 30, 905–922.

Who, C.O., 2020. World health organization. Air Quality Guidelines for Europe.

Witkowska, D., Słowik, J., Chilicka, K., 2021. Heavy metals and human health: possible exposure pathways and the competition for protein binding sites. Molecules 26.

Xie, J., Wu, S., Szadowski, H., Min, S., Yang, Y., Bowman, A.B., et al., 2023. Developmental Pb exposure increases AD risk via altered intracellular $Ca2^{+}$ homeostasis in hiPSC-derived cortical neurons. J. Biol. Chem. 299.

Xu, P., Liu, A., Li, F., Tinkov, A.A., Liu, L., Zhou, J.-C., 2021. Associations between metabolic syndrome and four heavy metals: a systematic review and meta-analysis. Environ. Pollut. 273, 116480.

Yates, P.L., Patil, A., Sun, X., Niceforo, A., Gill, R., Callahan, P., et al., 2021. A cellular approach to understanding and treating Gulf War Illness. Cell Mol. Life Sci. 78, 6941–6961.

Yavuz, I.H., Yavuz, G.O., Bilgili, S.G., Demir, H., Demir, C., 2018. Assessment of heavy metal and trace element levels in patients with telogen effluvium. Indian. J. Dermatol. 63, 246–250.

Ye, T., Duan, Y., Tsang, H.W.S., Xu, H., Chen, Y., Cao, H., et al., 2021. Efficient manipulation of gene dosage in human iPSCs using CRISPR/Cas9 nickases. Commun. Biol. 4, 195.

Yuan, Y., Tang, X., Bai, Y.F., Wang, S., An, J., Wu, Y., et al., 2018. Dopaminergic precursors differentiated from human blood-derived induced neural stem cells improve symptoms of a mouse Parkinson's disease model. Theranostics 8, 4679–4694.

Zaazaa, A.M., Abd El-Motelp, B.A., Ali, N.A., Youssef, A.M., Sayed, M.A., Mohamed, S.H., 2022. Stem cell-derived exosomes and copper sulfide nanoparticles attenuate the progression of neurodegenerative disorders induced by cadmium in rats. Heliyon 8, e08622.

Zakrzewski, W., Dobrzyński, M., Szymonowicz, M., Rybak, Z., 2019. Stem cells: past, present, and future. Stem Cell Res. Ther. 10, 68.

Zarei, S., Carr, K., Reiley, L., Diaz, K., Guerra, O., Altamirano, P.F., et al., 2015. A comprehensive review of amyotrophic lateral sclerosis. Surg. Neurol. Int. 6, 171.

Zhang, L., Wang, H., Abel, G.M., Storm, D.R., Xia, Z., 2020. The effects of gene-environment interactions between cadmium exposure and apolipoprotein E4 on memory in a mouse model of Alzheimer's disease. Toxicol. Sci. 173, 189–201.

Zhang, Y.X., Liu, Y.P., Miao, S.S., Liu, X.D., Ma, S.M., Qu, Z.Y., 2021. Exposure to persistent organic pollutants and thyroid cancer risk: a study protocol of systematic review and meta-analysis. BMJ Open 11, e048451.

Zhou, Y., Sun, M., Li, H., Yan, M., He, Z., Wang, W., et al., 2013. Recovery of behavioral symptoms in hemi-parkinsonian rhesus monkeys through combined gene and stem cell therapy. Cytotherapy 15, 467–480.

CHAPTER SEVEN

The use of human iPSC-derived neuronal cultures for the study of persistent neurotoxic effects

Anke M. Tukker and Aaron B. Bowman*
School of Health Sciences, Purdue University, West Lafayette, IN, United States
*Corresponding author. e-mail address: bowma117@purdue.edu

Contents

Abstract

Exposures can occur at any life stage. Durations of exposure vary from acute through chronic. In some cases, effects of exposure are reversed as soon as the toxicant is removed whereas in other cases, effects persist long after cessation of exposure. Persistent effects are changes in biological systems observed long after cessation of exposure regardless of exposure duration. These effects are often still present in

Advances in Neurotoxicology, Volume 12
ISSN 2468-7480, https://doi.org/10.1016/bs.ant.2024.07.003

absence of the toxicant and can occur following a latency period. Known cases of persistent effects in toxicological history include (a) the Minamata Bay disaster where people experienced a chronic methylmercury poisoning through their diet and (b) fetal alcohol disorder. In both cases, health effects were severe and the effects were observed long after the toxicant was removed from the body. Amongst the mechanisms involved in neuronal persistency are alterations in differentiation, development, migration and maturation of cell populations as well as neurotransmitter balance, epigenetic changes and somatic mosaicism. All these outcome measures and mechanisms can be studied using stem cell models. Additionally, stem cells can be differentiated to cells reflecting particular brain regions of interest and kept in culture for prolonged periods of time. Stem cells provide a platform to study effects of exposure from development, to maturation and all through the aging process and thus allows for sampling across most life stages.

1. Persistent toxicity

Toxicity can occur following different durations of exposure. Acute exposures generally last anywhere from a few seconds to several hours (Kimura et al., 2023; Dominah et al., 2017), whereas chronic exposures are defined as occurring over days to years (Jin et al., 2022; Du et al., 2023). Between acute and chronic toxicity, subacute effects as well as sub-chronic effects can be of interest and are studied as well. Exposure duration of these types of studies falls somewhere in between acute and chronic. The relevant duration of exposure depends on the model used and the question asked. Generally speaking, exposures in in vitro experimental settings tend to be shorter than those used in in vivo experiments regardless of whether acute or chronic exposure is studied. This discrepancy in durations is in part related to the speed with which the toxicant can reach its biological target. In in vitro systems the route to reach a target is more straightforward/direct than in vivo; as the compound does not first need to enter the bloodstream and travel through the circulation as it is immediately in the culture media. Additionally, in vitro models often lack relevant biological barriers to the 'target' cells, such as the blood-brain-barrier, and thus the toxicant will reach its target faster and at a higher at least initial concentration relative to the equivalent blood levels. Effects of acute and chronic exposures have typically been assessed in the scientific literature either in the continued presence of toxicant or shortly after its removal from the system (Fig. 1). An understudied type of effect following exposure is persistent toxicity, i.e. well after removal of the exposure and/or beyond the point of detectable levels of the toxicant in the system. Here we discuss the

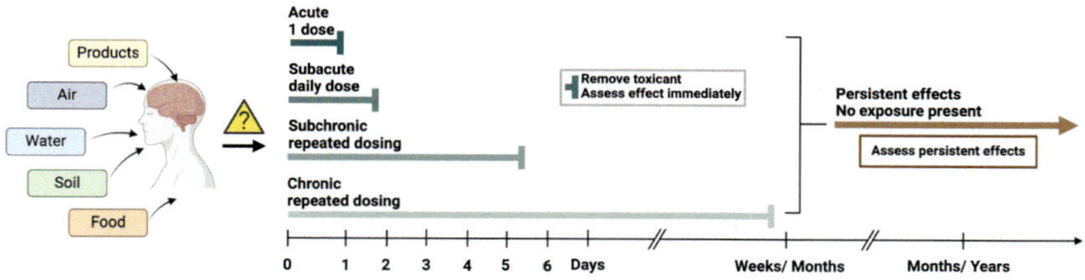

Fig. 1 Effects of exposures can be assessed following acute, subacute, sub-chronic or chronic exposure. These types of studies take place right after removal of exposure. Then, persistent effects can be assessed long after cessation of exposure. Persistent effects can occur regardless of exposure duration. *Image created with BioRender.com.*

potential and value of hiPSC model systems for studying the long-term impact of environmental chemical toxicants in the context of human health and disease.

1.1 What are persistent effects?

When studying persistent effects, the focus is really on changes in the biological model system that persist well after cessation of exposure and thus in absence of the toxicant. It does not require a specific exposure length. Exposure time can be short, aligning with acute exposure, or longer in line with chronic exposure (see Section 2 for more details). Moreover, effects of acute and chronic exposures can be studied and then by letting the model age, persistent effects can be investigated. An example of persistency following acute exposure is a study by Silva et al. They exposed mice to a single dose of methamphetamine and found depressive like behavior 49 days post-administration, long after the compound was removed from the body (Silva et al., 2014).

The underlying idea of persistent toxicity is that following cessation of exposure, the biological system can either fully recover from the insult after prolonged period of time, partially recover or no recovery is observed. Persistent effects are not necessarily irreversible. When persistent effects are irreversible, they become permanent effects (Neuwirth and Emenike, 2024). This recovery is toxicant specific and so is the time it takes to recover. For example, the inhibitory effect of perfluorooctanoic acid (PFOA) on the GABA-evoked current in an oocyte expressing a $GABA_A$-receptor is reversible within seconds of cessation of exposure while the effect of perfluorooctane sulfonate (PFOS) is not readily reversible (Tukker et al., 2020a). Additionally, short-term solvent exposure is reversible following cessation of exposure whereas effects caused by organophosphates and carbamates are reversible following reactivation of the target (Serafini et al., 2024).

Absence of the toxicant is key for persistency. This is in contrast to effects caused by toxicants that remain in the system such as cadmium (Cd) and lead (Pb). Both compounds will cause toxic effects for a long time, but these effects are at least plausibly due to being related more to a chronic exposure through the continued, or prolonged, toxicant presence; making it difficult to distinguish chronic effects from persistent effects in cases where the toxicant itself persists in the biological system. Cadmium exposure occurs via ingestion of contaminated food and water or through inhalation of cigarette smoke (Genchi et al., 2020). Once in the body, Cd mainly accumulates in the liver and kidneys leading to liver damage and kidney problems (Feng et al., 2022; Souza-Arroyo et al., 2022). Additionally, it affects DNA (Pizzino et al., 2014) and can increase cancer risk (Florez-Garcia et al., 2023). Due to a low excretion rate, Cd has a half-life time of more than 25–30 years in the human body (Satarug et al., 2010; Järup and Åkesson, 2009). It is thus unclear whether the effects observed for long time periods are persistent caused by earlier exposure or are the result of residual Cd. A similar issue occurs with the known neurotoxicant Pb for which exposure occurs via ingestion or inhalation. Although Pb is deposited in soft tissue, its main target is bone (Rodríguez and Mandalunis, 2018). Once in bone, the half-life time is 25–30 years (Agency for Toxic Substances and Disease Registry, 2020). Accumulated lead is released from the bone slowly during life. Pb is able to pass the blood brain barrier due to its ability to substitute for calcium ions (Sharma et al., 2015) and induces (developmental) neurotoxicity (Bjorklund et al., 2024). It is thought that many of the neurotoxic effects are related to this calcium-mimicking ability, but lead also inhibits calcium actions (Bressler and Goldstein, 1991). Additionally, Pb can alter the release of neurotransmitters at the presynaptic side (Bressler and Goldstein, 1991). But for Pb the same holds true as for Cd, due to the long half-life time and the slow release from bone, it is unclear whether effects are persistent or prompted by newly released Pb.

1.2 Persistent effects versus latent effects

Latent effects have a delay in onset of effects following exposure and can occur months to years after beginning (Vandivort and Eaton, 2014). Since most targets are not immediately accessible from the exposure site, most toxicants exhibit some degree of latency (Vandivort and Eaton, 2014). This is often the time the toxicant needs to be distributed throughout the body and reach the target site. The latency period is also related to the dose with a lower dose the longer the latent period (Pletz et al., 2016). The metal manganese (Mn) is an example of an environmental toxicant that induces latent effects (Kim et al., 2022). Following exposure, neurological effects are not observed for years and neurotoxicity may be asymptotic; however, structural alterations in the brain progress slowly (Lee et al., 2018a). Persistent effects could be latent in the sense that they do not occur immediately when exposure begins.

The main difference between persistent effects and latent effects is that persistent effects are maintained in absence of exposure, whereas latent effects start to occur long after exposure begins but whilst toxicant is still present. If the latent effect persists till long after cessation of exposure, it becomes a (latent) persistent effect. Persistent effects do not necessarily have a latency period, although this could be possible as is the case with exposure to methylmercury (MeHg; see Section 2).

2. Known cases of persistent toxicity throughout toxicological history

Throughout history, several cases of exposures capable of inducing persistent effects in humans became known. Some of those exposures had devastating adverse effects on human health. Exposures can for example come from environmental contaminants, licit drugs (e.g., alcohol and tobacco) or illicit drugs (e.g., cocaine, methamphetamine) (Derauf et al., 2009). In this section some known persistency case studies will be detailed. Additionally, maternal-fetal exposures as a source of persistent toxicity will be discussed. The case studies will emphasize the urge for studies of persistent toxicity.

2.1 Persistent toxicity following MeHg exposure

MeHg is an abundant environmental neurotoxin. Humans are exposed to this organic mercury via diet, especially through consumption of rice and fish. Once in the body, MeHg distributes via the bloodstream to the different organs. It can pass the blood-brain-barrier and can thus reach the brain. The biological half-life time of MeHg is approximately 50 days (Rand and Caito, 2019). Following exposure to MeHg, body burden can be reduced by chelating therapy. However this has limited effects on the improvement of neurological outcomes (Clarkson et al., 1981; Bakir et al., 1976; Nierenberg et al., 1998) though some chelating agents only redistribute MeHg (Kosnett, 2013). Failure in treatment of MeHg poisoning suggests that once brain damage occurs it cannot be reversed.

Neurons are the preferential target for MeHg and especially cortical glutamatergic (Ke et al., 2023; Feng et al., 2014; Kaur et al., 2007) and nigral dopaminergic neurons (Caito and Aschner, 2016; Kendricks and Newland, 2021; Daré et al., 2003; Ke et al., 2020). In the case of glutamatergic neurons, *ex vivo* studies showed that MeHg decreased glutamate uptake into mice cerebral cortical slices of both adults (Farina et al., 2003) and weanlings (Manfroi et al., 2004). This together with spontaneous release of glutamate induced by MeHg contribute to elevated extracellular glutamate levels and MeHg-induced excitotoxicity (Reynolds and Racz, 1987; Vendrell et al., 2007). MeHg inhibits synaptic glutamate recycling via redox and transporter-based mechanisms (Allen et al., 2001; Porciúncula et al., 2003). Exposure to MeHg also induces oxidative stress (Wei et al., 2023; Li et al., 2021). This affects glutamate homeostasis via inhibition of extracellular glutamate uptake and by alteration of the transcription of genes that are critical for glutamate cycling (Culbreth and Aschner, 2016). Thus, glutamate dyshomeostasis has a vital role in MeHg-induced neurotoxicity. When it comes to dopaminergic neurons, MeHg inhibits dopamine uptake and metabolism (Tiernan et al., 2015; Bonnet et al., 1994; Komulainen and Tuomisto, 1985). In rats, developmental MeHg exposure inhibits DA synthesis and turn-over (Bartolome et al., 1984). Additionally, a higher release of DA associated with lower dopamine transporter activity was found (Dreiem et al., 2009). Following *in utero* exposure, latent effects on dopamine recycling were identified (Castoldi et al., 2006). Disruptions in dopamine mediated behavior caused by developmental exposure, persist through adult life (Kendricks et al., 2020).

Exposure to MeHg can thus induce persistent effects. Subacute exposures to 25–40 ppm can cause neurological symptoms with a latent period of weeks or months in adults (Bakir et al., 1973; Weiss et al., 2002) and years in infants (Amin-Zaki et al., 1979). The long latency period has been confirmed in a low dose non-human primate study (Rice, 1996). Thus, the long-term health effects for those with elevated MeHg blood levels, for example by high fish diet, may not be unmasked clinically for decades (Weiss and Bellinger, 2006; Zheutlin et al., 2024). The decline in brain function due to MeHg exposure can last for decades as has been shown by longitudinal investigation of survival of patients in both Iraq and Japan decades post exposure (Davis et al., 1994; Yorifuji et al., 2011; Kinjo et al., 1993; Lam et al., 2013, 2012; Takaoka et al., 2018).

In 1971–72 in Iraq, wheat seeds that were meant for planting had been used mistakenly for human food production and the public ate MeHg contaminated bread. Signs of MeHg poisoning started to become apparent two to five weeks later and in December 1971 the first person was admitted to the hospital, reaching a total of 6148 admission in the following weeks (Skerfving and Copplestone, 1976). However, it is likely that not all patients consulted medical attention. Symptoms were sensory disturbances (including numbness of hands, feet, lips and tongue), motor disturbances (visible in impairment of gait) and visual disturbance (Skerfving and Copplestone, 1976). What is interesting about this Iraq case, is that most patients with mild symptoms recovered almost completely and those with moderate symptoms generally improved and sometimes fully recovered, and patients with severe symptoms either died or somewhat improved (Skerfving and Copplestone, 1976). This is different from the poisoning in Minamata Bay in Japan and could be due to the difference in exposure duration. Whereas exposure in Iraq was sub-chronic and occurred for approximately 4 months, exposure in Japan occurred over a much longer period of time and thus more chronically.

For more than 30 years, a chemical plant in Japan released waste water containing mercury in the Minamata Bay. Once mercury enters the water, microorganisms convert it into MeHg which then bioaccumulates in the food chain. The population living near the bay consumed polluted fish for extended period of time and started to exhibit symptoms of MeHg poisoning. Typical symptoms were the same as observed in Iraq: constriction of visual field, sensory disturbances and motor disturbances (Harada, 1995). Unlike the case in Iraq, symptoms were very persistent and could still be observed 50 years after the plant halted its waste disposal (Takaoka et al., 2018). Moreover, a stark latency period was noted. First symptoms kept on occurring until at least 50 years after exposure (Takaoka et al., 2018). What was also found was that latency period increased with a decreased exposure level (Takaoka et al., 2018). Additionally, it was found that individuals with Minamata Disease were more likely to suffer from forgetfulness, weakness and constriction of visual field as compared to age-matched controls and that their symptoms showed an accelerated decline (Kinjo et al., 1993). These studies were conducted with adults exposed to MeHg, the case for congenital MeHg exposure is different.

2.2 Persistent toxicity as a result of maternal-fetal exposures

The developing fetus is protected by the placenta during pregnancy. Although the placenta tightly regulates the exchange of exogenous and endogenous compounds between mother and fetus, some exposures are able to cross this barrier and will reach the fetal environment (Aylward et al., 2014; Needham et al., 2011). Exposure during development can have severe and persistent health effects as discussed in examples below.

2.2.1 Fetal Minamata bay disease

MeHg rapidly crosses the placental barrier and is thus able to reach the developing fetus. Besides being able to cross the barrier, MeHg seems to prefer the developing fetus over the mother (Rand and Caito, 2019). This became apparent during the Minamata bay disaster. Exposed expecting mothers often showed mild symptoms or no symptoms, whereas their babies did exhibit more severe symptoms than the mother (Harada, 1972). What was striking was that these mothers also showed light or mild symptoms as compared to other adults with similar mercury exposures (Dos Santos et al., 2018). Symptoms of fetal Minamata disease started to occur in infants at early ages, but were not apparent directly at birth. Thus, showing a latency period. Signs of disease included delayed movement, failure to follow visual stimuli, uncoordinated suckling and swallowing, mental retardation and cerebral palsy (Dos Santos et al., 2018; Gilbertson, 2009; Amin-Zaki et al., 1979; Eto et al., 1992). This congenital MeHg poisoning became known as fetal Minamata disease (Dos Santos et al., 2018; Fahrion et al., 2012).

A few autopsies of brains from fetal Minamata disease patients have been performed and the following neuropathological changes have been reported: decrease and disappearance of cortical nerve cells, cerebellar atrophy and hypoplasia, abnormal neural cytoarchitecture, hypoplasia of corpus callosum, dysmyelination of white matter and hydrocephalus (Dos Santos et al., 2018; Matsumoto et al., 1965). On a microscopic level, hypoplastic dendrites in Purkinje cells were observed as well as a horizontally oriented (disoriented position) (Chang, 1984). Neuronal loss mainly occurred in the granule cells in the cerebellum and pyramidal neurons in the cerebrum (Dos Santos et al., 2018). Additionally, there were strong indications of disrupted neuronal maturation, growth and migration (Choi et al., 1978; Choi, 1986). Disruptions in migration and cytoarchitecture can be the result of changes in neural cell adhesion molecules and microtubules (Dos Santos et al., 2018).

Fetal Minamata disease also made it apparent that timing of the exposure matters. The earlier during development the fetus or child was exposed, the more aberrant brain development and the more lesions could be observed. If exposure occurs during adult life, lesions are localized to certain regions of the cortex, in non-fetal Minamata disease lesions occur all over the cortical region and fetal exposure has more lesions all over the cortical area (Dos Santos et al., 2018).

Although the biological half-life of MeHg may be as high as in the range of months (Rand and Caito, 2019), effects of exposure are present long after removal of the metal. Therefore, congenital MeHg exposures result in long-lasting persistent effects.

2.2.2 Fetal alcohol spectrum disorders

Another example is the case of prenatal alcohol exposure. Alcohol can readily cross the placental barrier, a fact that became apparent in the 1900s (Brown et al., 2019) and accumulates in the amniotic fluid (Brien et al., 1983). However, it was not until 1973 that the effect of alcohol on the developing fetus gained widespread attention (Petrelli et al., 2018). Once in the fetus, it reaches fetal organs including the brain (Chung et al., 2021). This developmental exposure can cause fetal alcohol spectrum disorders (FASDs). FASDs is an umbrella term that covers all effects following prenatal alcohol exposure. Effects include physical impairments, neurological problems and behavioral development (Thomas et al., 2010). Symptoms and severity depend on the dose and timing of exposure and play an important role in the development of craniofacial abnormalities and in brain development (Petrelli et al., 2018). Alcohol affects the developing brain in the first half of the first trimester when organ development is rapid and in the third trimester when the brain growth spurt occurs (Petrelli et al., 2018).

Excessive neuron death has been seen as the cause for the damage and causing FASD symptoms (West et al., 1990) but late prenatal exposure is thought to block NMDA receptors and hyperactivate GABA transmission (Weiner and Valenzuela, 2006; Olney et al., 2002). Neuron depletion is not the only factor leading to FASD symptoms, some long-term consequences may be due to maladaptive plastic rearrangement of the remaining surviving neurons (Granato and Dering, 2018). The cerebral cortex is one of the most damaged parts of the brain by alcohol exposure with pyramidal neurons of layer 5 being more susceptible as compared to principal neurons in other layers of the cortex (Olney et al., 2002). GABAergic cortical interneuron populations do not seem to suffer naturally occurring cells death and some populations are increased in size following exposure to alcohol (Granato and Dering, 2018). Studies now seem to show that alcohol exposure delays the maturation process and prevents some neuron populations from undergoing cell decline (Schierle et al., 1997). Others point towards a change in migration speed resulting in higher number of interneurons in certain brain regions as compared to others (Cuzon et al., 2008). This increase in GABAergic interneurons can lead to a permanent shift of the excitatory-inhibitory balance towards the inhibitory side. (Skorput et al., 2015). Though above-described mechanisms may all result in apoptosis, some may not. Presence of GABA and glutamate neurotransmitters drive the migration of cortical neurons to the proper layer by acting as attractants (Behar et al., 1999, 2000). By inhibiting NMDA and activating GABA, receptor activation alcohol thus interferes with the migration of cortical neurons during brain development. Additionally, alcohol is known to interfere with mitogenic growth factors and growth inhibiting factors (Luo and Miller, 1998) thereby affecting the cell cycle.

Although breakdown of alcohol in the fetus is not as efficient as in adults and occurs at much slower pace, all will be removed relatively quickly, on the order of days to weeks. Many effects will thus be there in absence of continued exposure. Hence, alcohol exposure is persistently altering homeostasis with long-lasting consequences as result.

2.2.3 The case of thalidomide

Another well-known example of maternal-fetal exposure is the case of the drug thalidomide. During the 1950s, thalidomide (knowns as Softenon in Europe and Distaval in the UK and Australia) was widely prescribed to treat morning sickness in pregnant women. In the late 1950s, early 1960s it became apparent that this drug caused, besides peripheral neuropathy in mothers, severe birth defects and it became known as the biggest man-made medical disaster (Vargesson, 2015). This disaster changed the way drugs are tested before gaining market approval. Additionally, this case showed that inter-species differences do exist in drug responses as it was tested on mice and no issues were identified (Kim and Scialli, 2011). Thalidomide causes cell death in embryonic tissue, halted cell proliferation, induction of reactive oxygen species (ROS) and DNA oxidation (Wells et al., 2005; Knobloch et al., 2011). The results of all these immediate toxic effects are seen as developmental abnormalities especially in the upper limbs and hands (Mansour et al., 2019). However, the drug could affect almost all other tissues and organs in the developing fetus resulting in, amongst others, lower limb abnormalities, hearing problems, spinal issues and micro-ophthalmia (Vargesson, 2019; Mansour et al., 2019). These symptoms are collectively referred to as thalidomide embryopathy. It is hypothesized that thalidomide affects the upper limbs more than the lower limbs because these develop earlier (Vargesson, 2015). Thalidomide has a short half-life time, approximately 6 h (Teo et al., 2004), and thus a single dose early in pregnancy would affect upper limbs.

Considering all of this together, thalidomide can thusly be seen as a toxicant that causes persistent effects as effects persist after the drug is eliminated from the body. However, these are effects on an organismal level. Effects can be the result of the fact that cells were never made or are otherwise absent. This is thus very different from persistent cellular functional changes where cells were made and are still present in the organism, but have adapted their homeostasis in order to survive the toxic insult. Thus, persistency can be divided into persistency at the whole organism level or at a cellular level. In other words, a difference between structural persistency (presence or absence of cellular and organ system structure) and functional persistency (changes to the activity or functional characteristics of cellular and organ systems that are normally present).

3. Persistent effects and the developmental origins of health and disease hypothesis

The maternal-fetal exposures discussed in the previous section have been shown to cause persistent effects on either a cellular level or on whole organism level. Developmental (perinatal or congenital) exposures have been hypothesized to be environmental triggers for life-long persistent effects that can increase the risk for diseases, amongst which neurodegenerative diseases, later in life. This is the so-called Developmental Origins of Health and Disease (DOHaD) hypothesis. The DOHad hypothesis is an extension of the Barker hypothesis applied to early life toxin exposures leading to vulnerability for later-life chronic illnesses (Bousquet et al., 2015; Capra et al., 2013; Heindel and Vandenberg, 2015). Evidence in support of this hypothesis has been reported in rodent and iPSC models developmentally exposed to environmental toxicants, including metals and pesticides, showing persistent effects throughout adult life (Xie et al., 2023; Bousquet et al., 2015; Heindel and Vandenberg, 2015; Lee and Freeman, 2016; Lee et al., 2018b, Fredriksson et al., 1993). Additionally, these studies found an increased vulnerability to later-life exposures (Eriksson, 1997; Lucchini et al., 2018). Developmental exposure to Pb results in transcriptional and histopathological changes related to neurodegeneration and Alzheimer's disease pathology in adult brains of non-human primates, rodents and zebrafish (Bihaqi et al., 2014; Lee and Freeman, 2016; Wu et al., 2008). In a retrospective human study, a relationship between maternal-fetal Pb exposure and an alteration in genes and enzymes implicated in amyloid plague formation was identified (Mazumdar et al., 2012). Neonatal exposure to paraquat has been shown to result in permanent changes in dopamine homeostasis and behavior in mice (Fredriksson et al., 1993).

Thus maternal-fetal exposures can result in irreversible changes in the developing fetus. Irreversible processes causing toxic effects include irreversible receptor binding. However, in the case of developmental exposures causing persistent (neuro)toxic effects, this mechanism is unlikely. The biological process causing later-life effects must persist for decades, till well after any protein bound by toxicant would be turned over. The precise mechanisms by which environmental exposures (developmental or later in life) result in persistent effects remains unclear as well as how the persistency is maintained. However, some underlying mechanisms seem to be shared across different cases or are hypothesized to induce persistent effects.

4. Mechanisms of persistent effects and the potential for hiPSCs

From the above examples it is clear that there are many different mechanisms that can contribute to persistent effects. In the following section several mechanisms will be highlighted. Attention will be paid to how these mechanisms are or can be studied using stem cells. The highlighted mechanisms below are only a handful of mechanisms that could potentially be involved in persistency and are not meant as a complete list. Some examples of persistency mechanisms not discussed but that could be studied using stem cells are formation of reactive oxygen species (Neely et al., 2017; Brazdis et al., 2020; Pamies et al., 2022), disrupted neurotransmitter or calcium homeostasis (Hook et al., 2014; Xie et al., 2023) and neuronal network activity (Tukker et al., 2020b, Ronchi et al., 2021).

4.1 Persistent epigenetic changes induced by environmental exposures

Epigenetics play an important role during embryonal and fetal development by dividing the genome into transcriptionally active and quiescent parts that directly affect cell differentiation and later in life cell maturation (Ho et al., 2012). Over recent years, the impact of the environment on epigenetic regulation has received increased interest from researchers. Often, changes in gene expression induced by environmental exposures have been linked to altered DNA methylation patterns or changes in histone modifications such as acetylation (Feil and Fraga, 2012). In some cases, the effects of these alterations become immediately clear whereas in other cases, there is a latency period and effects surface later in life (Tang et al., 2012). Environmental exposures often have a nonlinear effect on epigenetic changes and severity of effects mainly depends on the window of exposure. A low dose exposure during fetal development will have a more severe impact than a high dose exposure during adult life (Ho et al., 2012). A study on the effect of polycyclic aromatic hydrocarbon (PAH) exposure found that DNA adducts increased in human cord blood, indicating PAH cross the placental barrier and altered the fetal epigenome (Perera et al., 2011). Additionally, researchers estimated that the fetal dose was approximately 10 times lower than the dose the mother received suggesting higher fetal sensitivity (Perera et al., 2011). A study testing samples of 406 seven year old children from the Seychelles that had been prenatally exposed to MeHg through maternal diet found altered DNA methylation of one GRIN2B (NMDA receptor subunit) CpG and two NR3C1 (glucocorticoid receptor) CPGs (Cediel Ulloa et al., 2021). These changes in methylation are hypothesized to lower gene expression which could result in adverse neurodevelopmental outcomes (Cediel Ulloa et al., 2021).

Epigenetic changes resulting from environmental exposures can also be studied using stem cell models. Spildrejorde et al exposed human embryonic stem cells during neuronal differentiation to maternal therapeutic doses of paracetamol and collected samples at different time points during exposure (Spildrejorde et al., 2023). Using single cell RNA-sequencing combined with assay for transposase-accessible chromatin (ATAC)-sequencing they identified paracetamol-induced chromatin changes linked to gene expression (Spildrejorde et al., 2023). Interestingly, genes they identified as altered, e.g., KCNE3, matched with differentially methylated genes identified in cord blood from paracetamol exposed samples (Spildrejorde et al., 2023; Gervin et al., 2017). This finding strongly suggests that stem cell-based models combined with high data content -omic approaches can be used to predict in vivo outcomes of environmental exposure.

It is important to keep in mind that, unlike somatic mosaicism (see Section 4.2), epigenetic changes do not cause alterations in the DNA sequence and can thus be reversible at least hypothetically if not in reality. They are though mitotically inheritable (D'urso and Brickner, 2014) and can thus be persistent for this reason. Nowadays, it is even hypothesized that epigenetic changes can be passed on to offspring via germline changes in epigenetic status (Bohacek and Mansuy, 2013).

4.2 Somatic mosaicism induced by environmental exposures

A mechanism that recently gained attention is environmental exposure induced somatic mosaicism (Amolegbe et al., 2022). If an exposure occurs during cell division, the risk on somatic mutations increases (Greenman et al., 2007). Somatic mutations occur after conception and cause alterations in somatic tissues (Jamuar et al., 2016). Due to the fact that these mutations do not affect the germline, they do not carry over to the next generation (Miles and Tadi, 2024). It has been speculated that somatic mutations during brain development contribute to neuronal diversity (Jamuar et al., 2016). When somatic mutations occur early in development, only a subset of cells harbor the mutation and mosaicism thus results due to this (Miles and Tadi, 2024). Depending on when the mutation happens during embryonic development, all cell types can be affected or only a few cell types (Jamuar et al., 2016). Mutations happening very late in development would potentially only affect a single neuron, whereas earlier mutations can affect brain tissue (D'gama and Walsh, 2018; Mohiuddin et al., 2022). When occurring in the brain, these mutations can contribute to normal and abnormal brain development depending on the number of cell(types) affected and the effect of the mutation on the cell (D'gama and Walsh, 2018). Severity of the effect thus depends heavily on timing. Somatic mosaicism can affect neuronal migration and focal brain overgrowth (D'gama and Walsh, 2018) and could result in neurodevelopmental and neuropsychiatric disorders when occurring in progenitor cells (Mohiuddin et al., 2022). An example of mosaic mutation with severe implications is a mutation in the mTOR pathway with dysmorphic and disorganized neurons and glia resulting in cortical malformation, which has been found to be an underlying cause of focal epilepsy (Kim et al., 2019).

Somatic mutations can be modeled in stem cell-based systems. Tuberous sclerosis is a classic example of brain mosaicism phenotype with cells in the brain containing mutations in one or both copies of the TSC1 or TSC2 gene creating patches called "tubers" (Wang et al., 2023). To gain better understanding in the effect of this mosaicism, human brain organoids were developed. One study suggests that dysfunctional astrocytes cause synapse disruption (Dooves et al., 2021), whereas another study showed that altered cytoskeletal dynamics alter neuronal development and contribute to the disease (Catlett et al., 2021). Baker et al compared the sensitivity of karyotyping, PCR-based methods and fluorescence in situ hybridization (FISH) for the study of mosaicism in hiPSCs (Baker et al., 2016). They showed that these methods have a limit of detection for mosaicism around 5%–10%. Somatic mosaicism can also be detected using single cell sequencing techniques (Bizzotto et al., 2021; Huang and Lee, 2022; Brazhnik et al., 2020). Single cells techniques have been extensively and successfully used in combination with hiPSCs (Tukker and Bowman, 2024). Thus, stem-cell based models combined with the available (single cell) methods can be used to study exposure induced mosaicism.

5. Opportunities in experimental designs for the study of persistent effects

To study persistent effects, the use of hiPSC-derived cultures can be a good option. Recently, an increasing number of protocols became available to differentiate stem cells into different neuronal subtypes or brain regions. Additionally, hiPSCs can be differentiated in 2D or in 3D to form organoids (Camp et al., 2015; Gantner et al., 2021; Perriot et al., 2021; Kim et al., 2021). The following section outlines opportunities regarding experimental design for the study of persistent neurotoxic effects.

5.1 Study the effect of exposure at different developmental timing windows

Human iPSCs can be used to detect changes in the mechanisms involved in persistency, but also because hiPSCs allow researchers to study induction of persistency at a multitude of different timepoints during development. As became clear from the Minamata Bay disaster, timing of exposure does matter for the severity of the effects. Thus, to gain insights in these temporal differences in sensitivity, being able to expose (h)iPSCs differentiating into neuronal cultures at different time points of development is greatly beneficial. Prince et al have shown that hiPSCs differentiating into cortical cultures using a dual SMAD inhibition method are suitable to detect exposure timing dependent differences (Prince et al., 2021). Human iPSCs were exposed to MeHg during different windows of development. Although the accumulation of MeHg was comparable between the exposure paradigms, significant differences in differentiation markers (FOXG1, TBR1 and CCND1) were observed (Prince et al., 2021). Neely et al used a similar experimental design in regard to differentiation and exposure windows and performed scRNA-seq experiments (Neely et al., 2021). They found that changes in gene expression were cell cluster and exposure paradigm specific (Neely et al., 2021). In Fig. 2, the exposure window is set at between day 5 to day 10, but such timing can be fully re-imagined, it can easily be adjusted in length and days can be changed to specific needs of the experimental questions being asked and suitable especially for studies on persistent and latent toxicity.

5.2 Study the temporal effects of persistency

Besides being able to study different exposure paradigms, using hiPSC stem cell models also allow for continuous monitoring of the same differentiation as those cells and structures mature. Following differentiation, part of the cells can be used for experiments whereas other splits of the cells can be used for monitoring their continued maturation. As long as cells are expanding or there are sufficient cells remaining, this process can be ongoing for weeks to months to even years potentially. Frazel et al differentiated hiPSCs into microglia and collected samples at several different days of maturation for single cell RNA-sequencing (Frazel et al., 2023). The resulting data allowed them to create

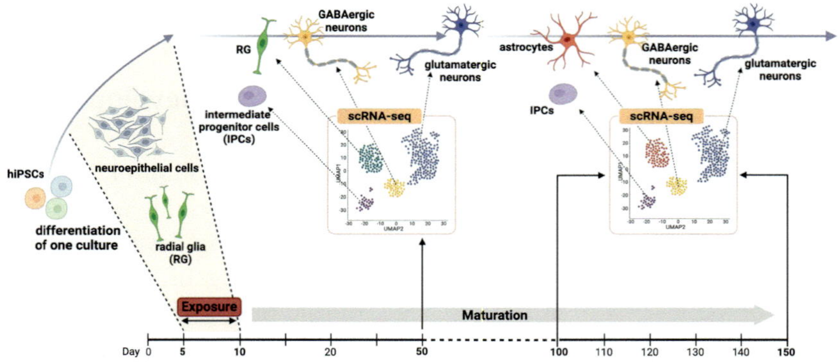

Fig. 2 Schematic overview of an example experimental design for the assessment of persistent toxicity. hiPSCs can be exposed during development into neuronal cultures to a toxicant of interest. During differentiation and maturation samples for e.g., scRNA-sequencing can be taken and the rest of the culture can be further matured. This allows for time comparisons within the same differentiation. Days in figure are meant as an example and can be shifted around to fit research question at hand. *Image created with BioRender.com.*

RNA velocity streamlines providing insights in the flow of cell states over time (Frazel et al., 2023). This helps understanding how cultures develop and from which progenitor cells mature cells stem. Ciceri et al used this approach to gain insights in the maturation rate of neurons (Ciceri et al., 2024). They identified the chromatin regulators EZH2, EHMT1, EHMT2 and DOT1L in the epigenetic barrier that are closely involved in the rate in which neuronal cultures mature. Additionally, using different markers for neuron function (e.g., but not exclusively SHANK1 and SHANK3 as post-synaptic markers, GRIN2A and GRIN2B as receptor markers, SCN1A and SCN2A as sodium channel markers) they showed the change of expression level of these markers in relation to culture maturation (Ciceri et al., 2024). Though the above studies were performed on 2D cultures, developmental trajectory analysis can also be performed using organoids. Kanton et al analyzed stem cell derived cerebral organoids to investigate human specific gene-regulatory changes (Kanton et al., 2019). They created differentiation trajectories over development starting at the stage of pluripotency to neuroectoderm, neuroepithelial stages and then differentiation in ventral forebrain, midbrain and hindbrain regions from different iPSC lines from different individuals. It was found that regional specific gene-expression profiles remained reproducible across different individuals (Kanton et al., 2019). This finding helps to assure us that stem cell work is reproducible and that findings do not solely rely on the iPSC line used.

The above-described types of analyses can be performed in absence or presence of prior exposure. In the context of the study of persistency it is interesting to see whether prior exposure affects the trajectory of development and maturation as well as the timing. Performing time course studies using hiPSC-derived cultures has as added benefit that this type of study does not necessarily have to be endpoint studies since the culture can continue. This is in contrast to animal experiments, where obtaining brain tissue often means an endpoint. Additionally maintaining a cell culture for an extended period of time is often more time efficient and affordable than keeping an in vivo study going for prolonged period of time. On top of that, using hiPSC-derived cultures also means that neurons sequenced later in time at mature stages come from the same pool as progenitors as the neurons analyzed at earlier time points. Thus, providing greater certainty that all cells have went through the same exposure regime/experience. Fig. 2 from above outlines some hypothetical days for single cell experiments and continuation of differentiation. These days can be adjusted to specific questions. Additionally, though this example uses single cell RNA-sequencing, the same holds true for other experiments.

5.3 Study regional differences in persistency

It is known that toxicants target specific brain regions. For example, a study in rats showed that fetal alcohol exposure persistently reduced the number of cerebellar Purkinje cells and granule neurons in the cerebellum and olfactory bulb, but did not affect granule cells in the dentate gyrus nor did it affect the total number of olfactory bulb mitral cells or hippocampal pyramidal cells (Maier and West, 2001). Stem cell research makes it possible to study specific brain regions by using targeted differentiation protocols. To illustrate, there are protocols for astrocyte differentiation (Tcw et al., 2017; Perriot et al., 2021), midbrain (Kim et al., 2021) and cortical cultures (Shi et al., 2012). Differentiating stem cells into specific brain regions is especially helpful to gain better understanding in the underlying mechanisms of persistency in affected areas.

5.4 Study individual differences in susceptibility to persistent toxic effects

When two individuals are exposed to the same level of toxicant, striking differences can occur in toxic response. As the example of fetal Minamata disease showed, sensitivity towards an exposure differs between age groups with the developing fetus being more susceptible. In other words, the developing brain has a higher sensitivity than the developed brain. Besides age related sensitivity differences, differences could occur from other reasons such as underlying interindividual genetic variations in susceptibility. Each person has a unique "toxicogenomic profile" that contributes to the way the body reacts (Nebert et al., 2013). Reactions are influenced by events in the genome, environmental factors (smoking, diet, etc.), endogenous factors (age, gender, fitness etc.) (Nebert et al., 2013). Often differences in susceptibility can be linked to an individual's ability to metabolize toxicants or to repair damage from exposure. The response will vary on the polymorphisms within critical genes related to metabolism (e.g., CYP and N-acetyltransferases), genes that are involved in detoxification (e.g., glutathione S-transferases) and genes that affect susceptibility to genotoxic damage (e.g., XRCC1 and XPD) (Miller et al., 2001). With the use of genomics techniques, insights in an individual's polymorphisms can be obtained. Comparing persistency in exposed iPSC-derived neuronal cultures from different individuals can help gaining insights in which polymorphisms contribute to underlying mechanisms and thus to gain insights in what makes one person more susceptible than another person.

Knowing genetic risk factors that contribute to rendering individuals more susceptible to persistent effects will help risk assessment where questions of sensitive populations arise.

5.5 Study persistent effects on population and individual level

Another interesting opportunity that comes with using stem cells is that they allow for assessment on an individual level as well as population level. When differentiating stem cells, this can be done with each line in a separate well or plate to create neuronal cultures that are individual specific. Though culturing different lines side by side can provide great insights on differences and similarities between individuals, it does not help to make estimates on risks for the general or sensitive populations. In order to gain insights in the effects exposures can have on the population, so called "cell villages" containing multiple different iPSC lines can be created. In such a cell village, different lines are cultured in a shared in vitro environment (Wells et al., 2023) and can be efficiently differentiated into neuronal cultures (Jerber et al., 2021). This in contrast to separate cultures, where each line has its own in vitro environment. Using single cell RNA-sequencing, insights of the effect of exposure on cell village level and thus population level can be gained. However, using individual specific information such as transcribed single nucleotide polymorphisms (SNPs), cells can be assigned back to individual lines (Wells et al., 2023) meaning that insights on both the individual and population level can be gained from the same data set. Thus, using this approach two types of questions can be answered with one run of single-cell based sequencing. This will significantly reduce the costs of single cell experiments and increase the output.

5.6 Toxicants of choice for the study of persistency

The mechanisms and impacts of persistent neurotoxic effects remain an area of active investigation for which our understanding is only just now beginning. Thus, the best toxicants to study for those seeking to gain understanding of how and when persistent effects occur, would be a toxicant that can definitively be shown to have left the model system. In other words, the toxicant is no longer present in the model system after a defined period time following exposure has ended. This time frame may vary based on the toxicokinetics of the chemical (uptake, storage, transport, metabolism, export), frequency of media changes, cell types present, composition of the culture media (e.g. protein content), the plasticware utilized to support the culture (e.g. for chemicals that may adhere to the plates), passaging of the cultures and other factors. It is known that the in vitro half life time of MeHg is approximately 2.5 days (data unpublished). This makes MeHg an ideal toxicant for persistency studies as the wait period for the chemical to leave the system is relatively short. The same holds true for ethanol. Other chemicals could be those of which it can be easily established that they are removed from the model by use of for example HPLC or LC-MS. Although lead (Pb) stays in the human body for a long time, it will leave in vitro systems and is thus a suitable toxicant for in vitro persistency studies.

5.7 Resilience and adaptation

It could very well be that the human body is able to cope with persistent effects caused by environmental exposures in such a way that homeostasis remains and the biological system is able to function properly. However, this could just be on the level of applied phenotypic measurements. Adaptation to the new balance may render the system less resilient and more vulnerable for future (or secondary) exposures. iPSCs models allow for the study of changes in sensitivity towards secondary exposures in a controlled environment (see Fig. 3). Indeed, it may be that many more toxicants than realized actually cause persistent effects. We are just not yet aware of this due to the resilience and adaption of the biological system.

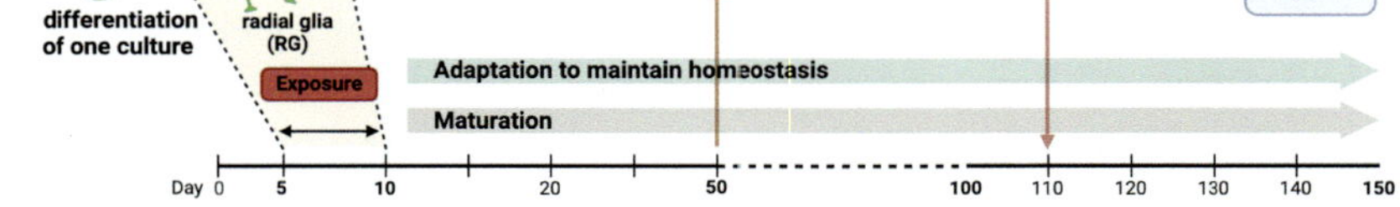

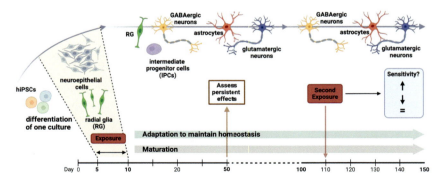

Fig. 3 Schematic overview of an experimental design that can be used for assessment of changes in sensitivity towards a secondary stressor. Following cessation of exposure and assessment of persistency, culture can be matured further. Then, a secondary stressor can be added and an assessment can be made whether the first (developmental) exposure rendered the culture more, less or equally sensitive to the second exposure. *Image created with BioRender.com.*

6. Concluding remarks

Persistent toxicity is till today an understudied form of toxicity. It is a multifaceted issue in the sense that it can occur in many forms following different exposure durations with multiple underlying mechanisms. Additionally, the definition of persistent toxicity can be quite complex. We pose a definition in which we see persistent toxicity as changes in the biological system following cessation of exposure and in absence of the toxicant, regardless of the duration and dose of exposure, or length of the cessation period. In our definition, effects can occur right after exposure and never reverse or occur after a latency period. Regardless of when effects occur or following what type of exposure, persistent toxicity can pose a serious threat to normal development and maturation, human health and healthy aging. However, we do want to point out the difference between persistent changes on a cellular level versus changes on an organism/structural level. In the latter, certain cell types are not made thereby creating persistent alterations to the system that are irreversible due to their structural basis. In our idea of cellular/functional persistent toxicity, we would not classify these alterations as persistent toxicity as these are persistent structural changes due to the absence of cells or other biological structures (e.g. cartilage, bones, etc). Rather, we focus here on persistent changes on a cellular level that could result in functional changes or an altered homeostatic state.

In this review, we showed the potential of developmental exposure to MeHg and alcohol to cause persistent toxicity with serious implications in later life. Even though there is not much known on the extent of environmental exposures to induce persistency (or a persistently altered homeostatic state), we posit with confidence that other chemicals are capable of doing this. Therefore, we urge further research into this form of toxicity to better understand underlying mechanisms and whether these are shared across (groups of) toxicants. We also showed that stem cells are a promising model for the study of persistent effect. We thus propose further research on mechanisms of persistency using stem cell derived models.

Conflict of interest

The authors declare no conflict of interest.

Funding

This work was supported by the National Institute of Environmental Health Sciences (NIEHS) under award numbers R01ES07331 and R01AG080917 (ABB) and K99ES036290 (AT). The content is solely the responsibility of the authors and does not necessarily represent the official views of the National Institutes of Health.

References

Allen, J.W., Mutkus, L.A., Aschner, M., 2001. Methylmercury-mediated inhibition of 3H-D-aspartate transport in cultured astrocytes is reversed by the antioxidant catalase. Brain Res. 902, 92–100.

Amin-Zaki, L., Majeed, M.A., Elhassani, S.B., Clarkson, T.W., Greenwood, M.R., Doherty, R.A., 1979. Prenatal methylmercury poisoning. Clinical observations over five years. Am. J. Dis. Child. 133, 172–177.

Amolegbe, S.M., Carlin, D.J., Henry, H.F., Heacock, M.L., Trottier, B.A., Suk, W.A., 2022. Understanding exposures and latent disease risk within the National Institute of Environmental Health Sciences Superfund Research Program. Exp. Biol. Med. (Maywood) 247, 529–537.

Agency for Toxic Substances and Disease Registry, 2020. Toxicological Profile for Lead. U. S. Department of Health & Human Services, Atlanta. Department of Health & Human Services Atlanta.

Aylward, L.L., Hays, S.M., Kirman, C.R., Marchitti, S.A., Kenneke, J.F., English, C., et al., 2014. Relationships of chemical concentrations in maternal and cord blood: a review of available data. J. Toxicol. Env. Health B Crit. Rev. 17, 175–203.

Baker, D., Hirst, A.J., Gokhale, P.J., Juarez, M.A., Williams, S., Wheeler, M., et al., 2016. Detecting genetic mosaicism in cultures of human pluripotent stem cells. Stem Cell Rep. 7, 998–1012.

Bakir, F., Al-Khalidi, A., Clarkson, T.W., Greenwood, R., 1976. Clinical observations on treatment of alkylmercury poisoning in hospital patients. Bull. World Health Organ. 53 (Suppl), 87–92.

Bakir, F., Damluji, S.F., Amin-Zaki, L., Murtadha, M., Khalidi, A., Al-Rawi, N.Y., et al., 1973. Methylmercury poisoning in Iraq. Science 181, 230–241.

Bartolome, J., Whitmore, W.L., Seidler, F.J., Slotkin, T.A., 1984. Exposure to methylmercury in utero: effects on biochemical development of catecholamine neurotransmitter systems. Life Sci. 35, 657–670.

Behar, T.N., Schaffner, A.E., Scott, C.A., Greene, C.L., Barker, J.L., 2000. GABA receptor antagonists modulate postmitotic cell migration in slice cultures of embryonic rat cortex. Cereb. Cortex 10, 899–909.

Behar, T.N., Scott, C.A., Greene, C.L., Wen, X., Smith, S.V., Maric, D., et al., 1999. Glutamate acting at NMDA receptors stimulates embryonic cortical neuronal migration. J. Neurosci. 19, 4449–4461.

Bihaqi, S.W., Bahmani, A., Subaiea, G.M., Zawia, N.H., 2014. Infantile exposure to lead and late-age cognitive decline: Relevance to AD. Alzheimer's Dement. 10, 187–195.

Bizzotto, S., Dou, Y., Ganz, J., Doan, R.N., Kwon, M., Bohrson, C.L., et al., 2021. Landmarks of human embryonic development inscribed in somatic mutations. Science 371, 1249–1253.

Bjorklund, G., Tippairote, T., Hangan, T., Chirumbolo, S., Peana, M., 2024. Early-life lead exposure: risks and neurotoxic consequences. Curr. Med. Chem. 31, 1620–1633.

Bohacek, J., Mansuy, I.M., 2013. Epigenetic inheritance of disease and disease risk. Neuropsychopharmacology 38, 220–236.

Bonnet, J.J., Benmansour, S., Amejdki-Chab, N., Costentin, J., 1994. Effect of CH3HgCl and several transition metals on the dopamine neuronal carrier; peculiar behaviour of Zn2+. Eur. J. Pharmacol. 266, 87–97.

Bousquet, J., Anto, J.M., Berkouk, K., Gergen, P., Antunes, J.P., Augé, P., et al., 2015. Developmental determinants in non-communicable chronic diseases and ageing. Thorax 70, 595–597.

Brazdis, R.-M., Alecu, J.E., Marsch, D., Dahms, A., Simmnacher, K., Lörentz, S., et al., 2020. Demonstration of brain region-specific neuronal vulnerability in human iPSC-based model of familial Parkinson's disease. Hum. Mol. Genet. 29, 1180–1191.

Brazhnik, K., Sun, S., Alani, O., Kinkhabwala, M., Wolkoff, A.W., Maslov, A.Y., et al., 2020. Single-cell analysis reveals different age-related somatic mutation profiles between stem and differentiated cells in human liver. Sci. Adv. 6, eaax2659.

Bressler, J.P., Goldstein, G.W., 1991. Mechanisms of lead neurotoxicity. Biochem. Pharmacol. 41, 479–484.

Brien, J.F., Loomis, C.W., Tranmer, J., Mcgrath, M., 1983. Disposition of ethanol in human maternal venous blood and amniotic fluid. Am. J. Obstet. Gynecol. 146, 181–186.

Brown, J.M., Bland, R., Jonsson, E., Greenshaw, A.J., 2019. A brief history of awareness of the link between alcohol and fetal alcohol spectrum disorder. Can. J. Psychiatry 64, 164–168.

Caito, S.W., Aschner, M., 2016. NAD+ supplementation attenuates methylmercury dopaminergic and mitochondrial toxicity in *Caenorhabditis elegans*. Toxicol. Sci. 151, 139–149.

Camp, J.G., Badsha, F., Florio, M., Kanton, S., Gerber, T., Wilsch-Bräuninger, M., et al., 2015. Human cerebral organoids recapitulate gene expression programs of fetal neocortex development. Proc. Natl Acad. Sci. U S A 112, 15672–15677.

Capra, L., Tezza, G., Mazzei, F., Boner, A.L., 2013. The origins of health and disease: the influence of maternal diseases and lifestyle during gestation. Ital. J. Pediatr. 39, 7.

Castoldi, A.F., Blandini, F., Randine, G., Samuele, A., Manzo, L., Coccini, T., 2006. Brain monoaminergic neurotransmission parameters in weanling rats after perinatal exposure to methylmercury and 2,2',4,4',5,5'-hexachlorobiphenyl (PCB153). Brain Res. 1112, 91–98.

Catlett, T.S., Onesto, M.M., Mccann, A.J., Rempel, S.K., Glass, J., Franz, D.N., et al., 2021. RHOA signaling defects result in impaired axon guidance in iPSC-derived neurons from patients with tuberous sclerosis complex. Nat. Commun. 12, 2589.

Cediel Ulloa, A., Gliga, A., Love, T.M., Pineda, D., Mruzek, D.W., Watson, G.E., et al., 2021. Prenatal methylmercury exposure and DNA methylation in seven-year-old children in the Seychelles Child Development Study. Environ. Int. 147, 106321.

Chang, L., 1984. Developmental toxicology of methylmercury. Toxicol. Newborn 175–197.

Choi, B.H., 1986. Methylmercury poisoning of the developing nervous system: I. Pattern of neuronal migration in the cerebral cortex. Neurotoxicology 7, 591–600.

Choi, B.H., Lapham, L.W., Amin-Zaki, L., Saleem, T., 1978. Abnormal neuronal migration, deranged cerebral cortical organization, and diffuse white matter astrocytosis of human fetal brain: a major effect of methylmercury poisoning in utero. J. Neuropathol. Exp. Neurol. 37, 719–733.

Chung, D.D., Pinson, M.R., Bhenderu, L.S., Lai, M.S., Patel, R.A., Miranda, R.C., 2021. Toxic and teratogenic effects of prenatal alcohol exposure on fetal development, adolescence, and adulthood. Int. J. Mol. Sci. 22.

Ciceri, G., Baggiolini, A., Cho, H.S., Kshirsagar, M., Benito-Kwiecinski, S., Walsh, R.M., et al., 2024. An epigenetic barrier sets the timing of human neuronal maturation. Nature 626, 881–890.

Clarkson, T.W., Magos, L., Cox, C., Greenwood, M.R., Amin-Zaki, L., Majeed, M.A., et al., 1981. Tests of efficacy of antidotes for removal of methylmercury in human poisoning during the Iraq outbreak. J. Pharmacol. Exp. Ther. 218, 74–83.

Culbreth, M., Aschner, M., 2016. Dysregulation of glutamate cycling mediates methylmercury-induced neurotoxicity. Adv. Neurobiol. 13, 295–305.

Cuzon, V.C., Yeh, P.W., Yanagawa, Y., Obata, K., Yeh, H.H., 2008. Ethanol consumption during early pregnancy alters the disposition of tangentially migrating GABAergic interneurons in the fetal cortex. J. Neurosci. 28, 1854–1864.

D'urso, A., Brickner, J.H., 2014. Mechanisms of epigenetic memory. Trends Genet. 30 (6), 230.

D'gama, A.M., Walsh, C.A., 2018. Somatic mosaicism and neurodevelopmental disease. Nat. Neurosci. 21, 1504–1514.

Daré, E., Fetissov, S., Hökfelt, T., Hall, H., Ogren, S.O., Ceccatelli, S., 2003. Effects of prenatal exposure to methylmercury on dopamine-mediated locomotor activity and dopamine D2 receptor binding. Naunyn Schmiedebergs Arch. Pharmacol. 367, 500–508.

Davis, L.E., Kornfeld, M., Mooney, H.S., Fiedler, K.J., Haaland, K.Y., Orrison, W.W., et al., 1994. Methylmercury poisoning: long-term clinical, radiological, toxicological, and pathological studies of an affected family. Ann. Neurol. 35, 680–688.

Derauf, C., Kekatpure, M., Neyzi, N., Lester, B., Kosofsky, B., 2009. Neuroimaging of children following prenatal drug exposure. Semin. Cell Develop. Biol. 20, 441–454.

Dominah, G.A., Mcminimy, R.A., Kallon, S., Kwakye, G.F., 2017. Acute exposure to chlorpyrifos caused NADPH oxidase mediated oxidative stress and neurotoxicity in a striatal cell model of Huntington's disease. Neurotoxicology 60, 54–69.

Dooves, S., Van Velthoven, A.J., Suciati, L.G., Heine, V.M., 2021. Neuron–glia interactions in tuberous sclerosis complex affect the synaptic balance in 2D and organoid cultures. Cells 10, 134.

Dos Santos, A.A., Chang, L.W., Liejun Guo, G., Aschner, M., 2018. Chapter 35 - Fetal minamata disease: a human episode of congenital methylmercury poisoning. In: Slikker, W., Paule, M.G., Wang, C. (Eds.), Handbook of Developmental Neurotoxicology, second ed. Academic Press.

Dreiem, A., Shan, M., Okoniewski, R.J., Sanchez-Morrissey, S., Seegal, R.F., 2009. Methylmercury inhibits dopaminergic function in rat pup synaptosomes in an age-dependent manner. Neurotoxicol Teratol. 31, 312–317.

Du, Z., Zang, Z., Luo, J., Liu, T., Yang, L., Cai, Y., et al., 2023. Chronic exposure to (2 R,6 R)-hydroxynorketamine induces developmental neurotoxicity in hESC-derived cerebral organoids. J. Hazard. Mater. 453, 131379.

Eriksson, P., 1997. Developmental neurotoxicity of environmental agents in the neonate. Neurotoxicology 18, 719–726.

Eto, K., Oyanagi, S., Itai, Y., Tokunaga, H., Takizawa, Y., Suda, I., 1992. A fetal type of Minamata disease. An. autopsy case Rep. Spec. Ref. Nerv. system. *Mol. Chem. Neuropathol,* 16, 171–186.

Fahrion, J.K., Komuro, Y., Li, Y., Ohno, N., Littner, Y., Raoult, E., et al., 2012. Rescue of neuronal migration deficits in a mouse model of fetal Minamata disease by increasing neuronal Ca2+ spike frequency. Proc. Natl Acad. Sci. U S A 109, 5057–5062.

Farina, M., Frizzo, M.E., Soares, F.A., Schwalm, F.D., Dietrich, M.O., Zeni, G., et al., 2003. Ebselen protects against methylmercury-induced inhibition of glutamate uptake by cortical slices from adult mice. Toxicol. Lett. 144, 351–357.

Feil, R., Fraga, M.F., 2012. Epigenetics and the environment: emerging patterns and implications. Nat. Rev. Genet. 13, 97–109.

Feng, S., Xu, Z., Liu, W., Li, Y., Deng, Y., Xu, B., 2014. Preventive effects of dextromethorphan on methylmercury-induced glutamate dyshomeostasis and oxidative damage in rat cerebral cortex. Biol. Trace Elem. Res. 159, 332–345.

Feng, X., Zhou, R., Jiang, Q., Wang, Y., Yu, C., 2022. Analysis of cadmium accumulation in community adults and its correlation with low-grade albuminuria. Sci. Total. Environ. 834, 155210.

Florez-Garcia, V.A., Guevara-Romero, E.C., Hawkins, M.M., Bautista, L.E., Jenson, T.E., Yu, J., et al., 2023. Cadmium exposure and risk of breast cancer: A meta-analysis. Env. Res. 219, 115109.

Frazel, P.W., Labib, D., Fisher, T., Brosh, R., Pirjanian, N., Marchildon, A., et al., 2023. Longitudinal scRNA-seq analysis in mouse and human informs optimization of rapid mouse astrocyte differentiation protocols. Nat. Neurosci. 26, 1726–1738.

Fredriksson, A., Fredriksson, M., Eriksson, P., 1993. Neonatal exposure to paraquat or MPTP induces permanent changes in striatum dopamine and behavior in adult mice. Toxicol. Appl. Pharmacol. 122, 258–264.

Gantner, C.W., Hunt, C.P.J., Niclis, J.C., Penna, V., Mcdougall, S.J., Thompson, L.H., et al., 2021. FGF-MAPK signaling regulates human deep-layer corticogenesis. Stem Cell Rep. 16, 1262–1275.

Genchi, G., Sinicropi, M.S., Lauria, G., Carocci, A., Catalano, A., 2020. The effects of cadmium toxicity. Int. J. Env. Res. Public. Health 17.

Gervin, K., Nordeng, H., Ystrom, E., Reichborn-Kjennerud, T., Lyle, R., 2017. Long-term prenatal exposure to paracetamol is associated with DNA methylation differences in children diagnosed with ADHD. Clin. Epigenetics 9, 77.

Gilbertson, M., 2009. Index of congenital Minamata disease in Canadian areas of concern in the Great Lakes: an eco-social epidemiological approach. J. Env. Sci. Health C. Env. Carcinog. Ecotoxicol. Rev. 27, 246–275.

Granato, A., Dering, B., 2018. Alcohol and the developing brain: Why neurons die and how survivors change. Int. J. Mol. Sci. 19.

Greenman, C., Stephens, P., Smith, R., Dalgliesh, G.L., Hunter, C., Bignell, G., et al., 2007. Patterns of somatic mutation in human cancer genomes. Nature 446, 153–158.

Harada, M., 1972. Clinical and epidemiological studies of Minamata disease. 16 years onset. *Shinkeikenkyuu no Shinpo,* 16, 870–880.

Harada, M., 1995. Minamata disease: methylmercury poisoning in Japan caused by environmental pollution. Crit. Rev. Toxicol. 25, 1–24.

Heindel, J.J., Vandenberg, L.N., 2015. Developmental origins of health and disease: a paradigm for understanding disease cause and prevention. Curr. Opin. Pediat. 27, 248–253.

Ho, S.-M., Johnson, A., Tarapore, P., Janakiram, V., Zhang, X., Leung, Y.-K., 2012. Environmental epigenetics and its implication on disease risk and health outcomes. ILAR J. 53, 289–305.

Hook, V., Brennand, K.J., Kim, Y., Toneff, T., Funkelstein, L., Lee, K.C., et al., 2014. Human iPSC neurons display activity-dependent neurotransmitter secretion: aberrant catecholamine levels in schizophrenia neurons. Stem Cell Rep. 3, 531–538.

Huang, A.Y., Lee, E.A., 2022. Identification of somatic mutations from bulk and single-cell sequencing data. Front. Aging 2.

Jamuar, S.S., D'gama, A.M., Walsh, C.A., 2016. Chapter 12 - Somatic mosaicism and neurological diseases. In: Lehner, T., Miller, B.L., State, M.W. (Eds.), Genomics, Circuits, and Pathways in Clinical Neuropsychiatry. Academic Press, San Diego.

Järup, L., Åkesson, A., 2009. Current status of cadmium as an environmental health problem. Toxicol. Appl. Pharmacol. 238, 201–208.

Jerber, J., Seaton, D.D., Cuomo, A.S.E., Kumasaka, N., Haldane, J., Steer, J., et al., 2021. Population-scale single-cell RNA-seq profiling across dopaminergic neuron differentiation. Nat. Genet. 53, 304–312.

Jin, H., Yang, C., Jiang, C., Li, L., Pan, M., Li, D., et al., 2022. Evaluation of neurotoxicity in BALB/c mice following chronic exposure to polystyrene microplastics. Env. Health Perspect. 130, 107002.

Kanton, S., Boyle, M.J., He, Z., Santel, M., Weigert, A., Sanchís-Calleja, F., et al., 2019. Organoid single-cell genomic atlas uncovers human-specific features of brain development. Nature 574, 418–422.

Kaur, P., Aschner, M., Syversen, T., 2007. Role of glutathione in determining the differential sensitivity between the cortical and cerebellar regions towards mercury-induced oxidative stress. Toxicology 230, 164–177.

Ke, T., Santamaria, A., Barbosa JR., F., Rocha, J.B.T., Skalny, A.V., Tinkov, A.A., et al., 2023. Developmental methylmercury exposure induced and age-dependent glutamatergic neurotoxicity in *Caenorhabditis elegans*. Neurochem. Res. 48, 920–928.

Ke, T., Tsatsakis, A., Santamaría, A., Antunes Soare, F.A., Tinkov, A.A., Docea, A.O., et al., 2020. Chronic exposure to methylmercury induces puncta formation in cephalic dopaminergic neurons in *Caenorhabditis elegans*. Neurotoxicology 77, 105–113.

Kendricks, D.R., Boomhower, S.R., Newland, M.C., 2020. Methylmercury, attention, and memory: baseline-dependent effects of adult d-amphetamine and marginal effects of adolescent methylmercury. Neurotoxicology 80, 130–139.

Kendricks, D.R., Newland, M.C., 2021. Selective dopaminergic effects on attention and memory in male mice exposed to Methylmercury during adolescence. Neurotoxicol Teratol. 87, 107016.

Kim, H., Harrison, F.E., Aschner, M., Bowman, A.B., 2022. Exposing the role of metals in neurological disorders: a focus on manganese. Trends Mol. Med. 28, 555–568.

Kim, J.H., Scialli, A.R., 2011. Thalidomide: the tragedy of birth defects and the effective treatment of disease. Toxicol. Sci. 122, 1–6.

Kim, J.K., Cho, J., Kim, S.H., Kang, H.-C., Kim, D.-S., Kim, V.N., et al., 2019. Brain somatic mutations in MTOR reveal translational dysregulations underlying intractable focal epilepsy. J. Clin. Invest. 129, 4207–4223.

Kim, T.W., Piao, J., Koo, S.Y., Kriks, S., Chung, S.Y., Betel, D., et al., 2021. Biphasic activation of WNT signaling facilitates the derivation of midbrain dopamine neurons from hESCs for translational use. Cell Stem Cell 28, 343–355.e5.

Kimura, M., Shoda, A., Murata, M., Hara, Y., Yonoichi, S., Ishida, Y., et al., 2023. Neurotoxicity and behavioral disorders induced in mice by acute exposure to the diamide insecticide chlorantraniliprole. J. Vet. Med. Sci. 85, 497–506.

Kinjo, Y., Higashi, H., Nakano, A., Sakamoto, M., Sakai, R., 1993. Profile of subjective complaints and activities of daily living among current patients with Minamata disease after 3 decades. Env. Res. 63, 241–251.

Knobloch, J., Jungck, D., Koch, A., 2011. Apoptosis induction by thalidomide: critical for limb teratogenicity but therapeutic potential in idiopathic pulmonary fibrosis? Curr. Mol. Pharmacol. 4, 26–61.

Komulainen, H., Tuomisto, J., 1985. 3H-dopamine uptake and 3H-haloperidol binding in striatum after administration of methyl mercury to rats. Arch. Toxicol. 57, 268–271.

Kosnett, M.J., 2013. The role of chelation in the treatment of arsenic and mercury poisoning. J. Med. Toxicol. 9, 347–354.

Lam, H.S., Fok, T.F., Ng, P.C., 2012. Long-term neurocognitive outcomes of children prenatally exposed to low-dose methylmercury. Hong. Kong Med. J. 18 (Suppl 6), 23–24.

Lam, H.S., Kwok, K.M., Chan, P.H., So, H.K., Li, A.M., Ng, P.C., et al., 2013. Long term neurocognitive impact of low dose prenatal methylmercury exposure in Hong Kong. Env. Int. 54, 59–64.

Lee, E.Y., Flynn, M.R., Lewis, M.M., Mailman, R.B., Huang, X., 2018a. Welding-related brain and functional changes in welders with chronic and low-level exposure. Neurotoxicology 64, 50–59.

Lee, J., Freeman, J.L., 2016. Embryonic exposure to 10 μg L(-1) lead results in female-specific expression changes in genes associated with nervous system development and function and Alzheimer's disease in aged adult zebrafish brain. Metallomics 8, 589–596.

Lee, J., Horzmann, K.A., Freeman, J.L., 2018b. An embryonic 100μg/L lead exposure results in sex-specific expression changes in genes associated with the neurological system in female or cancer in male adult zebrafish brains. Neurotoxicol Teratol. 65, 60–69.

Li, X., Pan, J., Wei, Y., Ni, L., Xu, B., Deng, Y., et al., 2021. Mechanisms of oxidative stress in methylmercury-induced neurodevelopmental toxicity. Neurotoxicology 85, 33–46.

Lucchini, R.G., Aschner, M., Landrigan, P.J., Cranmer, J.M., 2018. Neurotoxicity of manganese: Indications for future research and public health intervention from the Manganese 2016 conference. Neurotoxicology 64, 1–4.

Luo, J., Miller, M.W., 1998. Growth factor-mediated neural proliferation: target of ethanol toxicity. Brain 27, 157–167.

Maier, S.E., West, J.R., 2001. Regional differences in cell loss associated with binge-like alcohol exposure during the first two trimesters equivalent in the rat. Alcohol 23, 49–57.

Manfroi, C.B., Schwalm, F.D., Cereser, V., Abreu, F., Oliveira, A., Bizarro, L., et al., 2004. Maternal milk as methylmercury source for suckling mice: neurotoxic effects involved with the cerebellar glutamatergic system. Toxicol. Sci. 81, 172–178.

Mansour, S., Baple, E., Hall, C.M., 2019. A clinical review and introduction of the diagnostic algorithm for thalidomide embryopathy (DATE). J. Hand Surg. Eur. Vol. 44, 96–108.

Matsumoto, H., Koya, G., Takeuchi, T., 1965. Fetal Minamata disease: a neuropathological study of two cases of intrauterine intoxication by a methyl mercury compound. J. Neuropathol. Exp. Neurol. 24, 563–574.

Mazumdar, M., Xia, W., Hofmann, O., Gregas, M., Ho Sui, S., Hide, W., et al., 2012. Prenatal lead levels, plasma amyloid β levels, and gene expression in young adulthood. Env. Health Perspect. 120, 702–707.

Miles, B., Tadi, P., 2024. Genetics, somatic mutation. *StatPearls*. Treasure Island, FL: StatPearls Publishing Copyright © 2024, StatPearls Publishing LLC.

Miller, M.C., 3rd, Mohrenweiser, H.W., Bell, D.A., 2001. Genetic variability in susceptibility and response to toxicants. Toxicol. Lett. 120, 269–280.

Mohiuddin, M., Kooy, R.F., Pearson, C.E., 2022. De novo mutations, genetic mosaicism and human disease. Front. Genet. 13, 983668.

Nebert, D.W., Zhang, G., Vesell, E.S., 2013. Genetic risk prediction: individualized variability in susceptibility to toxicants. Annu. Rev. Pharmacol. Toxicol. 53, 355–375.

Needham, L.L., Grandjean, P., Heinzow, B., Jørgensen, P.J., Nielsen, F., Patterson JR., D.G., Sjödin, A., et al., 2011. Partition of environmental chemicals between maternal and fetal blood and tissues. Env. Sci. Technol. 45, 1121–1126.

Neely, M.D., Davison, C.A., Aschner, M., Bowman, A.B., 2017. From the cover: manganese and rotenone-induced oxidative stress signatures differ in iPSC-derived human dopamine neurons. Toxicol. Sci. 159, 366–379.

Neely, M.D., Xie, S., Prince, L.M., Kim, H., Tukker, A.M., Aschner, M., et al., 2021. Single cell RNA sequencing detects persistent cell type- and methylmercury exposure paradigm-specific effects in a human cortical neurodevelopmental model. Food Chem. Toxicol. 154, 112288.

Neuwirth, L.S., Emenike, B.U., 2024. Comment on "neurotoxicity and outcomes from developmental lead exposure: persistent or permanent?". Env. Health Perspect. 132, 48001.

Nierenberg, D.W., Nordgren, R.E., Chang, M.B., Siegler, R.W., Blayney, M.B., Hochberg, F., et al., 1998. Delayed cerebellar disease and death after accidental exposure to dimethylmercury. N. Engl. J. Med. 338, 1672–1676.

Olney, J.W., Tenkova, T., Dikranian, K., Qin, Y.-Q., Labruyere, J., Ikonomidou, C., 2002. Ethanol-induced apoptotic neurodegeneration in the developing C57BL/6 mouse brain. Develop. Brain Res. 133, 115–126.

Pamies, D., Wiersma, D., Katt, M.E., Zhao, L., Burtscher, J., Harris, G., et al., 2022. Human IPSC 3D brain model as a tool to study chemical-induced dopaminergic neuronal toxicity. Neurobiol. Dis. 169, 105719.

Perera, F.P., Wang, S., Vishnevetsky, J., Zhang, B., Cole, K.J., Tang, D., et al., 2011. Polycyclic aromatic hydrocarbons-aromatic DNA adducts in cord blood and behavior scores in New York city children. Env. Health Perspect. 119, 1176–1181.

Perriot, S., Canales, M., Mathias, A., Pasquier, R., D.U., 2021. Differentiation of functional astrocytes from human-induced pluripotent stem cells in chemically defined media. STAR. Protoc. 2, 100902.
Petrelli, B., Weinberg, J., Hicks, G.G., 2018. Effects of prenatal alcohol exposure (PAE): insights into FASD using mouse models of PAE. Biochem. Cell Biol. 96, 131–147.
Pizzino, G., Bitto, A., Interdonato, M., Galfo, F., Irrera, N., Mecchio, A., et al., 2014. Oxidative stress and DNA repair and detoxification gene expression in adolescents exposed to heavy metals living in the Milazzo-Valle del Mela area (Sicily, Italy). Redox Biol. 2, 686–693.
Pletz, J., Sánchez-Bayo, F., Tennekes, H.A., 2016. Dose-response analysis indicating time-dependent neurotoxicity caused by organic and inorganic mercury-Implications for toxic effects in the developing brain. Toxicology 347-349, 1–5.
Porciúncula, L.O., Rocha, J.B., Tavares, R.G., Ghisleni, G., Reis, M., Souza, D.O., 2003. Methylmercury inhibits glutamate uptake by synaptic vesicles from rat brain. Neuroreport 14, 577–580.
Prince, L.M., Neely, M.D., Warren, E.B., Thomas, M.G., Henley, M.R., Smith, K.K., et al., 2021. Environmentally relevant developmental methylmercury exposures alter neuronal differentiation in a human-induced pluripotent stem cell model. Food Chem. Toxicol. 152, 112178.
Rand, M.D., Caito, S.W., 2019. Variation in the biological half-life of methylmercury in humans: Methods, measurements and meaning. Biochimica et. Biophysica Acta (BBA) - Gen. Subj. 1863, 129301.
Reynolds, J.N., Racz, W.J., 1987. Effects of methylmercury on the spontaneous and potassium-evoked release of endogenous amino acids from mouse cerebellar slices. Can. J. Physiol. Pharmacol. 65, 791–798.
Rice, D.C., 1996. Evidence for delayed neurotoxicity produced by methylmercury. Neurotoxicology 17, 583–596.
Rodríguez, J., Mandalunis, P.M., 2018. A review of metal exposure and its effects on bone health. J. Toxicol. 2018, 4854152.
Ronchi, S., Buccino, A.P., Prack, G., Kumar, S.S., Schröter, M., Fiscella, M., et al., 2021. Electrophysiological phenotype characterization of human iPSC-derived neuronal cell lines by means of high-density microelectrode arrays. Advanced Biology 5, 2000223.
Satarug, S., Garrett, S.H., Sens, M.A., Sens, D.A., 2010. Cadmium, environmental exposure, and health outcomes. Env. Health Perspect. 118, 182–190.
Schierle, G.S., Gander, J.C., D'orlando, C., Ceilo, M.R., Vogt Weisenhorn, D.M., 1997. Calretinin-immunoreactivity during postnatal development of the rat isocortex: a qualitative and quantitative study. Cereb. Cortex 7, 130–142.
Serafini, M.M., Sepehri, S., Midali, M., Stinckens, M., Biesiekierska, M., Wolniakowska, A., et al., 2024. Recent advances and current challenges of new approach methodologies in developmental and adult neurotoxicity testing. Arch. Toxicol. 98, 1271–1295.
Sharma, P., Chambial, S., Shukla, K.K., 2015. Lead and neurotoxicity. Indian. J. Clin. Biochem. 30, 1–2.
Shi, Y., Kirwan, P., Livesey, F.J., 2012. Directed differentiation of human pluripotent stem cells to cerebral cortex neurons and neural networks. Nat. Protoc. 7, 1836–1846.
Silva, C.D., Neves, A.F., Dias, A.I., Freitas, H.J., Mendes, S.M., Pita, I., et al., 2014. A single neurotoxic dose of methamphetamine induces a long-lasting depressive-like behaviour in mice. Neurotox. Res. 25, 295–304.
Skerfving, S.B., Copplestone, J.F., 1976. Poisoning caused by the consumption of organomercury-dressed seed in Iraq. Bull. World Health Organ. 54, 101–112.
Skorput, A.G., Gupta, V.P., Yeh, P.W., Yeh, H.H., 2015. Persistent interneuronopathy in the prefrontal cortex of young adult offspring exposed to ethanol in utero. J. Neurosci. 35, 10977–10988.

Souza-Arroyo, V., Fabián, J.J., Bucio-Ortiz, L., Miranda-Labra, R.U., Gomez-Quiroz, L.E., Gutiérrez-Ruiz, M.C., 2022. The mechanism of the cadmium-induced toxicity and cellular response in the liver. Toxicology 480, 153339.

Spildrejorde, M., Samara, A., Sharma, A., Leithaug, M., Falck, M., Modafferi, S., et al., 2023. Multi-omics approach reveals dysregulated genes during hESCs neuronal differentiation exposure to paracetamol. iScience 26, 107755.

Takaoka, S., Fujino, T., Kawakami, Y., Shigeoka, S.I., Yorifuji, T., 2018. Survey of the extent of the persisting effects of methylmercury pollution on the inhabitants around the Shiranui Sea, Japan. Toxics 6.

Tang, W.Y., Morey, L.M., Cheung, Y.Y., Birch, L., Prins, G.S., Ho, S.M., 2012. Neonatal exposure to estradiol/bisphenol A alters promoter methylation and expression of Nsbp1 and Hpcal1 genes and transcriptional programs of Dnmt3a/b and Mbd2/4 in the rat prostate gland throughout life. Endocrinology 153, 42–55.

Tcw, J., Wang, M., Pimenova, A.A., Bowles, K.R., Hartley, B.J., Lacin, E., et al., 2017. An efficient platform for astrocyte differentiation from human induced pluripotent stem cells. Stem Cell Rep. 9, 600–614.

Teo, S.K., Colburn, W.A., Tracewell, W.G., Kook, K.A., Stirling, D.I., Jaworsky, M.S., et al., 2004. Clinical pharmacokinetics of thalidomide. Clin. Pharmacokinet. 43, 311–327.

Thomas, J.D., Warren, K.R., Hewitt, B.G., 2010. Fetal alcohol spectrum disorders: from research to policy. Alcohol. Res. Health 33, 118–126.

Tiernan, C.T., Edwin, E.A., Hawong, H.Y., Ríos-Cabanillas, M., Goudreau, J.L., Atchison, W.D., et al., 2015. Methylmercury impairs canonical dopamine metabolism in rat undifferentiated pheochromocytoma (PC12) cells by indirect inhibition of aldehyde dehydrogenase. Toxicol. Sci. 144, 347–356.

Tukker, A.M., Bouwman, L.M.S., Van Kleef, R., Hendriks, H.S., Legler, J., Westerink, R.H.S., 2020a. Perfluorooctane sulfonate (PFOS) and perfluorooctanoate (PFOA) acutely affect human α(1)β(2)γ(2L) GABA(A) receptor and spontaneous neuronal network function in vitro. Sci. Rep. 10, 5311.

Tukker, A.M., Bowman, A.B., 2024. Application of single cell gene expression technologies to neurotoxicology. Curr. Opin. Toxicol. 37, 100458.

Tukker, A.M., Van Kleef, R.G.D.M., Wijnolts, F.M.J., De Groot, A., Westerink, R.H.S., 2020b. Towards animal-free neurotoxicity screening: Applicability of hiPSC-derived neuronal models for in vitro seizure liability assessment. Altex 37, 121–135.

Vandivort, T.C., Eaton, D.L., 2014. Principles of toxicology. Reference Module in Biomedical Sciences. Elsevier.

Vargesson, N., 2015. Thalidomide-induced teratogenesis: history and mechanisms. Birth Defects Res. C. Embryo Today 105, 140–156.

Vargesson, N., 2019. The teratogenic effects of thalidomide on limbs. J. Hand Surg. Eur. Vol. 44, 88–95.

Vendrell, I., Carrascal, M., Vilaró, M.T., Abián, J., Rodríguez-Farré, E., Suñol, C., 2007. Cell viability and proteomic analysis in cultured neurons exposed to methylmercury. Hum. Exp. Toxicol. 26, 263–272.

Wang, L., Owusu-Hammond, C., Sievert, D., Gleeson, J.G., 2023. Stem cell–based organoid models of neurodevelopmental disorders. Biol. Psych. 93, 622–631.

Wei, Y., Ni, L., Pan, J., Li, X., Deng, Y., Xu, B., et al., 2023. Methylmercury promotes oxidative stress and autophagy in rat cerebral cortex: Involvement of PI3K/AKT/mTOR or AMPK/TSC2/mTOR pathways and attenuation by N-acetyl-L-cysteine. Neurotoxicol. Teratol. 95, 107137.

Weiner, J.L., Valenzuela, C.F., 2006. Ethanol modulation of GABAergic transmission: The view from the slice. Pharmacol. Therap. 111, 533–554.

Weiss, B., Bellinger, D.C., 2006. Social ecology of children's vulnerability to environmental pollutants. Env. Health Perspect. 114, 1479–1485.

Weiss, B., Clarkson, T.W., Simon, W., 2002. Silent latency periods in methylmercury poisoning and in neurodegenerative disease. Env. Health Perspect. 110 (Suppl 5), 851–854.

Wells, M.F., Nemesh, J., Ghosh, S., Mitchell, J.M., Salick, M.R., Mello, C.J., et al., 2023. Natural variation in gene expression and viral susceptibility revealed by neural progenitor cell villages. Cell Stem Cell 30 (312-332), e13.

Wells, P.G., Bhuller, Y., Chen, C.S., Jeng, W., Kasapinovic, S., Kennedy, J.C., et al., 2005. Molecular and biochemical mechanisms in teratogenesis involving reactive oxygen species. Toxicol. Appl. Pharmacol. 207, 354–366.

West, J.R., Goodlett, C.R., Bonthius, D.J., Hamre, K.M., Marcussen, B.L., 1990. Cell population depletion associated with fetal alcohol brain damage: mechanisms of BAC-dependent cell loss. Alcohol. Clin. Exp. Res. 14, 813–818.

Wu, J., Basha, M.R., Brock, B., Cox, D.P., Cardozo-Pelaez, F., Mcpherson, C.A., et al., 2008. Alzheimer's disease (AD)-like pathology in aged monkeys after infantile exposure to environmental metal lead (Pb): evidence for a developmental origin and environmental link for AD. J. Neurosci. 28, 3–9.

Xie, J., Wu, S., Szadowski, H., Min, S., Yang, Y., Bowman, A.B., et al., 2023. Developmental Pb exposure increases AD risk via altered intracellular Ca2+ homeostasis in hiPSerived cortical neurons. J. Biol. Chem. 299.

Yorifuji, T., Tsuda, T., Inoue, S., Takao, S., Harada, M., 2011. Long-term exposure to methylmercury and psychiatric symptoms in residents of Minamata, Japan. Env. Int. 37, 907–913.

Zheutlin, A.R., Sharareh, N., Guadamuz, J.S., Berchie, R.O., Derington, C.G., Jacobs, J.A., et al., 2024. Association between pharmacy proximity with cardiovascular medication use and risk factor control in the United States. J. Am. Heart Assoc. 13, e031717.

CHAPTER EIGHT

Simple and reproducible directed differentiation of cortical neural cells from hiPSCs in chemically defined media for toxicological studies

Hyunjin Kim, David Yi, and Aaron B. Bowman*
School of Health Sciences, Purdue University, West Lafayette, IN, United States
*Corresponding author. e-mail address: bowma117@purdue.edu

Contents

Abstract

Human induced pluripotent stem cell (hiPSC)-derived neural cell models serve as powerful tools for studying human brain health and disease and their interactions with toxicological exposures. Here, we provide a simple and reproducible protocol for

Advances in Neurotoxicology, Volume 12
ISSN 2468-7480, https://doi.org/10.1016/bs.ant.2024.08.002

the directed differentiation and derivation of cortical neural cultures comprising of neurons and glia. Through the use of chemically defined media and carefully timed passages at optimized densities, we circumvent the need to perform extensive laborious and time-consuming purification steps to obtain high quality cortical cultures and offer greater flexibility to design and implement experiments for toxicological research.

1. Introduction

1.1 Background

During embryonic development, the cerebral cortex develops from the telencephalon which, in turn, emerges from the neuroepithelium of the anterior neuroectoderm. As development proceeds, neuroepithelial cells (NECs) give rise to radial glia (RG) that serve as the principal neural stem cells of the developing cortex. In its simplest form, RG undergo a series of self-renewal to expand the neural stem cell pool and enter a neurogenic phase of symmetric or asymmetric cell divisions to either generate neurons or intermediate progenitor cells (IPCs) which subsequently undergo neurogenesis (Hansen et al., 2010). Throughout corticogenesis, developing neuroblasts migrate radially along the RG scaffold to enter the cortical plate in a characteristic 'inside-out' manner such that early-born neurons become deep-layer neurons and those born later become upper-layer neurons (reviewed in Di Bella et al., 2024; Lui et al., 2011; Molyneaux et al., 2007). Following neurogenesis, RG transition to a gliogenic fate (known as "gliogenic switch") to sequentially give rise to astrocytes and then to oligodendrocytes (Lanjewar and Sloan, 2021). These important events that occur during in vivo cortical development are elegantly recapitulated in the differentiation of human pluripotent stem cells (hPSCs) into cortical neural cells as demonstrated by pioneering studies (Chambers et al., 2009; Eiraku et al., 2008; Espuny-Camacho et al., 2013; Shi et al., 2012b). Advances in hPSC-based neural cell models have thus been instrumental in enhancing our knowledge on human-specific features of brain development and their implication in health and disease (Zhou et al., 2024). Moreover, the ability to generate various neural cell types in a manner that follows in vivo development from patient-specific induced pluripotent stem cells (iPSCs) offers numerous opportunities to dissect the underlying genetic/cellular basis of individual susceptibility to environmental insults and its contribution to disease risk (Kumar et al., 2012; Neely et al., 2021).

In this chapter, we present a directed differentiation protocol for the derivation of cortical neurons and glia from hiPSCs that can be further utilized for toxicological research.

1.2 Rationale

To date, most directed differentiation protocols for hiPSC-derived cortical neurons involve an initial dual SMAD inhibition (dSMADi) step to enrich in neuroectodermal cells and generally follow a protocol previously developed by Shi et al. with some modifications (Chambers et al., 2009; Shi et al., 2012a). While this strategy has been extremely successful and served as the basis for numerous studies that followed, it also has some critical limitations that pose significant hurdles for performing downstream assays.

One of the most frequently cited challenges is the collapse of the neuroepithelial sheet and massive cell death during the early stages of neural induction. Although this is not restricted to cortical differentiations and seems to be attributable to the use of 2D adherent culture environment for neural differentiation, several reports on such cell losses can be found throughout literature (Dias et al., 2022; García-León et al., 2020; Kim et al., 2022; Muratore et al., 2014; Neaverson et al., 2023; Shi et al., 2012a; Song et al., 2020). Indeed, we have consistently observed cell death typically between Days 6–7 of cortical differentiation (Fig. 1). In our experience, these events occur without overt signs of cell detachment and

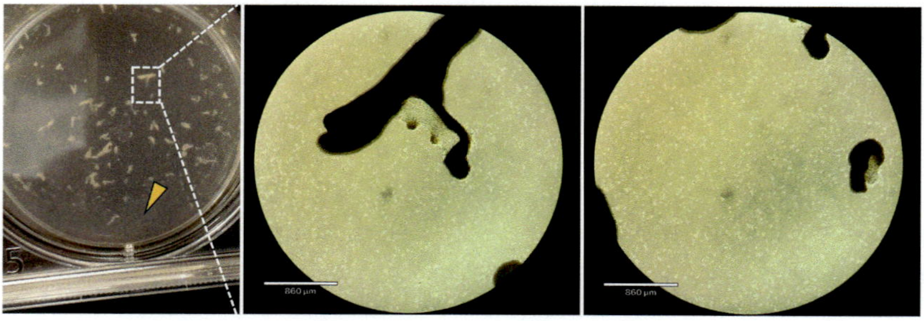

Fig. 1 Sample images of cell death during early stages of differentiation. Left: Photographic image of well undergoing spontaneous cell death during the night of Day 6 of cortical differentiation, just 7 h following media change. Note that only cells at the edges retain neuroepithelial sheet structure (yellow arrow). Center and Right: Bright field images capturing regions of the plate that are empty due to widespread cell loss and patches of remaining cells that are curled up. Dotted area corresponds to zoomed-in image. A side-by-side differentiation done using the same cell line (AD3–2) under the current protocol was successful and is represented in Fig. 5B. Scale bars, 860 μm.

progress rapidly overnight. While there are some line-dependent differences, we identified that more than half the cells within a well are gone as early as 7 h following the last media change, making it considerably difficult to capture the incident in time to passage the cells before the entire culture is lost.

Another significant obstacle that has been reported is the appearance of contaminating neural crest derivatives and/or non-neural cells. Akin to other studies, we have seen such cells appear primarily when, (1) only a small portion of cells were salvaged and replated following cell death in Days 6–7 and, (2) approximately Day 50 of differentiation in cultures where cells migrate and clump to form large aggregates resulting in empty gaps (Dias et al., 2022). On such occasions, non-specific cells exhibiting a flat morphology start to appear and proliferate to gradually take over the culture at the expense of neurons, suggesting a role for cell-to-cell contact in differentiation as well as the potential presence of a competitive mechanism for adhering to culture surface.

While the precise reason underlying these phenomena is unclear, several reports suggest, as do we, that the large-scale cell death in the early stages may be attributable to, at least in part, excessive cell growth and the concomitant depletion of nutrients or limited access to thereof, accumulation of metabolic waste products, and acidification of culture media (Bell et al., 2019; Dannert et al., 2023; García-León et al., 2020; Kim et al., 2022; Mucci et al., 2022; Muratore et al., 2014). Moreover, the practice of increasing media volume to accommodate cell growth throughout differentiation and our observation that this slightly mitigates cell death (though it did not prevent it) also support this notion (Bell et al., 2019; Dannert et al., 2023; Gantner et al., 2020; Nolbrant et al., 2017). Additionally, as cell detachment is not restricted to the edges of the well and instead occurs throughout the neuroepithelial sheet, the physical stress of media change cannot be the sole causative factor. Based on these observations and reports, we reasoned that excessive proliferation and low cell density can both lead to aberrant cortical differentiation in culture by causing cell death or deviating cells from their trajectory of becoming cortical progenitors. Thus, we hypothesized that the timely passaging of cells at densities permitting sufficient cell-to-cell contact without risking premature overgrowth will alleviate these challenges and promote successful cortical neural differentiation. Finally, in developing our protocol, we observed that cells proliferate at a higher rate under culture media containing knockout serum replacement (KSR) and thus further hypothesized that the use of chemically defined media will provide a more permissive environment by attenuating excessive growth and minimize potential batch-to-batch variability and unexpected interference with endogenous signaling pathways that are associated with complex reagents such as KSR (Blauwkamp et al., 2012; Kim et al., 2021; Tchieu et al., 2017; Zimmer et al., 2016).

2. Overview of protocol

To address the challenges described above, we introduced a defined number of critical passaging points at optimized cell densities throughout the differentiation period as described in detail in subsequent sections. Importantly, this standardized single cell passaging scheme offers enhanced flexibility in planning experimental designs, especially those involving early developmental exposures, and is considerably simpler and more cost-/time-efficient than other previously reported methods. Notable examples include multiple rounds of lengthy, labor-intensive neuroepithelium/rosette purification via dispase or manual mechanical isolation/dissociation using syringes or Pasteur pipettes, decreased Accutase incubation period to detach neurons from flat non-neural cells on the presumed basis of differential adherence properties, and spot-based differentiations, all of which require extensive additional training/preparation and are more prone to technical variability. Fig. 2 provides a general overview and description of the overall protocol described herein.

- Stage 1: Initial seeding of hiPSCs (Day −4 to Day 0).
- Stage 2: Neuralization via dSMADi (Day 0–Day 10). Passaging on Day 6.
- Stage 3: Neural progenitor cell (NPC) expansion and early neurogenesis (Day 11–Day 24). Passaging on Day 13.

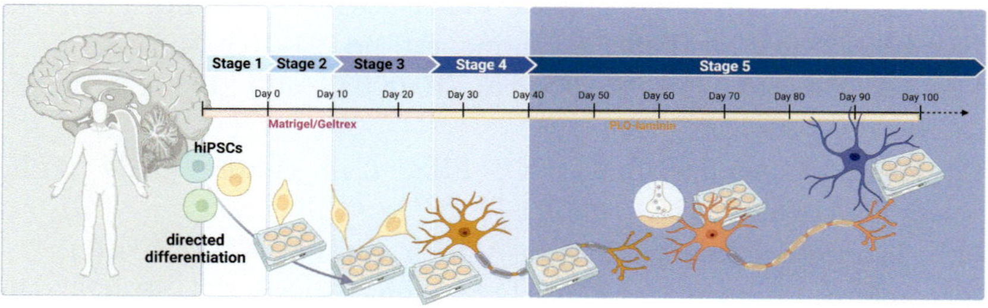

Fig. 2 Schematic representation of overall protocol. *Figure created via BioRender.com.*

- Stage 4: Immature post-mitotic neurons (Day 25–Day 39). Passaging on Day 25.
- Stage 5: Long-term culture for neuronal maturation and astrogliogenesis. Passaging on Day 40 and on need-basis thereafter.

3. Materials and reagent setup

3.1 Chemicals, peptides, and recombinant proteins

- 2-Mercaptoethanol (BME) (Sigma-Aldrich, cat. no. M3148-100 mL)
- Accutase (Fisher/Innovative Cell Technologies, cat. no. NC9839010/AT-104)
- B-27 supplement (50X), minus vitamin A (Thermo Scientific, cat. no. 12587010)
- B-27 supplement (50X), serum free (Thermo Fisher Scientific, cat. no. 17504044)
- DMEM/F-12 (Fisher Scientific, cat. no. 11-320-082)
- DMEM/F-12, GlutaMAX supplement (Fisher Scientific, cat. no. 10-565-042)
- Geltrex™ (Thermo Fisher Scientific, cat. no. A1413302)
- GlutaMAX Supplement (Fisher Scientific, cat. no. 35-050-061)
- HyClone™ Water, Cell Culture Grade (Endotoxin-Free) (Thermo Fisher Scientific, cat. no. SH3052903)
- Laminin (Sigma-Aldrich, cat. no. L2020-1MG)
- LDN-193189 (REPROCELL, cat. no. 04-0074)
- Matrigel™ (Thermo Fisher Scientific, cat. no. 08-774-552)
- MEM Non-Essential Amino Acids (NEAA) Solution (100X) (Fisher Scientific, cat. no. 11-140-050)
- mTeSR1 (STEMCELL Technologies, cat. no. 85850)
- mTeSR1 5X Supplement (STEMCELL Technologies, cat. no. 85850)
- N-2 Supplement (100X) (Thermo Fisher Scientific, cat. no. 17502048)
- Neurobasal medium (Fisher Scientific, cat. no. 21-103-049)
- Penicillin Streptomycin (Pen/Strep) (10,000 U/mL) (Fisher Scientific, cat. no. 15-140-122)
- Phosphate Buffered Saline (1X); PBS, Cell culture grade (Fisher Scientific, cat. no. MT21040CV)
- Poly-L-ornithine solution (Sigma-Aldrich, cat. no. P4957-50 mL)
- Recombinant human FGF2 (R&D Systems, cat. no. 233-FB-025/CF)
- RevitaCell Supplement (100X) (Thermo Scientific, cat. no. A2644501)

- SB431542 (REPROCELL, cat. no. 04-0010-10)
- XAV-939 (Tocris, cat. no. 3748)
- Y-27632 2HCl (Selleck Chemicals, cat. no. S1049)

3.2 Cell culture media recipes

- *Cortical neural induction (CNi) media* (Table 1)
- *Neural progenitor cell (NPC) media* (Table 2)
- *Neural maintenance (SHI CTX) media* (Table 3)

3.3 Primary antibodies for immunofluorescence

See Table 4 for detail.

3.4 Coating for cell culture plates

- Matrigel™ & Geltrex™ coating
  - Thaw Matrigel/Geltrex vials on ice at 4 °C overnight.
  - Distribute aliquots into 1.5 mL microcentrifuge tubes prechilled in −20 °C for 20 min. Store resulting aliquots in −80 °C until further use.
  - For coating, dilute with 24 mL of cold DMEM/F-12 per aliquot. Store working solution in 4 °C and use within 2 weeks.
  - Add 1 mL to each well of a 6-well plate (50–60 μL for 96-well plates) and incubate plates in 37 °C for at least 30 min prior to use.
- Poly-L-Ornithine (PLO) laminin coating

PLO.

- Stock concentration: 0.1 mg/mL (resuspend 10 mg in 100 mL sterile HyClone cell culture grade water).
- Dilute stock 1:1 (final 50 μg/mL) in sterile HyClone cell culture grade water for coating.
- Incubate plates with 1 mL of PLO solution per well for at least 2 h at room temperature (RT).
- Seal plates with parafilm and store at RT overnight.

Laminin.

- Original concentration: 1 mg/mL.
- Dilute laminin 1:100 in PBS for coating.
- Aspirate PLO, making sure not to touch the coated surface, and rinse 3 times with sterile PBS.
- Add 1 mL of laminin solution per well and incubate at RT for at least 2 h.
- Seal plates with parafilm and store PLO-laminin-coated plates at 4 °C for up to 2 weeks.
- Incubate at 37 °C for 30 min prior to seeding cells.

Note: A single, combined PLO-laminin solution can be used in lieu of the two-step coating procedure detailed above. Dilute laminin (1:100) using the PLO solution (50 μg/mL) instead of PBS and apply directly to the plate. Incubate at 37 °C for at least 2 h before plating cells.

3.5 hiPSC lines used in this study

Images and protocols described in this chapter have been performed in our own lab across several different hiPSC lines and include both healthy subjects and neurological disease-associated subjects. Example images in this chapter come from hiPSC lines referred to in our own research as

Table 1 Cortical neural induction (CNi) media.

Component	Volume
DMEM/F-12	235 mL
B-27 supplement (50X), minus vitamin A	5 mL
N-2 supplement (100X)	2.5 mL
GlutaMax (100X)	2.5 mL
NEAA (100X)	2.5 mL
Pen/Strep (10,000 U/mL)	2.5 mL
BME (14.3 M)	0.965 μL

Table 2 Neural progenitor cell (NPC) media.

Component	Volume
DMEM/F-12	118.75 mL
Neurobasal medium	118.75 mL
B-27 supplement (50X), minus vitamin A	5 mL
N-2 supplement (100X)	2.5 mL
GlutaMax (100X)	2.5 mL
Pen/Strep (10,000 U/mL)	2.5 mL
BME (14.3 M)	0.965 μL

Table 3 Neural maintenance media (SHI CTX).

Component	Volume
DMEM/F-12, GlutaMax	241.25 mL
Neurobasal medium	241.25 mL
B-27 supplement (50X), serum free	5 mL
N-2 supplement (100X)	2.5 mL
GlutaMax (100X)	2.5 mL
NEAA (100X)	2.5 mL
Pen/Strep (10,000 U/mL)	5 mL
BME (14.3 M)	1.75 μL

Table 4 Primary antibodies.

Antibody	Dilution	Manufacturer	Cat. no.
Mouse anti-beta-III tubulin	1:500	Thermo Fisher Scientific	MA1-19187
Rat anti-CTIP2	1:200	Abcam	ab18465
Rabbit anti-GFAP	1:1000	Abcam	ab7260
Chicken anti-MAP2	1:1000	Abcam	ab5392
Mouse anti-MAP2	1:500	Thermo Fisher Scientific	13-1500
Goat anti-Nanog	1:20	R&D Systems	AF1997
Rabbit anti-Nestin	1:50	Thermo Fisher Scientific	19483-1-AP
Mouse anti-Oct3/4	1:100	BD Transduction Laboratories	611202
Rabbit anti-PAX6	1:200	BioLegend	901301
Mouse anti-PSD95	1:500	Thermo Scientific	MA1-046
Goat anti-SOX1	1:100	R&D Systems	AF3369
Rabbit anti-Synapsin I	1:1000	Sigma Aldrich	S193-10UG
Rabbit anti-vGLUT1	1:300	Abcam	ab77822

AD2–2, AD3–2, AD4–2, CM-1, PSEN1, and AD-1s; and come from a variety of sources, reflecting both male and female subjects as well. More details about these lines can be found in the funding section of this chapter.

4. Procedure

This section provides a detailed step-by-step description of cortical neural differentiation from hiPSCs involving dual SMAD inhibition (dSMADi) to obtain mixed cultures of neurons and radial/astroglia. The procedure described here is optimized for a standard 6-well tissue culture plate (culture surface area = 9.6 cm^2), but the protocol may be adjusted accordingly to accommodate other formats.

4.1 Stage 1: Initial seeding of hiPSCs

Ensure that hiPSC cultures are of high quality and void of spontaneous differentiations around colony edges (or below 5%). We recommend seeding two wells per line for neural induction when hiPSCs are at 70–80% confluence.

1. Rinse hiPSCs once with mTeSR1 and add 1 mL Accutase to each well.
2. Incubate for 15–20 min at 37 °C and occasionally check the plate to avoid prolonged exposure to Accutase.
3. Add 3 mL replating media (mTeSR1 supplemented with 10 μM Y-27632) to each well and transfer detached cells to a 50 mL conical tube.
4. Centrifuge in standard room temperature (RT) for 5 min at 200 × *g*.
5. Aspirate supernatant and gently resuspend cells in 2.5–3 mL of replating media to obtain single cell suspension.
6. Determine cell concentration using an automated cell counter or hemocytometer and prepare a diluted solution of 0.1×10^6 cells/mL for seeding.
7. Aspirate Matrigel/Geltrex and disperse 2 mL of cell suspension to each well (resulting in a starting density of 200,000 cells/well or 20,833 cells/cm^2).
8. Gently shake the plate by alternating between up/down and left/right movement to ensure the cells are evenly distributed (do not swirl the plate to mix as this will result in uneven distribution).
9. Incubate overnight at 37 °C.

10. On the following day, rinse hiPSCs with mTeSR1 to remove cell debris and replenish each well with 2.5 mL of mTeSR1.
11. Change mTeSR1 everyday until the cells form a confluent monolayer. This typically takes four days from the initial replating (e.g., if replated on Friday, cells should be near 100% confluent by Tuesday).

Note: Increase media volume in 0.5 mL increments each day to account for cell proliferation.

4.2 Stage 2: Neuralization via dSMADi

The following steps describe the process of cortical neural differentiation via dSMADi using SB431542 (SB4) and LDN-193189 (LDN) with an initial passaging of neuroepithelial cells on Day 6. In addition, the WNT signaling inhibitor XAV-939 (XAV) is added to promote an anterior neuroectodermal identity. Table 5 below provides an overall summary of media composition for each day of this stage.

1. Ensure that the cells are 90–100% confluent at this point as large gaps may lead to aberrant differentiation.
2. Aspirate mTeSR1 and rinse cells with DMEM/F-12.

Table 5 Media composition for differentiation (Days 0–10).

Day	Daily media setup
0	CNi medium + 10 μM SB4 + 500 nM LDN + 5 μM XAV
1	CNi medium + 10 μM SB4 + 500 nM LDN + 5 μM XAV
2	CNi medium + 10 μM SB4 + 500 nM LDN + 5 μM XAV
3	CNi medium + 10 μM SB4 + 500 nM LDN + 5 μM XAV
4	CNi medium + 10 μM SB4 + 500 nM LDN + 5 μM XAV
5	CNi medium + 10 μM SB4 + 500 nM LDN
6	CNi medium + 10 μM SB4 + 500 nM LDN + RevitaCell (1:100)
7	CNi medium + 10 μM SB4 + 500 nM LDN + RevitaCell (1:100)
8	CNi medium + 10 μM SB4 + 500 nM LDN + RevitaCell (1:100)
9	NPC medium + 10 μM SB4 + 500 nM LDN + RevitaCell (1:100, optional)
10	NPC medium + 10 μM SB4 + 500 nM LDN + FGF2 (20 ng/mL)

3. Commence cortical neural differentiation by adding 4 mL of CNi media supplemented with SB4, LDN, and XAV. This is considered Day 0.
4. Days 1–2: Aspirate exhausted media. Rinse cells with DMEM/F-12 and add 4 mL of Day 1/2 media.
5. Days 3–5: Aspirate exhausted media. Rinse cells with DMEM/F-12 and 5 mL of Day 3/4/5 media.

   Note 1: A lot of cell debris may be present during these early stages. This is normal and should not affect downstream processes.

   Note 2: Occasionally, depending on the hiPSC line, cell death and/or detachment may be observed. In contrast to previous reports, under this protocol, the cells at the center of the wells will remain intact. As cells are highly proliferative at these stages, we have consistently been able to expand cultures on Day 6 and continue with the differentiation without difficulty. Should this occur, perform media change with extra caution (volume may be reduced to 3–4 mL depending on the number of surviving cells).
6. Day 6 passaging:
   i. Rinse cells with DMEM/F-12 and add 1 mL Accutase to each well.
   ii. Incubate for 17–20 min at 37 °C. Incubation time may be reduced to 10–15 min for wells in which cell death/detachment have occurred.
   iii. Add 3 mL replating media (Day 6 media supplemented with RevitaCell) to each well and transfer detached cells to a 50 mL conical tube.
   iv. Centrifuge in RT for 5 min at 200 × *g*.
   v. Aspirate supernatant and gently resuspend cells in replating media to obtain single cell suspension. The volume of media required may vary (3–5 mL) depending on the size of the pellet.
   vi. Determine cell concentration and prepare a diluted solution of 1.0×10^6 cells/mL for seeding.
   vii. Aspirate Matrigel/Geltrex and disperse 2 mL of cell suspension to each well.
   viii. Gently shake the plate in up/down and left/right motion to evenly distribute cells and incubate overnight at 37 °C.

   Note 1: At this stage, a yield of 12–20 million cells from the starting two wells can be expected. As cells will remain highly proliferative until the next two passages, massive culture expansion is not required, thus we recommend expanding to 3–4 wells per line at this passage.

   Note 2: A subset of cells can be replated into 96-well plates for initial quality control via immunofluorescence. For this, plate cells at

0.5×10^6 cells/mL in 100 μL volume per well. Cells can be fixed 24–48 h following replating.

7. Days 7–8: Aspirate exhausted media and thoroughly rinse cells 2–3 times with DMEM/F-12 to remove cell debris from replating. Failure to do so will affect the cells directly below the swarm of dead cells (this can be morphologically observed under a standard tissue culture microscope). See *4.2.1 Expected Outcomes* for representative images following passaging on Day 6.

   IMPORTANT: It is critical to use RevitaCell supplement (not Y-27632) during the Day 6 split as well as replenishing it in Days 7 and 8. Not doing so will result in progressive cell death and non-specific differentiation. RevitaCell treatment may be extended to Day 9 depending on culture morphology.
8. Days 9–10: Aspirate exhausted media. Rinse cells with DMEM/F-12 and add 4 mL of Day 9/10 media.

4.2.1 Expected outcomes

As shown in Fig. 3A, by Day 8 of differentiation, cells display robust expression of the definitive neuroectodermal markers PAX6, SOX1, and the neural progenitor marker NES. Conversely, pluripotency markers such as NANOG and OCT4 should not be detected at this stage. We recommend performing such immunostaining analyses as an early quality control measure to ensure successful differentiation, especially during the initial implementation stages of the protocol as well as when attempting to differentiate a new hiPSC line.

4.3 Stage 3: NPC expansion and early neurogenesis

Following Day 10 of differentiation, this stage involves an additional 2-day treatment of FGF2 to promote NPC propagation and a second standard passaging on Day 13 to prevent overgrowth of culture. Early neurogenesis will start to occur approximately at Day 21 of differentiation and subsequent passaging at Day 25 will allow investigators to obtain and expand mixed cultures of cortical neurons and proliferating progenitors for further maturation.

1. Days 11–12: Perform daily media change with NPC media supplemented with 20 ng/mL FGF2 at 4–5 mL volume per well. Cells need not be rinsed prior to media change provided that there is no cell debris present.

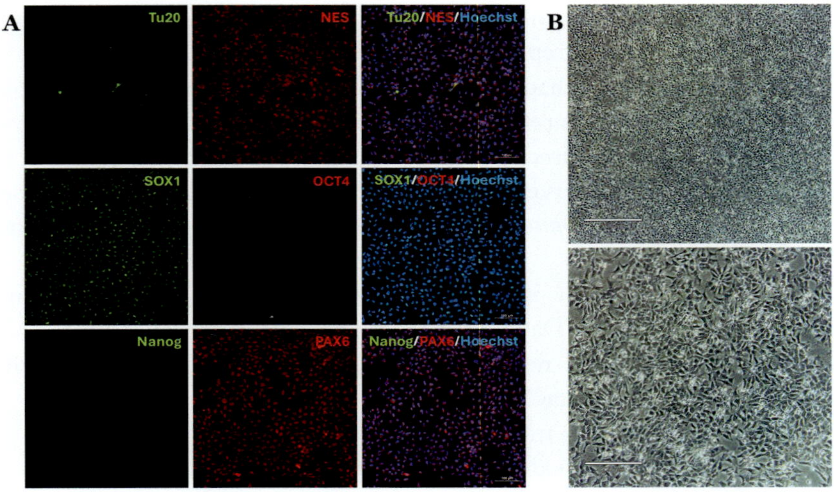

Fig. 3 Expected immunocytochemical and morphological properties of cortical differentiation cultures on Days 8 and 7, respectively. (A) Immunofluorescence staining on Day 8 demonstrating the expression of cortical progenitor markers PAX6, SOX1, and NES with concomitant loss of pluripotency markers NANOG and OCT4. Scale bars, 100 µm. (Top, middle: AD2–2, bottom: AD4–2) (B) Representative bright field images of cells on Day 7; 24 h after replating on Day 6 (AD-1s). Scale bars 560 µm (top), 220 µm (bottom). (A) and (B) from two independent differentiations.

2. Day 13 passaging: Passage cells in NPC media at 1.0×10^6 cells/mL in 2 mL volume per well following the procedure described in the previous section under 'Day 6 passaging'. At this point and onwards, cells may be passaged using either RevitaCell or Y-27632 without the need for replenishing the next day. Alternatively, Day 13 passaging can be postponed to Day 15 to accommodate specific experimental plans.

   Note: Similar to the split on Day 6, there is no need for large-scale culture expansion at this stage as cells still retain proliferative activity. A range of 18–27 million cells can be expected from 3 wells that had been replated on Day 6. Refer to Fig. 4 under *4.3.1 Expected Outcomes* below for representative images following replating.
3. Day 14: Aspirate exhausted media and rinse cells with DMEM/F-12 to remove cell debris from replating.
4. Days 15–24: Perform daily media change with NPC medium. From this point onwards, only aspirate half the existing media and add 3–4 mL of fresh media per well.

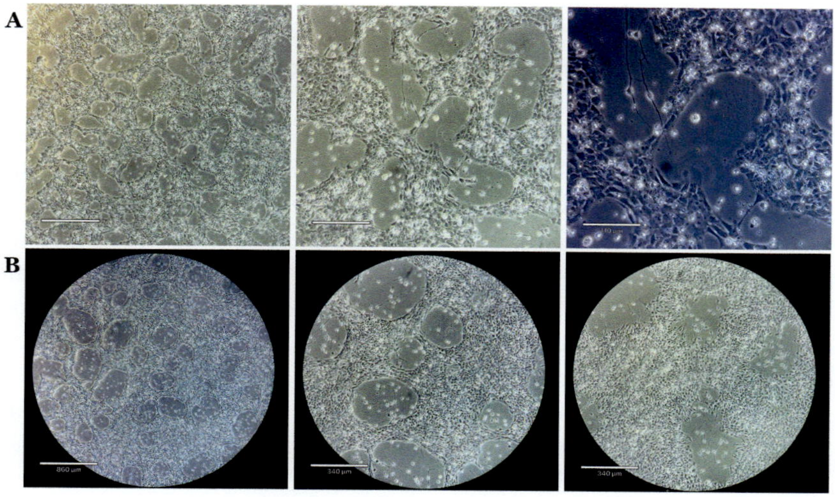

Fig. 4 Sample bright field images following 24 h after replating on Day 13 or Day 15. (A) Representative images of Day 14 PSEN1 hiPSC-derived cortical progenitors following splitting on Day 13. Scale bar 560 μm (left), 220 μm (center), 110 μm (right). (B) Sample images of Day 16 AD2–2 hiPSC-derived cortical progenitors 24 h (left, center) and 48 h (right) after replating on Day 15. Scale bars 860 μm (left), 340 μm (center, right). Differentiations from (A) and (B) are independent of each other.

4.3.1 Expected Outcomes

See Fig. 4 and corresponding legend for detail.

4.4 Stage 4: Immature post-mitotic neurons

Starting from this stage, differentiated cultures will be comprised of early-born immature post-mitotic cortical neurons and proliferating progenitors. Cells are cultured in neural maintenance media previously formulated by Shi and colleagues with minor modifications and is referred to here as SHI CTX media (Shi et al., 2012a).

1. Day 25 passaging:
  - **i.** Rinse cells with DMEM/F-12 and add 1 mL Accutase to each well.
  - **ii.** Extend the incubation period to 30–45 min at 37 °C and occasionally monitor the progress of enzymatic digestion.
  - **iii.** Add 3 mL replating media (SHI CTX media supplemented with Y-27632 or RevitaCell) to each well and transfer detached cells to a 50 mL conical tube.
  - **iv.** Centrifuge in RT for 5 min at 200 × *g*.

v. Aspirate supernatant and initially resuspend the cells in 3 mL of replating media as this will facilitate disaggregating large pellets.

vi. Pass the cells through a 40 μm cell strainer to remove residual cell aggregates. This step may be omitted, but we recommend the use of strainers here as the cultures are highly confluent at this point. Further, straining the cells will also facilitate uniform seeding.

vii. Initially add 2–3 mL more media to take a preliminary cell count and dilute further accordingly to obtain a cell suspension that permits accurate estimation of the cell concentration. This can range from a final volume of 7–12 mL of media.

viii. Prepare cell suspensions of 2.0–2.25 × 10^6 cells/mL for seeding.

Note: We also recommend replating a subset of cells for immunostaining to determine successful neuronal differentiation. Cells can be fixed after 3–5 days following replating at this stage.

ix. Aspirate PLO-laminin solution from plate and rinse once with DMEM/F-12 or PBS before seeding cells at 2 mL volume per well.

x. Gently shake the plate up/down and left/right to evenly distribute cells and incubate overnight at 37 °C.

Note: This step allows researchers to obtain multiple plates of cortical neural cultures for downstream assays. Here, 3 wells prepared from the last split on Day 13 can yield 30–50 million cells. See Fig. 5A in *4.4.1 Expected Outcomes* below for culture morphology before and after passaging.

2. Day 26: Rinse the cells with DMEM/F-12 to remove cell debris and add 3 mL of SHI CTX media. From this point until the next passage on Day 40, cells can be fed every other day by removing 50% of consumed media and adding 3 mL of fresh media.

4.4.1 Expected outcomes

As shown in Fig. 5B, early-born Tu20$^+$ neurons can be visualized throughout the culture along with the expression of the deep-layer cortical marker CTIP2, mimicking laminar formation of the developing human cortex in vivo.

4.5 Stage 5: Long-term culture for neuronal maturation and astrogliogenesis

In this stage, we provide a general description on how to maintain cells further to promote neuronal maturation and/or subsequently astrogliogenesis.

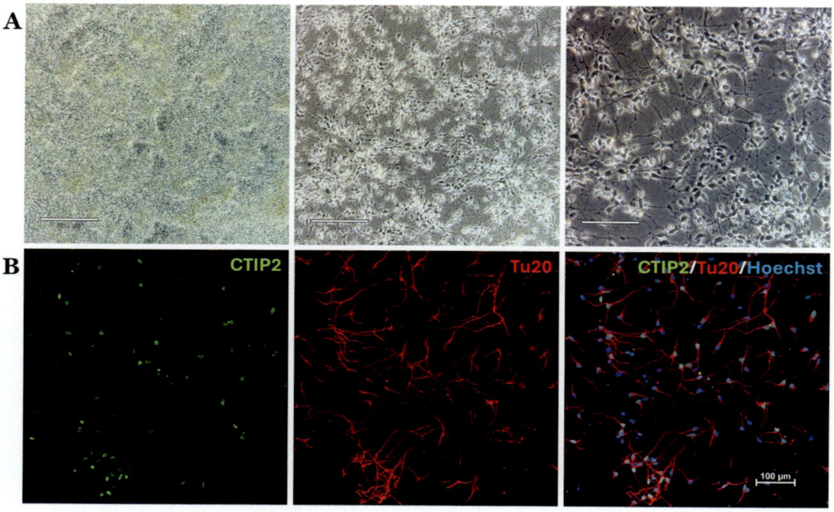

Fig. 5 Morphological and immunocytochemical characterization of early cortical neurons. (A) Left: Morphology of D24 AD3–2 hiPSC-derived cortical cultures prior to passaging on Day 25. Center, right: Cellular morphology of corresponding culture on Day 26 (i.e., 24 h after replating). Scale bars, 560 µm (left), 220 µm (center), 110 µm (right). (B) Immunocytochemical staining of Day 29 AD3–2 hiPSC-derived early cortical neurons expressing deep-layer cortical marker CTIP2. Scale bar, 100 µm. (A) and (B) from two independent differentiations.

1. Following the procedure described for Day 25 splits, passage the cells between Days 40 and 45 for long-term culture. Fig. 6 below depicts anticipated culture morphology on Day 39 to serve as a reference point.

   Note: Unless required for immediate to short-term downstream assays, for extended culture, we do not recommend seeding cells at densities below 2.0×10^6 cells/mL.
2. Feed the cells every 72 h by aspirating 50% of consumed media and replenishing it with 3 mL of fresh media.

   Note: Alternatively, cells can also be cultured in the commonly used NB/B27 media.
3. Cell proliferation will gradually slow down and cultures will not need to be replated as frequently. Depending on the cell line, cultures can be passaged every 30–50 days, with longer intervals as cultures mature.

   Note 1: Regularly monitor the cultures to ensure that cells do not cluster to form large aggregates, detach, or become overconfluent.

   Note 2: We recommend characterizing cultures monthly (e.g., Day 70, Day 100) via immunostaining to accommodate downstream assays. Some notable aspects to assess include timepoints of neuronal maturation or astrogliogenesis. See *4.5.1 Expected Outcomes* below for examples.

4. After approximately Day 150, cells can be maintained in the same plate for months without risking overgrowth.

4.5.1 Expected outcomes

Cortical neurons exhibit signs of maturation in a progressive manner with extended culture. As shown in Fig. 7A, expression of pre- and

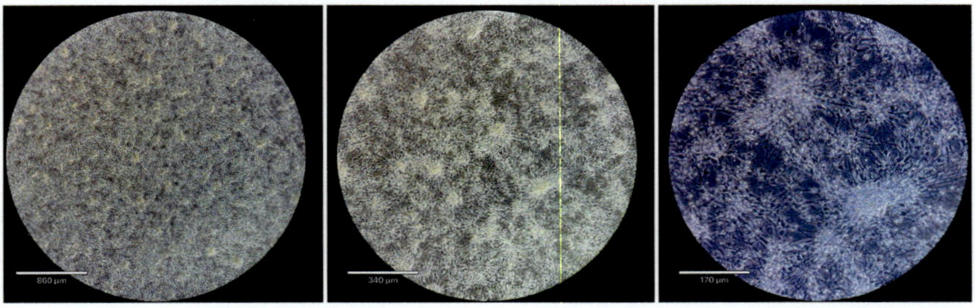

Fig. 6 Bright field images of cortical cultures prior to being passaged for long-term maintenance. Image from Day 39 cortical cultures derived from AD3–2 hiPSC line. Neuronal processes can be observed throughout the culture. Scale bars, 860 µm (left), 340 µm (center), 170 µm (right).

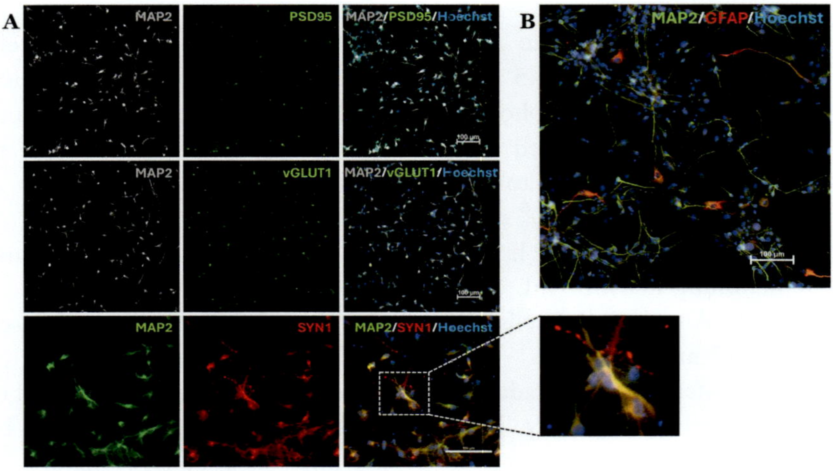

Fig. 7 Cortial neural cultures gradually progress to maturity. (A) Immunofluorescence detection of neuronal and synaptic markers in Day 62 AD3–2 hiPSC-derived cortical neurons. Dotted area corresponds to zoomed-in image. Scale bars, 100 µm. (B) Confirmation of astrocytic development as determined by immunocytochemical detection of GFAP in Day 105 CM-1 hiPSC-derived cortical cultures. Scale bar, 100 µm.

post-synaptic proteins vGLUT1, SYN1, and PSD95 can be detected (in this case, on Day 62) at discernable levels along neuronal processes and in the cell body, indicating that the cells are gradually progressing to maturity. Following prolonged culture, GFAP-positive astrocytes start to emerge as well as shown in Fig. 7B. As the process of astrogliogenesis under this protocol relies on endogenous spontaneous development, the rate or time of astrocyte development may vary depending on the cell line and requires validation by the researcher. Additional functional outcome measures assessing electrophysiological activity, neurotransmitter release/uptake, response to inflammatory stimulus, or calcium imaging can provide further validation of functional maturation.

5. Conclusions

This protocol describes a dSMADi-based directed differentiation method of generating cortical neural cultures comprising of neurons, neural progenitors, and subsequently astrocytes from hiPSCs using chemically defined media. Immunocytochemical analyses demonstrated that cells exit pluripotency and efficiently differentiate towards a dorsal telencephalic identity as confirmed by the decrease in NANOG, OCT3/4 and strong expression of PAX6, SOX1, and NES on Day 8 of differentiation. In addition, neurons resulting from the first wave of neurogenesis expressed the deep-layer cortical marker CTIP2, recapitulating the 'inside-out' patterning of laminar fate specification in corticogenesis. Furthermore, cells gradually adopt a mature phenotype as shown by the cellular distribution of pre-/post-synaptic markers along neuronal processes and the soma. Astrogliogenesis occurred following prolonged culture maintenance, mimicking the temporal sequence of neurogenesis and astrocyte development in vivo. Importantly, astrocytes generated under this protocol were confirmed to respond to inflammatory stimuli by increasing inflammatory gene expression, displaying nuclear translocation of nuclear factor kappa B, and undergoing morphological shift to exhibit a reactive phenotype (unpublished data). Finally, as described thus far, this method has been applied to multiple cell lines ($n = 6$) to bolster confidence and has been tested by multiple members of our group to ensure reproducibility.

6. Troubleshooting

- *Problem*: hiPSC lines are not ready for neural induction on the same day.
    - *Solution*: Some lines may grow at a slower rate. We recommend seeding slow-growing lines at a density of 0.125×10^6 cells/mL. We also recommend starting the differentiation 1 day apart for slower lines rather than initiating the process when the cultures are less than 90% confluent.
- *Problem*: Cells start to gradually die following replating on Day 6.
    - *Solution*: It is critical to replenish RevitaCell within 18–24 h (i.e., do not let cultures go beyond 24 h without media change and RevitaCell replenishing). As shown in Table 5, RevitaCell treatment can be extended to Day 9, depending on culture morphology. Also make sure cultures are thoroughly rinsed to remove cell debris. It is imperative to prevent cell death at these stages as it may lead to the appearance of non-neural cells.

Funding

This work was supported by the National Institute Of Environmental Health Sciences (NIEHS) under award numbers R01ES031401, R01ES07331, R01ES010563, and R01AG080917 (ABB). Samples from the National Centralized Repository for Alzheimer's Disease and Related Dementias (NCRAD), which receives government support under a cooperative agreement grant (U24 AG21886) awarded by the National Institute on Aging (NIA), were used in this study. We thank contributors who collected samples used in this study, as well as patients and their families, whose help and participation made this work possible. Some of the iPSC lines from NCRAD include lines obtained from other investigators with support from the iPSC Initiative (1 RF1 AG048083–01) (PI Lawrence Goldstein, PhD) and herein referred to as CM-1, deposited as IUGB 55.1; and from the LEADS study funded by NIA grants (R56 AG057195) and (U01 AG057195) and herein referred to as AD2–2, AD3–2, and AD4–2. Other lines were obtained from the Coriell Institute for Medical Research and herein referred to as PSEN1 and AD-1s corresponding to lines AG25367 and GM24666. The content is solely the responsibility of the authors and does not necessarily represent the official views of the National Institutes of Health.

Conflict of interest

The authors declare no conflict of interest.

References

Bell, S., Hettige, N., Silveira, H., Peng, H., Wu, H., Jefri, M., et al., 2019. Differentiation of human induced pluripotent stem cells (iPSCs) into an effective model of forebrain neural progenitor cells and mature neurons. Bio. Protoc. 9. https://doi.org/10.21769/BioProtoc.3188.

Blauwkamp, T.A., Nigam, S., Ardehali, R., Weissman, I.L., Nusse, R., 2012. Endogenous Wnt signalling in human embryonic stem cells generates an equilibrium of distinct lineage-specified progenitors. Nat. Commun. 3, 1070. https://doi.org/10.1038/ncomms2064.

Chambers, S.M., Fasano, C.A., Papapetrou, E.P., Tomishima, M., Sadelain, M., Studer, L., 2009. Highly efficient neural conversion of human ES and iPS cells by dual inhibition of SMAD signaling. Nat. Biotechnol. 27, 275–280. https://doi.org/10.1038/nbt.1529.

Dannert, A., Klimmt, J., Cardoso Gonçalves, C., Crusius, D., Paquet, D., 2023. Reproducible and scalable differentiation of highly pure cortical neurons from human induced pluripotent stem cells. STAR Protoc. 4, 102266. https://doi.org/10.1016/j.xpro.2023.102266.

Di Bella, D.J., Domínguez-Iturza, N., Brown, J.R., Arlotta, P., 2024. Making Ramón y Cajal proud: development of cell identity and diversity in the cerebral cortex. Neuron 112, 2091–2111. https://doi.org/10.1016/j.neuron.2024.04.021.

Dias, C., Nita, E., Faktor, J., Hernychova, L., Kunath, T., Ball, K.L., 2022. Generation of a CHIP isogenic human iPSC-derived cortical neuron model for functional proteomics. STAR Protoc. 3, 101247. https://doi.org/10.1016/j.xpro.2022.101247.

Eiraku, M., Watanabe, K., Matsuo-Takasaki, M., Kawada, M., Yonemura, S., Matsumura, M., et al., 2008. Self-organized formation of polarized cortical tissues from ESCs and its active manipulation by extrinsic signals. Cell Stem Cell 3, 519–532. https://doi.org/10.1016/j.stem.2008.09.002.

Espuny-Camacho, I., Michelsen, K.A., Gall, D., Linaro, D., Hasche, A., Bonnefont, J., et al., 2013. Pyramidal neurons derived from human pluripotent stem cells integrate efficiently into mouse brain circuits in vivo. Neuron 77, 440–456. https://doi.org/10.1016/j.neuron.2012.12.011.

Gantner, C.W., Cota-Coronado, A., Thompson, L.H., Parish, C.L., 2020. An optimized protocol for the generation of midbrain dopamine neurons under defined conditions. STAR Protoc. 1, 100065. https://doi.org/10.1016/j.xpro.2020.100065.

García-León, J.A., García-Díaz, B., Eggermont, K., Cáceres-Palomo, L., Neyrinck, K., Madeiro da Costa, R., et al., 2020. Generation of oligodendrocytes and establishment of an all-human myelinating platform from human pluripotent stem cells. Nat. Protoc. 15, 3716–3744. https://doi.org/10.1038/s41596-020-0395-4.

Hansen, D.V., Lui, J.H., Parker, P.R.L., Kriegstein, A.R., 2010. Neurogenic radial glia in the outer subventricular zone of human neocortex. Nature 464, 554–561. https://doi.org/10.1038/nature08845.

Kim, J., Jeon, J., Song, B., Lee, N., Ko, S., Cha, Y., et al., 2022. Spotting-based differentiation of functional dopaminergic progenitors from human pluripotent stem cells. Nat. Protoc. 17, 890–909. https://doi.org/10.1038/s41596-021-00673-4.

Kim, T.W., Piao, J., Koo, S.Y., Kriks, S., Chung, S.Y., Betel, D., et al., 2021. Biphasic activation of WNT signaling facilitates the derivation of midbrain dopamine neurons from hESCs for translational use. Cell Stem Cell 28, 343–355.e5. https://doi.org/10.1016/j.stem.2021.01.005.

Kumar, K.K., Aboud, A.A., Bowman, A.B., 2012. The potential of induced pluripotent stem cells as a translational model for neurotoxicological risk. Neurotoxicology 33, 518–529. https://doi.org/10.1016/j.neuro.2012.02.005.

Lanjewar, S.N., Sloan, S.A., 2021. Growing glia: cultivating human stem cell models of gliogenesis in health and disease. Front. Cell Dev. Biol. 9. https://doi.org/10.3389/fcell.2021.649538.

Lui, J.H., Hansen, D.V., Kriegstein, A.R., 2011. Development and evolution of the human neocortex. Cell 146, 18–36. https://doi.org/10.1016/j.cell.2011.06.030.

Molyneaux, B.J., Arlotta, P., Menezes, J.R.L., Macklis, J.D., 2007. Neuronal subtype specification in the cerebral cortex. Nat. Rev. Neurosci. 8, 427–437. https://doi.org/10.1038/nrn2151.

Mucci, S., Rodriguez-Varela, M.S., Isaja, L., Ferriol-Laffouillere, S.L., Sevlever, G.E., Scassa, M.E., et al., 2022. Protocol for morphometric analysis of neurons derived from human pluripotent stem cells. STAR Protoc. 3, 101487. https://doi.org/10.1016/j.xpro.2022.101487.

Muratore, C.R., Srikanth, P., Callahan, D.G., Young-Pearse, T.L., 2014. Comparison and optimization of hiPSC forebrain cortical differentiation protocols. PLoS One 9, e105807. https://doi.org/10.1371/journal.pone.0105807.

Neaverson, A., Andersson, M.H.L., Arshad, O.A., Foulser, L., Goodwin-Trotman, M., Hunter, A., et al., 2023. Differentiation of human induced pluripotent stem cells into cortical neural stem cells. Front. Cell Dev. Biol. 10. https://doi.org/10.3389/fcell.2022.1023340.

Neely, M.D., Xie, S., Prince, L.M., Kim, H., Tukker, A.M., Aschner, M., et al., 2021. Single cell RNA sequencing detects persistent cell type- and methylmercury exposure paradigm-specific effects in a human cortical neurodevelopmental model. Food Chem. Toxicol. 154, 112288. https://doi.org/10.1016/j.fct.2021.112288.

Nolbrant, S., Heuer, A., Parmar, M., Kirkeby, A., 2017. Generation of high-purity human ventral midbrain dopaminergic progenitors for in vitro maturation and intracerebral transplantation. Nat. Protoc. 12, 1962–1979. https://doi.org/10.1038/nprot.2017.078.

Shi, Y., Kirwan, P., Livesey, F.J., 2012a. Directed differentiation of human pluripotent stem cells to cerebral cortex neurons and neural networks. Nat. Protoc. 7, 1836–1846. https://doi.org/10.1038/nprot.2012.116.

Shi, Y., Kirwan, P., Smith, J., Robinson, H.P.C., Livesey, F.J., 2012b. Human cerebral cortex development from pluripotent stem cells to functional excitatory synapses. Nat. Neurosci. 15, 477–486. https://doi.org/10.1038/nn.3041.

Song, B., Cha, Y., Ko, S., Jeon, J., Lee, N., Seo, H., et al., 2020. Human autologous iPSC–derived dopaminergic progenitors restore motor function in Parkinson's disease models. J. Clin. Investig. 130, 904–920. https://doi.org/10.1172/JCI130767.

Tchieu, J., Zimmer, B., Fattahi, F., Amin, S., Zeltner, N., Chen, S., et al., 2017. A modular platform for differentiation of human PSCs into all major ectodermal lineages. Cell Stem Cell 21, 399–410.e7. https://doi.org/10.1016/j.stem.2017.08.015.

Zhou, Y., Song, H., Ming, G., 2024. Genetics of human brain development. Nat. Rev. Genet. 25, 26–45. https://doi.org/10.1038/s41576-023-00626-5.

Zimmer, B., Piao, J., Ramnarine, K., Tomishima, M.J., Tabar, V., Studer, L., 2016. Derivation of diverse hormone-releasing pituitary cells from human pluripotent stem cells. Stem Cell Reports 6, 858–872. https://doi.org/10.1016/j.stemcr.2016.05.005.

Printed in the United States
by Baker & Taylor Publisher Services